WHEAT PRODUCTION AND UTILIZATION
Systems, quality and the environment

To our wives and children
Kaye and Philip
Janice, Nick and Suzy

Wheat Production and Utilization

Systems, Quality and the Environment

by

MIKE J. GOODING

University of Reading, UK

and

W. PAUL DAVIES

Royal Agricultural College, Cirencester, UK

CAB INTERNATIONAL

CAB INTERNATIONAL
Wallingford
Oxon OX10 8DE
UK

Tel: +44 (0)1491 832111
Fax: +44 (0)1491 833508
E-mail: cabi@cabi.org

CAB INTERNATIONAL
198 Madison Avenue
New York, NY 10016-4341
USA

Tel: +1 212 726 6490
Fax: +1 212 686 7993
E-mail: cabi-nao@cabi.org

A catalogue record for this book is available from the British Library, London, UK.
A catalogue record for this book is available from the Library of Congress, Washington DC, USA.

ISBN 0 85199 155 6

Typeset in Palatino by Advance Typesetting, Oxford
Printed and bound in the UK at the University Press, Cambridge

Contents

1

An Introduction to the Utilization, Development and Production of Wheat

Wheat production and its utilization has been intimately linked with the development of both agriculture and civilization over at least the last 12,000 years. Some archaeological evidence suggests even earlier utilization, around 15,000 BC (Harlan, 1981). The domestication of wheat, more than any other plant, has allowed food to be produced in sufficient quantities to support community settlement, religious/cultural development, and continuing population growth. The impact of wheat exploitation has been global, and not limited to particular regions or ethnic preferences. Its tolerance of a very wide range of growing conditions has ensured successful production throughout the world.

Wheat provides a major source of energy, protein and dietary fibre in human nutrition. Primitive unleavened bread is first thought to have been made in the Neolithic era, and since early Egyptian times wheat has been processed through fermentation into bread of various types. These products continue to form a very significant part of modern human diets. Bread continues to be the 'staff of life' and is probably the world's oldest convenience food (Dukes *et al.*, 1995). Recognition of the important contribution of complex carbohydrates in cereal grains to a healthy diet has increased wheat consumption in the West generally, and North America in particular (Faridi and Faubion, 1995). Wheat can provide more than half of the calorie requirements in a healthy daily diet (Dukes *et al.*, 1995) and new mineral enriched varieties (International Food Policy Research Institute, 1995) could reinforce the nutritional benefits in developing countries in particular.

Wheat is the world's single most important food crop in terms of tonnes of grain produced each year (Fig. 1.1), and the wheat trade represents a significant component of the balance of trade of national economies. Globally, this trade is a major issue in political and economic relationships between governments. Historically, wheat supply and the price of bread has underpinned democracy and has supported or broken governments. The most famous example probably being the French Revolution initiated by bread riots in 1789. The increase in wheat production, more than any other crop, has allowed food supply to keep pace with world population growth. For example, total world production increased at an average rate of 9.5 Mt per annum between 1946 and 1995, and this allowed global stocks to remain at levels of between 20 and 35% of consumption (Fig. 1.2). Increases since 1974 have been most dramatic in Asia (Fig. 1.3). Higher production has been achieved through increased yields per unit area (Fig. 1.4) rather than by changes in the area under wheat cultivation, which has remained relatively static between the 1970s and 1990s (Fig. 1.5). Maintaining yield growth rates is considered critical for securing future food supplies in developing countries (Agcaoili and Rosegrant, 1994) as the world demand for wheat and other cereals continues to increase substantially. Continuing investment in agricultural development to ensure yield growth rates, at least to the level achieved since 1988, will be required to meet basic requirements for cereals in some parts of the world in the early part of the next century.

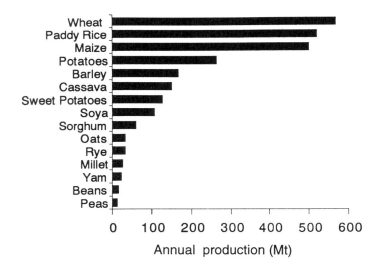

Fig. 1.1. Annual production from the world's food crops excluding sugar and vegetable crops (three year means from 1989 to 1991). Source: FAO.

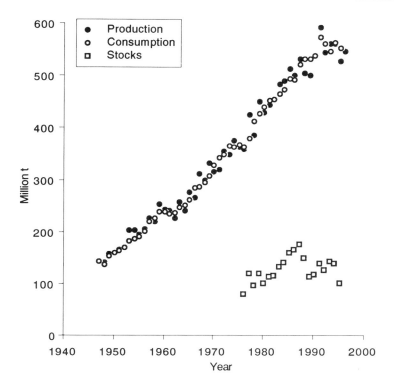

Fig. 1.2. Total world wheat production, consumption and stocks. Source: International Wheat Council.

UTILIZATION

Of all the wheat grain consumed, it has been estimated that about 65% is used directly as food for humans, 21% as a feed for livestock, 8% as seed, and 6% for other uses including industrial raw material (Orth and Shellenberger, 1988). In addition, much of the stem and leaf is exploited either as straw, or less commonly as fresh forage. Because of the high total tonnage of wheat production, all of these sectors are of great importance in terms of production systems and utilization.

Human Consumption

The most important dietary significance of wheat is as an energy source, and direct consumption of wheat products is thought to make up about 20% of the energy supplied in the worldwide total human diet (Dukes *et al.*, 1995). This is an underestimate of total contribution, however, as

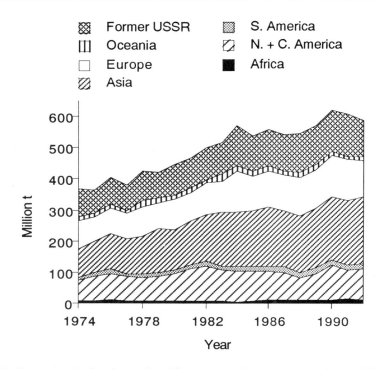

Fig. 1.3. The total production of wheat from different regions from 1974 to 1992. Source: FAO.

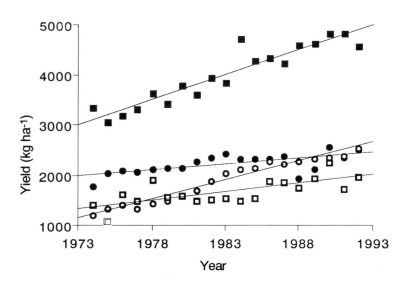

Fig. 1.4. Yields of wheat from the major wheat producing regions of the world including Europe (■), North and Central America (●), Asia (○) and former USSR (□). Source: FAO.

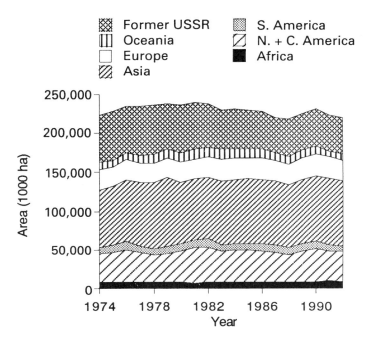

Fig. 1.5. The area of wheat grown in different regions from 1974 to 1992. Source: FAO.

wheat energy is utilized in livestock enterprises and thereby also supplied indirectly. As well as energy, wheat is also a major source of protein compared with other foodstuffs (Table 1.1) and contributes over 25% of the protein consumed in the human diet.

Wheat is utilized and processed for a multitude of products, reflecting the large quantities produced by peoples of diverse cultures and socio-economic groups (Faridi and Faubion, 1995). The global success of wheat as a food crop not only derives from its geographical range of climate and soil tolerance, but also the adaptability of grain and derived flours as a source material for many different food products, thousands of which are produced from wheat worldwide (Faridi and Faubion, 1995).

Different approaches to wheat use can be illustrated by comparing utilization in North America and neighbouring Central American countries. In both cases a wide variety of products have been developed, some of which are clearly similar (Tables 1.2 and 1.3). Soft wheats have been exploited in particular in the USA and Canada to produce a wide range of snacks from crackers to pretzels, puff pastries, doughnuts and refrigerated doughs for affluent, sophisticated consumers. Breakfast cereals in North America have also become a significant cultural market, often linked to healthy diet interests. In North America plant breeders have focused

Table 1.1. Relative energy and nutrient contents of some common foodstuffs per unit fresh weight of product (wheat = 100; adapted from McCance and Widdowson, 1960).

	Energy	Water	Protein	Fat	Carbohydrate
Whole wheat flour	100	100	100	100	100
White bread	71	255	57	56	76
Rice, raw	106	78	45	40	125
Milk, fresh, whole	19	580	25	148	6
Butter	233	92	2	3404	0
Cheese, Cheddar	125	246	186	1380	0
Beef steak, fried	80	379	150	816	0
Haddock, fried	51	434	150	332	5
Potatoes, raw	20	533	18	0	23
Peas, canned	25	484	43	0	23
Cabbage, boiled	2	638	9	0	1
Orange, with peel	8	432	4	0	9
Apple	13	560	2	0	17
White sugar	116	0	0	0	151

Table 1.2. Generic types of wheat food products in the USA and Canada (Faridi and Faubion, 1995).

Hard wheat derived
  White pan breads
  White speciality breads
  Wheat breads

Wheat – other mixtures
  Rye and pumpernickel breads
  Mixed grain breads

Soft wheat
  Crackers
  Cookies
  Cakes
  Pie crusts
  Pretzels
  Puff pastries
  Doughnuts
  Refrigerated dough

Durum wheat
  Pasta products

Wheat breakfast cereals
  Traditional hot cereals
  Instant traditional hot cereals
  Ready-to-eat cereals
  Miscellaneous cereals

Table 1.3. Generic wheat food products in Mexico and Central America (Pena, 1995).

Hard wheat derived
    Pan blanco (popular white breads)
    Tortilla de harina (flour tortillas)
    Flauta and michita (Panama, white breads)
    Sweet breads
    Semita (El Salvador, sweet breads)
    Champurrada (Guatemala, sweet flat bread)

Durum wheat
    Pasta products

Soft wheat
    Cookies
    Sweet breads (El Salvador)
    Pan de piceblo (Mexico)
    Tortilla de harina (Guatemala)
    Pastel de masa amarilla cakes (Guatemala)
    Quesadilla (El Salvador, cheese cake)

Other products
    Atole (Guatemala, beverage)
    Cafe de trigo (Guatemala, wheat coffee)

on distinctive combinations of grain protein and hardness for particular market outlets. The construction of sophisticated milling facilities and investment in larger bread and baking companies has also contributed to the development of a wide range of foods. There are reported to be at least 150 pasta products produced from durum wheat, all of which have a viable niche in the North American market place.

In Mexico and other Central American countries, in contrast, there is a much wider range of milling facilities and baking outlets from small village bakers to an increasing number of larger bulk baking plants. Countries such as El Salvador and Guatemala have mostly small and medium-sized bakers producing, from labour-intensive preparations, a diverse range of non-uniform products. In contrast to North America, sweet breads with a higher sugar content and longer shelf-life are popular (Pena, 1995).

The concept of end-use 'quality' of grain and characters required to meet highly specific criteria has not, until more recently, been a major factor in the breeding objectives and development of new local wheat varieties. The tendency has been more towards the production of high yielding wheats of low protein and, therefore, inferior 'quality' grain has tended to predominate. For these reasons, large increases in the import of

higher quality wheats from the USA into Mexico, in particular, is antici-
pated as a result of the recent North America Free Trade Agreement
(NAFTA) – which could displace more local wheats in some baking
processes.

Fewer products are derived from durum wheat in Central America,
which has to be imported into some countries for pasta production. However, food products from local maize in Central America are much more
popular traditionally than in the USA, except perhaps in those areas in the
southern states bordering Mexico where people have migrated and settled.

Wheat is pre-eminent amongst cereals in the ability of doughs produced from the flours of most bread wheat varieties to trap carbon
dioxide liberated from fermentation. This allows the production of
leavened bakery products, the diverse forms of bread being by far the
most important. Breads have been classified into three groups with
respect to their specific volume: (i) those with a high specific volume such
as pan breads; (ii) those with medium specific volume such as French
bread; and (iii) those with low specific volume such as flat breads (Qarani
et al., 1992). Many types of bread have been developed in different
locations and become associated with their country or region of origin.
Particular examples are the dark breads of Russia; light crusty rolls of
Vienna bread made from matured Hungarian flour with relatively large
amounts of yeast, which is baked quickly in a hot oven with steam; and
French bread where 45 cm long sticks of dough are baked without sugar
or fat to produce a crusty bread with very little crumb (Orth and
Shellenberger, 1988). These types of bread are often classified as hearth
breads to distinguish them from those produced in baking pans, or pan
bread.

Some breads consist of wheat mixes with other cereals, such as the
wheat–rye breads of Central Europe (Meuser *et al.*, 1994). The most common bread consumed in Germany is reported to be a 70% : 30% wheat–rye
mix. Elsewhere in Central Europe other breads consist of wheat blended
with oats, sorghum, barley and rice.

Much of the white-sliced loaves predominant in the UK are produced
with the Chorleywood Baking Process which allows the use of relatively
low protein content flours by applying high-speed mechanical mixing,
the use of oxidizing improvers, special fats and higher levels of yeast
(Blackman and Payne, 1987).

Because of the major importance of bread in diets, many countries
have legislation requiring a minimum content of specified nutrients to be
present. For example in the UK each 100 g of flour must contain no less
than 0.24 mg of vitamin B_1, 1.6 mg of nicotinic acid, 1.65 mg of iron and
235 mg of calcium carbonate. In the UK, therefore, bread contributes 17%
of the energy, 17% of the protein, 16% of the calcium, 20% of vitamin B_1,
19% of nicotinic acid and 20% of iron in an average diet. As well as bread,

wheat flour is baked to produce a variety of cakes, biscuits, crackers and pastries.

In China, for at least the last 3000 years, wheat has been utilized to produce rolls of fermented dough which are steamed rather than baked to produce a product with a dense crumb and thin white skin, rather than the brown crust of traditional western bread. This so-called steamed bread is the staple food of northern China, where it is eaten at all meals. Steamed bread is also popular in southern China, Korea, Japan and most Southeast Asian countries (Lin *et al.*, 1990). Chinese flat bread (Lao Bing) is also popular in East Asia (Nagao, 1995).

Unleavened bread in the form of chapatis is the predominant form in which wheat is consumed in India and Pakistan (Chaudhri and Muller, 1970). Chapatis are made by placing thin sheets of worked dough on a hot plate. No yeast is used, so the rising that does occur is solely a result of the steam produced during cooking. Other so-called flat breads, which may or may not require at least some yeast fermentation, are known as nan in North India, and in the rest of India as roti or rotta. In Ethiopia flat bread is known as injera, and in Arabian countries as khoubs or pitta (Lathia and Koch, 1989). Depending on location they can also include barley, sorghum and pearl millet. In Central and South America products which are more commonly made from maize, such as tortillas and tacos, can also be partly or wholly made from wheat. Flat breads are reported to be increasing in popularity in the USA in particular (Qarani *et al.*, 1992). They are regarded as a good source of dietary fibre and are highly flexible for incorporating foods such as meat and vegetables for ethnic meals. The variety of flat breads from different countries and their relationship is illustrated in Fig. 1.6.

In many Asian countries, noodles from bread wheat flour are made by cutting sheeted doughs into thin strips which are then boiled and dried ready for subsequent cooking. In certain countries this may account for up to 50% of flour utilization. There are two main types of Asian noodles: (i) the cream white soft noodles of Japan, Korea and Northern China made by mixing flour with salt water; and (ii) the yellow noodles of Malaysia, Singapore, Indonesia, Thailand and Southern China made by mixing flour with lye (alkaline) water (Miskelly and Moss, 1985). Colour, good surface appearance, favourable texture, minimal cooking loss, and high noodle yield are important noodle quality factors sought in the wheat flour (Nagao, 1995). Off colours and visible specks in noodles are, for example, considered highly undesirable in Japan. Instant noodles, produced as fried or non-fried products have become a particularly important convenience food.

Pastes produced from the semolina of durum wheat are extruded into different types of pasta such as spaghetti, lasagne and macaroni. Alternatively semolina grains can be steamed in a couscousière to

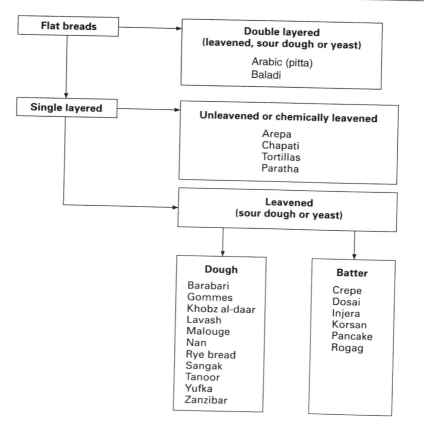

Fig. 1.6. Classification of flat breads.

produce couscous, an important indigenous dish of North Africa. In the Middle East durum wheat is often used to make flat bread (Tessemma, 1987), or can be harvested before full maturity (usually during dough development) and either boiled to produce burghul or parched by roasting through ignition without burning to produce frekeh. Total production of frekeh in the Middle East and North Africa is thought to be in the range of 200,000–300,000 tonnes (t), with additional similar products produced from durum wheat in Germany known as Gruenkern (Takruri *et al.*, 1990).

Other processes lead to the multitude of wheat forms used in breakfast cereals. Alternatively, some wheat is simply prepared by soaking for use in porridge, broth or pudding. Similarly, bulgur and wurld wheat can be prepared by removing part or all of the bran respectively from whole grain to be used in a similar way to rice. In Thailand, enzymes from wheat sprouts are mixed with rice to make glucose for sweet production, while crushed roasted wheat is fermented with soya bean to prepare soy sauce and miso (Chandhanamutta, 1985).

As a source of nutrients, however, the value of the wheat depends greatly on the parts of the wheat plant which are exploited. Human consumption of wheat is dominated by products milled from the grain or caryopsis (Fig. 1.7). The starchy endosperm represents about 80% of the dry matter and 72% of the protein in the grain and makes up about 99%

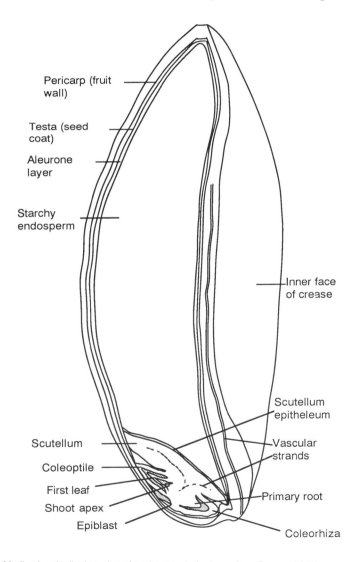

Fig. 1.7. Median longitudinal section of a wheat grain (redrawn from Barnes, 1989).

of the flour used to produce white bread. It is also relatively rich in pantothenic acid, riboflavin and some minerals (Orth and Shellenberger, 1988). The bran consists of the pericarp, testa and aleurone. Together these constitute about 15% of the dry matter and 20% of the protein. The bran is also a source of dietary fibre, potassium, phosphorus, magnesium, calcium, niacin, phytic acid and some other minerals. This is, however, often disregarded in many flour products, as are the scutellum and embryo which contain about 8% of the protein, fats, lipids, B vitamins and sugars. Not only is total protein reduced for white flour, but the quality of protein in the starchy endosperm is also relatively poor in essential amino-acids. Since the discarded bran layer is often fed to livestock, however, much of these may still be consumed by humans in certain cultures, and the production of high bran content flours can give storage problems as bran and germ oils can become rancid. Nonetheless, whole wheat flour, cracked wheat and flaked wheat products contain the bran and germ giving improved nutritional value in terms of mineral, fibre and protein content (Table 1.4). It is evident, therefore, that the nutritional value of wheat grain depends greatly on which milling process and flour products are used (Table 1.4).

Modern milling processes involve a complex array of grinding, separating and mixing procedures to produce a multitude of flour grades. Common systems rely first on breaking the grain by passing it between grooved cylinders rotating at differential speeds. This produces some relatively fine flour ('first break flour'), coarse nodules of flour ('semolina'), and relatively large grain fragments with the bran layers still attached. These fractions are separated by sieving and fed to successive break rolls to release more semolina from the bran. Bran fractions may be separated by sieving and various other processes including air purifiers. Further flour may be scraped from the bran with specialized 'scratch' rolls. The semolinas are ground to fine flour by a series of 'reduction' rolls. Each level of grinding produces streams of different grain materials which vary

Table 1.4. Typical ranges in composition of wheat and white flour (%).

	Wheat	Flour
Moisture	9–18	13.0–15.5
Starch	60–68	65–70
Protein	8–15	8–13
Cellulose (fibre)	2.0–2.5	trace (0.2)
Fat	1.5–2.0	0.8–1.5
Sugars	2–3	1.5–2.0
Mineral matter	1.5–2.0	0.3–0.6

in particle size and composition such that the final flour is often the product of 20 or more streams. Different grades of flour can be produced by varying the relative contributions of the different streams. In typical white flour production, about 72% of the grain is utilized. The remainder are milling by-products commonly ascribed as bran (14%), middlings (12.6%), shorts (0.3%) and red dog (1.1%). Alternatively the bran and middlings are combined as mill run or pollards (Boucqué and Fiems, 1988).

Starch carbohydrate in wheat grain can, like in other cereals, be fermented for use in alcoholic beverages. Traditional methods of brewing beer using malted wheat employ temperature programmed mashing followed by wort separation. The wort is fermented with yeast and the final product can be bottled or casked as a 'wheat beer' (Seaton, 1987; Dale *et al.*, 1987). Low nitrogen content soft wheats can give a high spirit yield for grain whisky production (Rifkin *et al.*, 1990) or be used in the barley brewing industry to aid head retention in pasteurized beers (Blackman and Payne, 1987). Wheat for alcohol production can be very important in certain regions. For example, in the 1988/89 season, approximately 60% (450,000 t) of the total Scottish wheat crop was required for grain distilling (Rifkin *et al.*, 1990). On a worldwide basis, however, wheat is a relatively minor source of alcohol production compared with barley and maize.

Livestock Consumption

Wheat takes second place to maize on a worldwide basis as feed grain for livestock. However, wheat grain can equal maize in energy value and often betters maize with respect to protein concentration (Table 1.5). In

Table 1.5. Relative digestible crude protein (DCP) and metabolizable energy (ME), both per unit dry matter, of cereal grains for different non-ruminant livestock enterprises (wheat = 100; adapted from Williams and Chesson, 1989).

	Pigs		Poultry	
	DCP	ME	DCP	ME
Wheat	100	100	100	100
Maize	97	64	108	89
Rice (polished)	112		96	
Barley	92	100	88	112
Rye	97		80	
Oats	89	74	78	91
Grain sorghum	105	76	100	98
Pearl millet	90	93	92	121

1985 in the UK wheat accounted for 95% of the cereal component of compound feeds for poultry and 80% for pigs (Williams and Chesson, 1989). As well as processed grain, by-products of the milling industry can also be utilized by livestock. Wheat bran, for example, can provide 170 g of crude protein, and provide 10.1 and 7.4 MJ of metabolizable energy (ME) per kilogram of dry matter for ruminants and poultry, respectively. Comparative protein, fat and fibre compositions of other milling by-products are shown in Table 1.6.

The livestock sector is often the recipient of wheat which has been rejected for human consumption. As a result of which the feed quality is variable, and nutritional problems sometimes occur leading to poor performance. More attention is now being paid to the quality criteria for feeds, in particular for pig and poultry rations (Wiseman, 1993).

In comparison with barley and oat straw, wheat straw is low in both crude protein content (about 30 g kg^{-1} dry matter (DM)) and ME (about 5.7 MJ ME kg^{-1} DM for ruminants). This negates wheat straw being used as a major nutritive source in relatively productive livestock systems. As with most straws, digestibility can be improved by treatment to solubilize the lignin, for example with sodium hydroxide, but increased utilization still requires an additional nitrogen supplement such as urea. Wheat straw is also widely used as an animal litter.

An alternative approach to utilizing straw and grain products separately after harvest is to harvest the wheat before full grain maturation as a whole crop cereal. In some situations it has been calculated that entire wheat plants harvested at 50% DM and preserved in a clamp with the addition of 4% urea can produce higher yields of DM with comparable or higher levels of crude protein and ME in one harvest than grass silage systems grown with the same amount of nitrogen and harvested in three cuts per year (Pain *et al.*, 1993).

Winter wheat, grown primarily for grain production, can also be grazed before true stem elongation. This often occurs in the southern Great Plains region of the USA in late autumn and early winter, whilst in the UK, grazing has been more commonly conducted in the early spring

Table 1.6. Characteristics of by-products from wheat flour milling (after Church, 1984).

By-product	Minimum protein (%)	Minimum fat (%)	Maximum fibre (%)
Bran	13.5–15	2.5	12.0
Middlings	10–14	3.0	9.5
Mill run	14–16	2.0	9.5
Shorts	14–16	3.5	7.0
Red dog	13.5–15	2.0	4.0

when alternative forage is in short supply. Cereals in the vegetative state can give a useful forage with a crude protein content between 20 and 30% DM and a digestibility of over 80% (Cosser *et al.*, 1994b). As well as producing additional forage at times of the year when grass production is limited, grazing can also potentially benefit the wheat crop in stimulating tillering, reducing weed competition by removing growing points from plants with an erect growth habit, reducing subsequent lodging risk by shortening straw length and increasing lower internode strength. The removal of foliage through grazing can also remove proud winter growth which may be susceptible to frost damage. If forage is severely limiting and/or the forage is more valuable than grain, cereal can be grazed until it matures but, as with grasses, digestibility declines as the stem becomes increasingly lignified.

Industrial and Non-Food Uses

Wheat has the potential to be used as a non-food raw material in many industrial processes. For example, starch can be extracted and used as a stiffening or surface coating agent in the manufacture of paper and board; as an adhesive in the manufacture of corrugated boxes (Jones, 1987a); as a loading agent in the production of resins and plastics; as a fermentation substrate in the production of antibiotics, vitamins and hormones; as a gelling agent or emulsifier in paint; as well as a host of other more minor uses (Leygue, 1993).

Wheat grains can also be used in the production of bioethanol for use as fuel. One hundred kilograms of wheat can yield 35 l of ethanol after a process of grinding; liquefaction and saccharification with enzymes; fermentation with yeasts; distillation and finally dehydration (Leygue, 1993). Wheat straw can also be combusted for the production of heat and electricity. Other uses of straw include construction as thatch and panels, use in craft industries, and as a pulp for paper manufacture.

WHEAT CLASSIFICATION AND EVOLUTION

The name wheat is given to a number of species within the genus *Triticum* (family *Poaceae*, syn. *Gramineae*). This genus is classified within the tribe of *Triticeae* Dumort (syn. *Hordeae* Benth.) which also contains the rye (*Secale*) and barley (*Hordeum*) genera. The tribe is further divided into subtribes and *Triticum, Secale, Aegilops* (goat grasses), *Agropyron, Eremopyron* and *Haynaldia* constitute the subtribe *Triticinae*.

The cultivated species within *Triticum* are often grouped on whether they have 14 (diploid), 28 (tetraploid) or 42 (hexaploid) chromosomes.

There are four different sets of 14 chromosomes that can be present in a wheat species, designated A, B, D or G. The genome formulae and names used in this book conform to the nomenclature of Miller (1987) (Table 1.7) which, although not taxonomically based, is aimed at reducing confusion and relating more easily to previous classifications. Synonyms with earlier systems are given in Miller (1987). Examples of ears are shown in Fig. 1.8. The cultivated diploid, tetraploid and hexaploid wheats commonly contain AA, AABB and AABBDD allopolyploid groups respectively. The exceptions being *T. timopheevi* Zhuk. (AAGG) and *T. zhukovskyi* Men and Er. (AAAAGG) which are cultivated forms found in western Georgia.

The evolution of wheat as a cultivated crop is thought to have started with selection from both the wild diploid *T. boeoticum* ssp. *aegilopoides* to produce the cultivated form of einkorn (*T. monococcum*, L.), and the wild tetraploid *T. dicoccoides* to produce the cultivated form of emmer

Table 1.7. Ploidy levels, scientific and common names, and genome formulae of different wheats (after Miller, 1987).

Ploidy level	Scientific name	Common name(s)	Genome formulae
Diploid	*T. uratu* Tum.	Wild einkorn	AA
	T. boeoticum Bioss.	Wild einkorn	
	ssp. *aegilopoides*		AA
	ssp. *thaoudar*		AA
	T. monococcum L.	Cultivated einkorn	AA
	T. sinskajae A. Filat and Kurk.	Cultivated einkorn	AA
Tetraploid	*T. dicoccoides* (Korn) Schweinf.	Wild emmer	AABB
	T. dicoccum (Schrank.) Schulb.	Cultivated emmer	AABB
	T. paleocolchicum Men.		AABB
	T. carthlicum Nevski	Persian wheat	AABB
	T. turgidum L.	Rivet or cone wheat	AABB
	T. polonicum L.	Polish wheat	AABB
	T. durum Desf.	Durum or macaroni wheat	AABB
	T. turanicum Jakubz	Khorasan wheat	AABB
	T. araraticum Jakobz.	Wild emmer	AAGG
	T. timopheevi Zhuk.		AAGG
Hexaploid	*T. spelta* L.	Spelt or dinkel	AABBDD
	T. vavilovi (Tum.) Jakobz.	Spelt	AABBDD
	T. macha Dek and Men.	Spelt	AABBDD
	T. sphaerococcum Perc.	Indian dwarf or shot wheat	AABBDD
	T. compactum Host.	Club wheat	AABBDD
	T. aestivum L.	Bread or common wheat	AABBDD
	T. zhukovskyi Men. and Er.		AAAAGG

Fig. 1.8. Ears of different wheats. (a) Wild einkorn, *T. boeoticum* ssp. *aegilopoides*. (b) Cultivated einkorn, *T. monococcum*. (c) Wild emmer, *T. dicoccoides*. (d) Cultivated emmer, *T. dicoccum*. *Continued over.*

Fig. 1.8. *Continued from previous page.* (e) Rivet wheat, *T. turgidum.* (f) Polish wheat, *T. polonicum.* (g) Durum wheat, *T. durum.* (h) Spelt, *T. spelta.*

Fig. 1.8. *Continued from previous page.* (i) Spelt, *T. macha.* (j) Shot wheat, *T. sphaerococcum.* (k) Club wheat, *T. compactum.* (l) Bread or common wheat, *T. aestivum* (non-bearded variety).

(*T. dicoccum*, Schulb.). It is generally agreed that this occurred within the 'fertile crescent' of mountain chains flanking the plains of Mesopotamia and the Syrian desert, as well as in areas of Anatolia and the Balkans. Earliest remains of cultivated forms of both einkorn and emmer from Ali Khosh, Iran, have been dated as being 9500 years old (Evans, 1993). Today, cultivated einkorn can still be found in remote areas of Turkey and the Balkans while a free threshing form (*T. sinskajae* A. Filat and Kurk.) is cultivated in Daghestan, west of the Caspian Sea. Emmer, except for very small amounts grown in the USA and the former USSR, has almost entirely been replaced as a cultivated species. The AABB grouping, however, is present as the much more widely grown species of durum wheat (syn. macaroni wheat) (*T. durum* Desf.) which covers about 8% of the land under wheat production and constitutes 5% of total wheat production (Anon., 1992a). Durum wheat is relatively tolerant of hot and dry conditions and nearly 45% of durum wheat cultivation is in West Asia and North Africa (Porceddu and Srivastava, 1990). It is also widely grown in Mediterranean countries, parts of eastern Europe, the north central and southwestern USA, Mexico, and areas in and surrounding Saskatchewan in Canada. In South America cultivation is practised in Chile, Argentina and Andean regions. Polish wheat (*T. polonicum* L.) and rivet wheat (*T. turgidum* L.) are closely related to durum wheat and are found in similar Mediterranean and European locations but in much smaller amounts. Minor cultivated forms of the AABB genome in the Middle East and Asia include *T. paleocolchicum* Men. (western Georgia), the Persian wheat (*T. carthlicum* Nevski; Iraq, Iran, Turkey and Southern Transcaucasia), and the Khorasan wheat (*T. turanicum* Jakubz.; central Asia and northern Middle East).

The origin of the B genome has received a great deal of scientific investigation and debate without a definite conclusion. Morphological, geographical and some cytological evidence has supported *Aegilops speltoides* Tausch as the most likely candidate but this has not been confirmed by some chromosome banding, *in situ* hybridization, isoenzyme and chromosome-pairing studies undertaken in the 1970s (Miller, 1987). The origin of the D genome is almost certainly *Aegilops squarrosa* L. and its addition to the AABB tetraploid to form the AABBDD hexaploid had a major impact on the utilization and geographical distribution of wheat. Genes contained on the D genome, for example, have profound influences on the bread-making characteristics and dough rheology of wheat flour. *Aegilops squarrosa* also appears to have a much wider range of environmental adaptation than the proposed donators of the A and B genomes, as it is naturally distributed from the eastern edge of the Fertile Crescent to Khazakhstan. Its role in hexaploid wheats, therefore, has been a main contributory factor, together with the efforts of plant breeders, to the present wide geographical distribution of cultivated hexaploid wheats

(Zohary *et al.*, 1969). The most primitive of the cultivated hexaploid wheats is thought to be spelt or dinkel (*T. spelta* L.). The grain is not free threshing but the plant is hardier than other hexaploid wheats, particularly in the seedling stages, and is therefore more suited to mountain areas. Cultivation is practised in Spain and areas of central Europe as well as a small amount in the USA. Interest in spelt has recently increased in many areas of the European Union as it appears that it grows relatively well in production systems using lower levels of nitrogen fertilizer and crop protection chemicals (Ruegger *et al.*, 1993). This interest is also being driven by a demand for specialist breads and wheat beers that can be made from spelt. A further hexaploid is club wheat (*T. compactum* Host) which is grown principally in Washington, Oregon, Idaho, and California in the USA, with a very small amount grown in Canada. Hexaploid wheats cultivated in the Middle East and Asia include *T. vavilovi* (Tum.) Jakobz. (Armenia), *T. macha* Dek and Men. (western Georgia) and Indian dwarf or shot wheat (*T. sphaerococcum* Perc.; northwest India and Iran).

By far the most important wheat, however, is the hexaploid bread or common wheat (*T. aestivum* L. syn. *T. vulgare* Host). This includes the vast majority of varieties which themselves show great diversity in agroecological adaptation and utilization. Several groupings of varieties within bread wheat can be made on the basis of milling properties (hard or soft), dough rheology (strong or weak), bran colour (red or white) and vernalization requirement (spring or winter). For example, the USA has four commercial classes describing different variety groups principally within bread wheat, although club wheat varieties also contribute to the White Wheat classification:

1. Hard Red Spring Wheat – grown in the north central areas of the USA, particularly North Dakota, and also in Canada, the former USSR and Poland. These are areas where winters are too severe for winter wheat and the varieties and climate are particularly suited for the production of bread flour. In the USA this class has been further divided into three subclasses, namely Dark Northern, Northern and Red. Hard red spring varieties are also grown in the UK when spring sowing is particularly required for rotational reasons; when sowing conditions in autumn have been difficult; and/or when their high bread-making quality can compensate for reduced yields compared with winter wheats.

2. Hard Red Winter Wheat – found on the Great Plains of the USA such as in Kansas and Oklahoma where annual rainfall is less than 900 mm, southern areas of the former USSR, the Danube Valley of Europe, the UK and Argentina. Depending on variety, these can be good for producing bread flour and can be divided in to Dark Red, Hard and Yellow Hard subclasses.

3. Soft Red Winter Wheat – grown on the eastern side of the USA where rainfall is greater than 750 mm, and in Europe including the UK. The flour from these varieties is often used for the production of biscuits, cakes and pastries.

4. White Wheat – grown on the western side of the USA. This group also includes much of the club wheat (about 15% of USA white wheat) grown in the same area. White wheats of bread wheat are also grown in Michigan and New York, some areas of Europe, Australia, South Africa, western South America, and Asia. They are not usually suitable for the UK. This classification includes both spring and winter varieties and can be divided into the Hard White, Soft White, White Club and Western White subclasses.

DEVELOPMENT AND MORPHOLOGY OF *T. AESTIVUM*

The physiological development of wheat has been reviewed (e.g. Evans *et al.*, 1975) and will not be considered in detail here. In order to appreciate the biological properties of wheat, however, in the context of different production systems, it is first necessary to summarize the main features of wheat growth.

Seed Structure and Germination

Wheat grain consists not only of the true seed but also the pericarp or fruit. A more precise term for the grain or kernel of wheat is, therefore, the caryopsis. It is described as naked because it is threshed free of the lemma and palea which remain fused to the grain of, for example, oats and barley during harvest. The grain has two distinct sides, that which is dissected by a crease, known as the ventral side, and the rounded back or dorsal side of the grain (Fig. 1.9). One end of the grain is covered in a fine layer of hairs, known as the brush end, opposite to the germ end which encases the embryo. The dorsal regions immediately adjacent to the embryo area are known as the shoulders of the grain, while the bulges on either side of the crease on the ventral side are referred to as the cheeks. The dimensions of the grain can vary greatly depending on growing conditions, variety and position within the ear but typically lengths range between 3 and 8 mm, and maximum widths between 3 and 5 mm. Individual grain weights are often between 35 and 65 mg. Surface colour ranges from golden brown with a reddish hue to creamy wheat depending largely upon variety. The endosperm may either be glassy or white and floury depending on variety and growing conditions (Fig. 1.9).

On the imbibition of water, the grain can swell by over 40% at the start of germination. Minimum moisture content and temperature for

Fig. 1.9. Morphology and germination of wheat grain. (a) Grain placed on dorsal side to show crease and cheeks. (b) Grain placed on the ventral side showing the embryo end (left) and brush end (right). (c) Cross-sectional surface of grain to reveal glassy or vitreous endosperm (left) and floury endosperm (right). (d) Coleorhiza rupturing the seed coat. (e) First seminal root and two laterals emerging with plumule (top) extending.

Fig. 1.10. (*and opposite*) Vegetative development of wheat. (a and b) First leaf emerging from the coleoptile. (c) First leaf unfolding and seminal root development. (d) Leaf emergence.

Fig. 1.10. *Continued from previous page*
(e) Hairs on auricles at junction between the lamina and leaf sheath. (f) Tiller production. (g) Emergence of adventitious roots (left).

germination is 35–45% and 4°C, respectively, although the speed of germination is faster at higher moisture contents and temperatures (20–25°C is optimal). The coleorhiza is the first part of the embryo to emerge from the seed coat layers which are ruptured by the extension (Fig. 1.9). This is shortly followed by the appearance of the seminal or seedling roots. A primary root and then two pairs of lateral roots break through the coleorhiza followed by the plumule. The scutellum is the modified seedling leaf, or cotyledon, of the wheat plant. This remains in the germinating seed, protecting the developing seedling from the hydro-lytic reactions occurring in the endosperm. Germination is, therefore, hypogeal. The plumule elongates to the soil surface and the coleoptile, a sheath protecting the first true leaf, emerges as a single pale tube-like structure (Fig. 1.10). The apical meristem or growing point is subsequently raised to the soil surface by expansion of an underground stem (rhizome) as the subcrown internode lengthens. Coleoptile elongation generally stops when it is exposed to light. The first true leaf then emerges from a slit in the top of the coleoptile and is quickly followed by others (Fig. 1.10).

The subsequent growth of wheat is determinate, i.e. there are two dis-tinct phases of growth: the vegetative and reproductive stages.

Vegetative Growth

During this period the apical meristem or growing point remains close to the soil surface and gives rise to leaves, tillers (the branches of cereal plants) and roots.

Leaves

The apical meristem produces leaf primordia which are attached at nodes or joints on the stem. During the vegetative stage the stem is very short and, therefore, the nodes are compressed with very short internodes. The leaves are pushed upwards and consist of two main sections, the leaf sheath and the leaf blade or lamina. The leaf sheaths are usually hairy (non-glabrous). The leaf blades are narrow with about 12 veins. The junc-tion between the sheath and blade is known as the collar where the ligule and auricles can be found (Fig. 1.10). The auricles are commonly 1–3 mm long and are usually hairy. The ligule is blunt and between 2 and 4 mm long. New leaves emerge from within the leaf sheath of the next young-est leaf. The rate at which leaves are formed on the meristem, emerge, expand and unfold, as well as size and shape of the leaf lamina, depends on temperature, light intensity, day-length, nutrient availability and variety

(Evans *et al.*, 1975). The total number of leaves formed on the main stem depends greatly on the environment but commonly ranges between seven and nine (Bunting and Drennan, 1966). The number of green leaves on the stem at any one time, however, is often two or four depending on nutrition, moisture, and the incidence of pests and diseases. In the absence of drought, it is in the producer's interests to obtain a canopy which has an upper surface leaf area of four to six times the area of ground that the crop is growing on (i.e. a leaf area index of 4–6). This amount of leaf appears to be necessary for maximal levels of light interception, photosynthesis and DM accumulation (Evans *et al.*, 1975; Biscoe and Gallagher, 1978). The arrangement of the leaves, however, is also important. For example, varieties of wheat can often be distinguished on the angle of the leaves relative to the stem (Fig. 1.11). When crop leaf density and solar radiation intensity

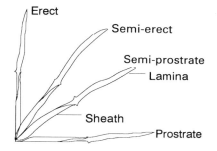

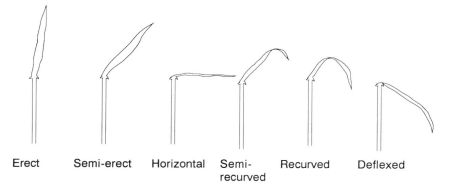

Fig. 1.11. Names given to different types of leaf angles and attitudes for cereals (adapted and redrawn from Jarman and Pickett, 1994).

are high, there are benefits to having leaves, particularly towards the top of the canopy, which are erect rather than horizontal or lax. Under high light intensities horizontal leaves can quickly become light saturated, i.e. the rate of photosynthesis stops increasing with increasing light energy. At the same time, however, they shade leaves lower down in the canopy making them less efficient at accumulating DM. More erect upper leaves, however, allow more light to penetrate down into the canopy. They themselves also become saturated less frequently. Many modern varieties designed for high yielding intensive production systems have, therefore, more erect leaves than their predecessors. During the early stages of crop growth, when ground cover is low, there are still benefits from having more horizontal leaves to reduce the amount of light reaching the soil and not being intercepted by the crop. Light intensities are often lower at this time of the year, and leaf saturation is less frequent.

Tillers

Nodes can also give rise to side shoots or tillers (Fig. 1.10). These develop from buds in the axils where the leaf sheath joins the stem. During initial growth the tiller is entirely enclosed in the leaf sheaths of the parent shoot and is dependent on the parent shoot for nutrition. Independency does not normally occur until the tiller has three mature leaves of its own. Tillering allows the plant to expand and to produce more ears and, therefore, more grains per plant when interplant competition is low. Tillers, thereby permit the wheat crop to compensate partly for adverse conditions such as poor germination and establishment, frost or hail injury, pest attack and grazing damage. Tillering is increased with increasing light and nitrogen availability during the vegetative phase and also depends greatly upon variety (Bunting and Drennan, 1966; Evans *et al.*, 1975). It is delayed and reduced by deep planting and is generally less in spring wheats than winter wheats. Tillering from a single plant in a fertile, non-competitive environment can be substantial. Not only can the main stem give rise to primary tillers, but primary tillers can also produce secondary tillers and, in some cases, even tertiary tillers may be created. Excessive tillering, however, is undesirable. Maturity of the late produced tillers will lag behind that of the main stem resulting in an uneven crop, which is difficult to manage and harvest at optimum times. Not all tillers produced will be fertile and reach maturity, and differences between varieties may become smaller towards flowering when the number of tillers is largely a function of environmental conditions. Tillers which do not produce grain may therefore be seen as a waste of resources. In certain conditions, however, carbohydrates and other nutrients can move between adjacent shoots so that late, infertile tillers may have a role in ensuring continuous nutrition to grain bearing stems (Evans *et al.*, 1975).

Roots

As well as the seminal roots, roots also derive from the coleoptile node or lower nodes of the main stem and tillers (Fig. 1.10). These adventitious or nodal roots develop about 1–2 months after germination and generally overtake the seminal roots, occupying over 90–95% of the total root volume of a fully grown crop. The adventitious roots are generally thicker (0.3–0.7 mm) than seminal roots (0.2–0.4 mm) and, if permitted, will commonly extend to 2 m depth at ear emergence.

At a later date, the apical meristem stops producing the initials of the vegetative organs, and starts giving rise to the structures of the ear. It eventually rises to emerge above the leaves. This is known as the reproductive stage.

Reproductive Growth

When this stage begins depends largely upon the prevailing climate and variety. Some varieties pass into the reproductive stage more effectively if they have experienced a cold period (vernalization) or nights getting shorter (long day photoperiodism). Vernalization appears to occur most effectively at 3°C, but very slowly below 0°C. This requirement helps prevent untimely flowering, especially in early autumn sown crops leading into a relatively harsh winter. The length of time required at low temperatures will depend upon variety, resulting in differences between winter wheats in their latest safe sowing dates in autumn and winter. Varieties also differ in their response to photoperiod. Exposure to long days will generally speed up the time to flowering, although most varieties will initiate inflorescences eventually even in short days. Like vernalization, this response helps the plant to synchronize its development with appropriate seasons. However, it also restricts the use of a particular variety to specific cropping periods and regions. In certain cases, therefore, plant breeders have generated varieties with reduced photoperiod sensitivity so that they are more adaptable for use in a wider range of cropping systems and areas (Borlaug *et al.*, 1964; Villareal *et al.*, 1985).

The first evidence of commencement of the reproductive stage is the elongation of the shoot apex. The production of leaf primordia stops, and the development of leaf primordia already formed towards the top of the apex will be halted. Growth of the tissue occurs between ridges on the apex and it is at this stage that primordia of ear structures first become visible, commonly known as the double ridge stage. These ridges then differentiate into the primordia of structures found in the ear, namely the glumes, lemmas and florets. The glumes encase the wheat spikelet which consists of several florets each surrounded by a lemma and a palea. The end of spikelet formation is marked by the initiation of the terminal spikelet. After this stage it is not possible to increase further the number of spikelets on the shoot, although the number of florets per spikelet can be increased. The number of florets formed determines the maximum numbers of grain per ear. This is obviously an important component of yield and processes influencing the number of grains per ear need to be understood. Spikelet number is increased by high light intensities and nitrogen availability, particularly before inflorescence initiation. Some compensation for thin plant and/or tiller densities can therefore be achieved with the production of longer ears containing more spikelets, as each shoot is less shaded by its neighbour. Increased nitrogen availability can increase the number of grains per spikelet as can high light levels between initiation and anthesis.

During inflorescence formation the external appearance of the plant changes as the stem elongates, pushing the developing ear to emerge finally from the flag leaf sheath. The first noticeable change is that the plant assumes a more upright posture, known as the pseudo-stem erect stage. This is more accurately defined as when the leaf sheath on the main shoot is more than 5 cm long. Later, it is possible to feel nodes above the base of the stem (Fig. 1.12). The first node detectable growth stage is defined when there is an internode of 1 cm or more present. Second and subsequent nodes are counted when the internode below them exceeds 2 cm. Unfortunately, the external structure of the plant is not always closely related to the stage of inflorescence development. Wibberley (1989), for example, cites work showing that the terminal spikelet stage was reached more than a month before the second node was detectable in one variety whilst in another variety the gap was less than a week.

Shortly before the ear emerges above the collar of the flag leaf the position of the ear can be seen as it expands and extends the leaf sheath. This is known as the booting stage. The structure of the ear is shown in Fig. 1.13. A major variation amongst varieties is the length of the awns deriving from the lemmas. The majority of wheat varieties have long awns and are known as bearded wheats, although non-bearded varieties predominate in the UK. Awns appear to increase yield potential for wheat grown in relatively warm climates. In cooler areas, however, there appears to be little difference between the yields of awned and awnless varieties. A further major distinction amongst wheat varieties, particularly evident by this stage, is plant height. Final plant height can range from 0.3–1.5 m. Shortening of the stem has been a major factor in the development of modern wheat production systems (Figs. 1.14 and 1.15) and contributed to what has been termed the 'Wheat Green Revolution' (MacKenzie, 1982). Shorter stemmed plants often allow a greater response to nitrogen as they are less likely to fall over, or lodge, in very fertile conditions. Figure 1.16 shows the effect of applying different amounts of nitrogen on the lodging

Fig. 1.12. Leaf sheath displaced to reveal node (joint) on stem.

severity of two varieties of winter wheat assessed one week before harvest. This clearly indicates the potential hazards of applying large amounts of nitrogen to taller varieties. The worldwide use of dwarfing genes in breeding programmes started when occupation forces of the USA sent a collection of Japanese wheat germplasm to Dr O.A. Vogel, USDA scientist at Washington State University. Japan had had a long history of dwarf wheat breeding, first reported in the USA in 1873 (Hanson *et al.*, 1982). Vogel crossed the Japanese semi-dwarf wheat

Fig. 1.13. (*and opposite*) Structure of the wheat ear. (a) An individual spikelet. (b) The upper spikelets of a wheat ear showing terminal spikelet face on and the others side on. (c) Structures of the spikelet dissected, from left to right, glume, lemma, grain, palea, lemma, grain, palea, infertile floret (top centre), palea, grain, lemma, palea, grain, lemma, glume. Bottom centre is rachis internode, smallest division on rule is 1 mm.

Norin 10 with the locally adapted cv. Brevor. Material from this cross was made available to other plant breeders, most notably Dr N.E. Borlaug in Mexico who first made crosses to improve rust resistance in the shorter wheats. With colleagues at the International Maize and Wheat Improvement Centre (CIMMYT) in Mexico he later developed a range of semi-dwarf genotypes, which were successfully exploited in many parts of the world. The effect of including the genes responsible for height reduction (*Rht* genes) is shown in Fig. 1.17.

After full ear emergence anthesis (flowering) begins. Not all florets reaching this stage will set grain. Low numbers of grain may be the result of excessive temperatures and, in particular, water stress during anthesis. Fertilization in some varieties is reported to be optimal between 18 and 24°C; at its lowest at 10°C and maximum at 32°C (Evans *et al.*, 1975). After fertilization, grain filling proceeds rapidly at first and then slows down as the mature grain weight is approached. This is clearly seen in Fig. 1.18 which shows the grain filling, nitrogen accumulation and drying of field grown Avalon winter wheat on sandy loam soil in the UK receiving 200 kg N ha^{-1} divided equally between the pseudo-stem erect and first node detectable growth stages. The carbohydrate within the grain derives largely from photosynthesis occurring after anthesis. The final weight per grain, therefore, is largely a function of green leaf survival late in the season,

Fig. 1.13. *Continued from previous page*
(d) Detail of glume showing nerve on left culminating in a short beak. (e) Detail of lemma with nerve extending to produce the awn. (f) Detail of palea.

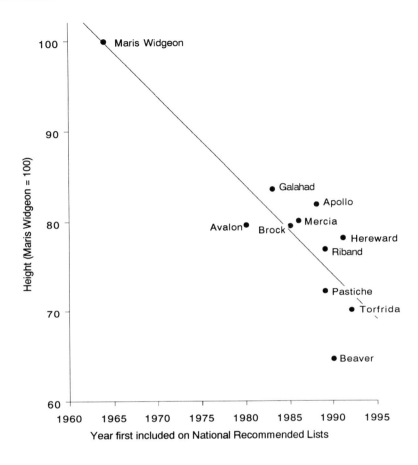

Fig. 1.14. Heights of wheat varieties recommended at different times for use in the UK when grown in the same field experiment, in the absence of nitrogen fertilizer (Thompson *et al.*, 1992).

particularly of the flag leaf. Also important, however, is temperature. Lower temperatures allow grain filling to proceed for longer and thereby result in larger grain (Evans *et al.*, 1975). As well as carbohydrate, the grain rapidly accumulates nitrogen (Fig. 1.18). In the most rapid phase of grain filling, though, the rate of nitrogen accumulation is outstripped by carbo-hydrate deposition such that nitrogen concentration falls. As grain filling slows, however, nitrogen accumulation continues such that the concentration of nitrogen recovers and can continue increasing until shortly before harvest. This pattern is shown in Fig. 1.18 and has also been found by others (Johnson *et al.*, 1967; Li *et al.*, 1991). It is indicative of situations where nitrogen uptake can continue late into grain filling such that half of

Fig. 1.15. Comparison between the height of a modern variety (cv. Hereward, foreground) and an older variety (cv. Maris Widgeon, background).

the grain protein may derive from nitrogen taken up from the soil after anthesis. This often occurs in many crops receiving greater and later applications of nitrogen fertilizers (Evans *et al.*, 1975). Under conditions of low nitrogen availability, nearly all nitrogen in the grain is derived from remobilization from the stems and leaves.

The amount of water in grain does not decline until late into grain filling (Fig. 1.18). However, as carbohydrate is accumulating moisture concentration declines, particularly as it falls below 50%. The accumulation of carbohydrate and grain drying can be appraised by simple tests on the grain to assess grain maturity. The watery-ripe growth stage occurs soon after anthesis when a watery, transparent to translucent droplet is

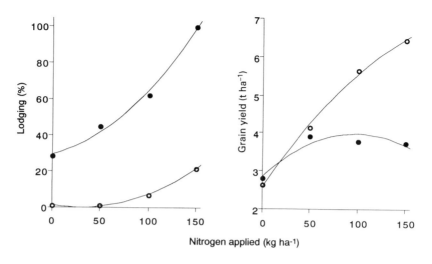

Fig. 1.16. The effect of nitrogen fertilizer application on the lodging severity and yield response of a relatively tall (Maris Widgeon, ●) and a shorter, stiffer strawed variety (Maris Ranger, ○) of winter wheat (after Holmes, 1980).

exuded as the caryopsis is squeezed. Later, a similar but white-coloured droplet signifies milky-ripeness. If no droplet is exuded then the dough development stage has started. If the dorsal side will not retain a thumb-nail impression the grain is said to be at early-dough, while if it does it is at mid-dough. If the grain cannot be split along the crease with a thumb nail, hard dough has been reached and from this stage harvest can proceed. This depends on the weather and the grain being at least below 20% moisture, and preferably below 16%.

Final grain yield will be a function of the total DM accumulated by the crop multiplied by the proportion of the total DM that is partitioned to the grain. The percentage of the above ground DM making up the final grain yield of the crop is designated the harvest index. Harvest index has been significantly increased by the adoption of the semi-dwarf varieties (Gallagher and Biscoe, 1978). A 50% yield increase between 1900 and 1980 in the performance of UK wheats has been attributed to *Rht* gene influences on harvest index (Austin *et al.*, 1980). It can also be influenced by temperature and moisture availability during grain growth.

Growth Stage Scores

Many physiological and agronomic responses in wheat are dependent on the growth stage of the crop. It is often necessary, therefore, to time

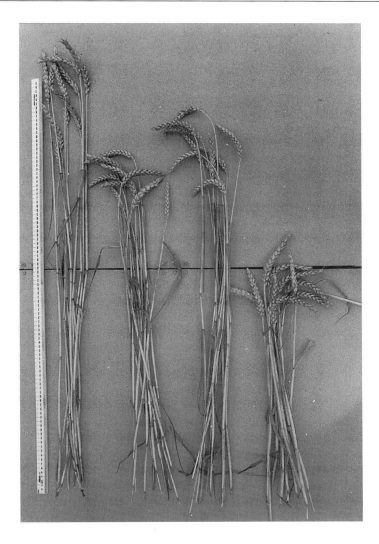

Fig. 1.17. Isogenic lines of Maris Widgeon. From left to right, with no *Rht* dwarfing genes, with *Rht1*, with *Rht2*, and with *Rht1* plus *Rht2*. Total rule length is 1 m.

applications of agrochemicals and fertilizers as accurately as possible to particular growth stages of the crop to maximize crop response and/or to prevent crop damage. Researchers also need to define growth stages so that results can be interpreted and related to the work of others. Several authors have, therefore, devised keys relating specific, defined periods in the crop's development to a numbered score. The first to gain widespread acceptance was the Feekes scale (Large, 1954) which was originally

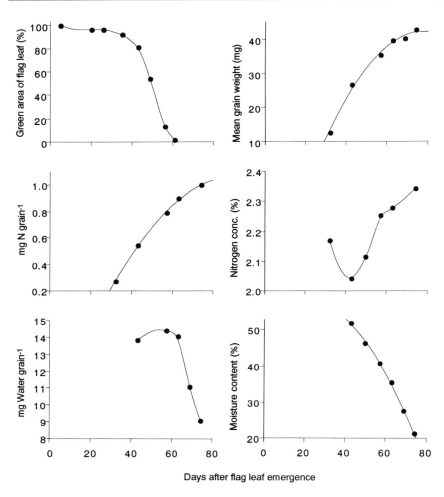

Fig. 1.18. Changes in the composition of developing Avalon wheat grain in relation to flag leaf senescence (Gooding, 1988).

devised to interpret disease severity by accounting for the developmental stages of the plant. It was subsequently adopted in many wheat growing areas in the 1970s to help the accurate timing of fungicides, herbicides and plant growth regulators. The Feekes scale was, however, considered to be cumbersome and imprecise, running from 1 to 11.4 (Wibberley, 1989) and has commonly been replaced by a score introduced by Zadoks *et al.* (1974). This is known as the decimal growth stage (GS) score, running as it does from 0 to 9 principal phases of development. Each of these phases is divided into more minor stages. A summary of the Zadoks *et al.* (1974) system is shown in Fig. 1.19 which also shows equivalent stages on the

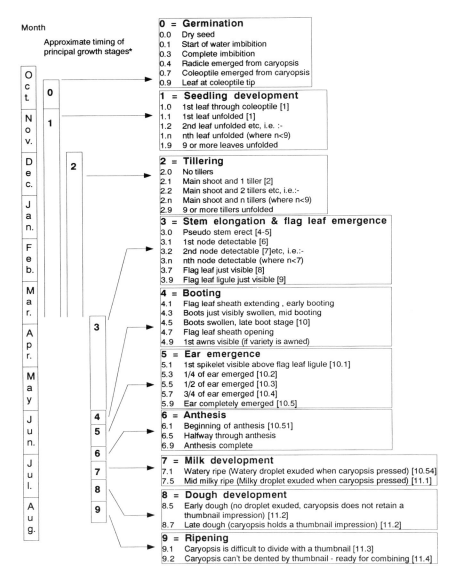

Fig. 1.19. The decimal growth stage score for cereal development (Zadoks *et al.*, 1974) and the approximate timings(*) and durations for principal growth stages for a winter wheat crop sown in October in the centre of England. Figures in square brackets denote equivalent scores on the Feekes scale. (*) Growth stage timings are greatly dependent on temperature and variety.

Feekes scale. It also shows approximate timings of the growth stages by calender month for an English-grown crop of winter wheat. It should be remembered, however, that one of the main reasons for using a growth stage score is to eliminate errors caused by differential speeds of development due to changes in weather patterns and variety from one year to the next. Plant appearance at selected growth stages is illustrated in Fig. 1.20.

PRODUCTION SYSTEMS

Wheat is extremely adaptable. It is grown from the Arctic Circle to the equator, from sea level to 3000 m, and in areas with between 250 and 1800 mm of rainfall. Wheat is, however, best suited to areas between 30° and 50°N, and 25° and 40°S latitude (Stoskopf, 1992) partly because wheat is a C_3 plant and therefore photorespiration limits DM assimilation in conditions of high light intensity, high temperatures and/or water shortage (Hall and Rao, 1994). Nonetheless, significant production extends through large areas of temperate, subtropical and tropical highland regions which, together with associated agronomic and system factors, have been grouped into the six following agroecological situations (Fischer, 1981):

1. Latitude over 45°; very cold winter with semi-arid conditions, e.g. some areas of Canada and central parts of the former USSR where photoperiod sensitive, rainfed spring wheats are sown in wheat–fallow rotations.
2. Latitude between 35° and 50°; cold winter and semi-arid conditions, e.g. some areas of the USA, western parts of the former USSR, Turkey and Iran where winter wheats with a strong vernalization requirement are sown in the autumn in rainfed wheat–fallow rotations.
3. Latitude between 40° and 60°; cold winter and humid conditions, e.g. some areas of western and eastern Europe where wheats with a strong vernalization requirement are sown in the autumn in rainfed, wheat–annual crop rotations.
4. Latitude between 25° and 40°; mild winter with semi-arid conditions, e.g. some areas of the Mediterranean, Australia, Argentina and China where wheats with little or no vernalization or photoperiod response are sown in early winter in rainfed wheat–fallow–grazed pasture rotations.
5. Latitude between 15° and 35°; mild winter with semi-arid conditions, e.g. in many less developed countries including parts of India, Pakistan, Mexico, Egypt and areas of China where wheats with little or no vernalization or photoperiod response are sown in early winter in irrigated wheat–annual crop rotations.
6. Latitude between 0° and 15°; cool summer with humid conditions, typical of upland areas of Kenya, Ethiopia, Colombia and Ecuador where

Fig. 1.20. Wheat plants at various growth stages (GS). From left to right, GS 1.1 (first leaf unfolded), GS 2.2 (main stem and two tillers), GS 3.2 (two nodes detectable), GS 3.9 (flag leaf emerged), GS 5.5 (50% of ear emerged), 9.2 (ripe).

wheats with little or no vernalization or photoperiod response are sown in early summer in rainfed wheat–annual crop rotations.

CIMMYT has further subdivided wheat production into environments to characterize wheat breeding objectives for the developing world (Figs 1.21 and 1.22). This classification does not specifically include the wheat production environments for the major wheat producing areas

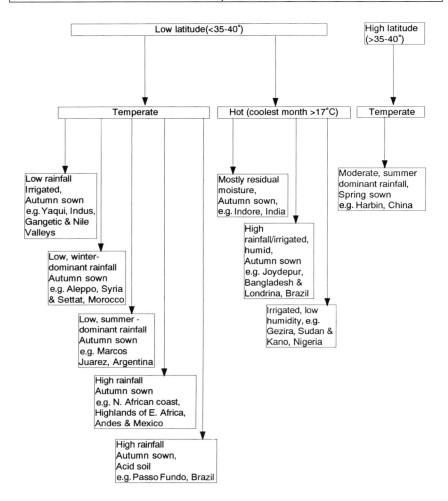

Fig. 1.21. Description of environments used by the CIMMYT Spring Wheat Programme. High and low rainfall refers to > 500 mm and < 500 mm just before or during the crop cycle (adapted from Fischer and Varughese, 1994).

in the developed world but does, nonetheless, highlight the diversity of wheat adaptation. Wheat is also increasingly being sown in warmer areas where it is grown in rotations with soya bean (e.g. Southern Brazil), rice (e.g. Southeast Asia), cotton, or less commonly with sugarcane, maize and sorghum (Wall *et al.*, 1991). Often in these associations, wheat is not the main economic crop but is grown to fill a void in the year for which there is no other more profitable or suitable crop. The sale of wheat, therefore, provides income during establishment of the major crop and provides food at a time of shortage.

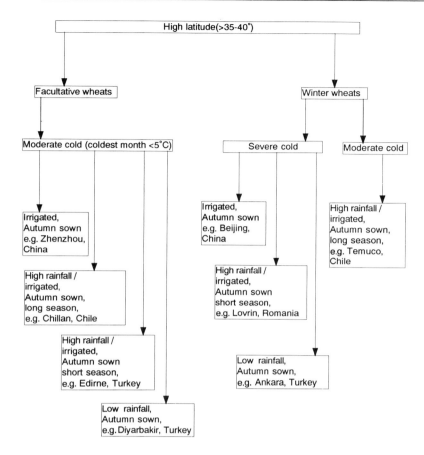

Fig. 1.22. Description of environments used by the CIMMYT Winter and Facultative Wheat Programme. High and low rainfall refers to > 500 mm and < 500 mm just before or during the crop cycle (adapted from Fischer and Varughese, 1994).

The Intensification of Wheat Production

Fertilizer and agrochemicals

The adoption of shorter, more nitrogen responsive varieties has had many implications for wheat production. The increase in nitrogen fertilizer renders the crop more susceptible to most diseases due to changes in leaf nutrient status, canopy morphology and microclimate. Those particularly affected include wind dispersed pathogens such as powdery mildew (*Erysiphe graminis*) and the rusts (*Puccinia* spp.) which have increased rates of sporulation as the nitrogen content of the leaves from which they extract nutrients is boosted (Last, 1954). Some insects, particularly certain aphid species, can also increase with nitrogen applications although effects are less consistent compared with diseases (Honek, 1991; Zhou and Carter, 1991). Increased pest and disease pressures, together with the greater crop yield potential encourages the use of more crop protection measures, particularly the application of fungicides. Increased nitrogen availability can also increase the risk of lodging (Gooding *et al.*, 1986b). Excessive tillering increases interstem competition causing them to grow taller and weaker, making the application of plant growth regulators to shorten and strengthen stems more necessary. These modern trends in production are well illustrated by the changing practices of UK wheat production, where the temperate maritime climate and price support mechanisms of the Common Agricultural Policy of the European Union have encouraged the adoption of high yielding, intensive wheat production systems since the 1970s. Wheat varieties grown in the mid-1990s were commonly 30% shorter than those grown three decades earlier (Fig. 1.14) during which time national UK yields have increased from 4.87 to 7.25 tonne ha^{-1}. Nitrogen fertilizer applications to wheat rose dramatically during the late 1970s and early 1980s (Fig. 1.23), while the total number of crop protection active ingredient applications per hectare increased three-fold (Table 1.8). Increases were particularly marked in the application of

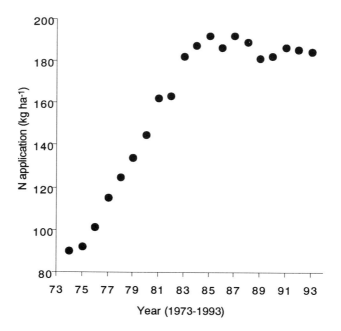

Fig. 1.23. The average amount of nitrogen fertilizer applied to winter wheat in England and Wales between 1973 and 1993.

Table 1.8. The use of crop protection active ingredients (a.i.) on wheat in the UK between 1977 and 1992.

	1977	1988	1990	1992	% Change 1977–92
Number of a.i. applications ha^{-1} (including repeats of the same a.i.)					
Insecticide + molluscicide	0.5	1.0	1.2	0.9	+76
Fungicide	0.4	2.4	4.3	5.1	+1316
Herbicide	1.6	2.7	3.4	3.6	+125
Plant growth regulator	0.2	0.7	0.9	1.1	+587
Total	2.6	6.8	9.8	10.7	+306
Average rate of each a.i. application (kg ha^{-1})					
Insecticide + molluscicide	0.31	0.21	0.11	0.11	−64
Fungicide	0.75	0.68	0.34	0.26	−65
Herbicide	1.73	1.28	0.86	0.60	−65
Plant growth regulator	1.31	1.01	1.02	0.82	−36
Average	1.29	0.88	0.56	0.42	−67
Total amount of a.i. (kg ha^{-1})					
Insecticide + molluscicide	0.16	0.21	0.13	0.1	−37
Fungicide	0.27	1.62	1.49	1.33	+392
Herbicide	2.77	3.45	2.94	2.17	−21
Plant growth regulator	0.21	0.71	0.92	0.91	+333
Total	3.41	5.99	5.48	4.51	+32
Fertilizers (kg ha^{-1})					
Nitrogen (kg N ha^{-1})	115	189	182	184	+60
Phosphate (kg P_2O_5 ha^{-1})	40	54	49	50	+25
Potassium (kg K_2O ha^{-1})	33	53	50	50	+51
Yields (ha^{-1})					
Grain yield (tonne DM)	4.19	5.28	5.94	6.23	+49
Yield of nitrogen in grain (kg)	88	111	130	139	+58
Environmental indicators					
N harvested – N applied (kg ha^{-1})	−27	−78	−52	−45	−67
Energy for fertilizers + a.i. (GJ ha^{-1})	10.3	16.8	16.1	16.0	+55
Fertilizer and a.i. energy used to produce 1 tonne grain DM (GJ)	2.5	3.2	2.7	2.6	+4

fungicides and plant growth regulators. Interestingly, the rate of each active ingredient application has declined for all classes of crop protection chemical. This reflects the combined consequences of the utilization of more active chemicals, reduced rates of individual active ingredients used in mixtures, and possibly better advice on optimum application rates in particular circumstances. Nonetheless, the total amount of fungicide and plant growth regulator applied in terms of kg a.i. ha^{-1} has increased by 4- and 3.3-fold, respectively. By 1992, on average, five fungicide active ingredients were applied as four products in either two or three (average 2.3) sprays. These were commonly applied at the start at the first node detectable growth stage to control stem base and foliar disease in the spring, at flag leaf emergence to protect the flag leaf from foliar disease and/or at ear emergence. Plant growth regulator was usually applied around the first node detectable growth stage, the most effective timing for reducing the risk of subsequent lodging. The 3.6 herbicide active ingredient applications were, on average, applied as 2.8 products in 2.1 sprays. Common timings included pre-emergence applications of soil-acting, residual herbicides in the autumn and/or sprays during stem extension. Other potential timings, however, included the late control of mature cleavers (*Galium aparine*) and chickweed (*Stellaria media*) with fluroxypyr and metsulphuron-methyl, respectively, until flag leaf emergence. Pre-harvest applications of glyphosate were also recommended for the control of common couch (*Agropyron repens* syn. *Elymus repens*). A single application of insecticide was often applied in the autumn or early summer to reduce virus transmission or direct damage by aphids, and/or a molluscicide to control slugs.

Another major feature of intensification of wheat in the UK has been the trend towards earlier sowing of winter wheat in autumn, with optimum yields being achieved by sowing in September rather than October (Fielder, 1988). Sowing earlier increases the risk of weeds, pests and diseases. As with the adoption of higher nitrogen fertilizer rates, therefore, earlier sowing has increased the requirement for agrochemicals.

Intensification, as evidenced by an increase in nitrogen fertilizer application rate in particular, has been observed in many of the major wheat growing areas of the world. However, many such areas have lower potential maximum yields due to constraints imposed by the climate, or have not perhaps received the same degree of price support for wheat grain. In the USA, for example, the average rate of nitrogen fertilizer on winter wheat doubled from the mid-1970s to the early 1990s. By 1992, however, the rate was still only estimated to be 72 kg N ha^{-1}, only 40% of that used in the UK. Nitrogen usage in other countries can be much lower. For example, nitrogen fertilizer application rates in Canada were only estimated to average 45 kg N ha^{-1} in 1993 (K. Isherwood, International Fertilizer Association, Paris, 1995, personal communication), while in Australia an average rate of between only 2 and 3 kg N ha^{-1} was reported in 1989 (McDonald, 1989). Of course, large regional variations in fertilizer practice can occur within individual countries.

Implications of intensification

Most of the substantial increases in global food production over the past 25 years have come from farming favourable areas more intensively, particularly in the developing world (CIMMYT, 1995). This intensification, for example in the Punjab of India, has been based largely on the increasing adoption of higher yielding, earlier maturing, modern varieties of wheat and rice; the greater availability and increasing use of chemical fertilizers (Borlaug and Dowsell, 1993); development of irrigation (Sidhu and Byerlee 1992); and the increasing adoption of new and more productive crop rotations. The exploitation of genetically improved varieties with increasing fertilizer use, for example, has provided essential prerequisites for the development of a new rice–wheat rotation in South Asia during the last 30 years. This now covers 13 million ha of the best agricultural land and feeds some 150 million people (CIMMYT, 1995). Not only has more food been produced but the real price of food for the rural poor has gone down, in particular for wheat consumers in India (Sidhu and Byerlee, 1992), as a result of this intensification. As well as the alleviation of hunger, poverty has declined in absolute terms – reducing the pressures on the natural resource base (CIMMYT, 1995). It has been argued that 40 million more hectares of farmland would be required to feed the population of India if farmers were still growing varieties of wheat from the early 1960s (CIMMYT, 1995). Land-saving and poverty alleviation resulting from modern wheat variety adoption has wide-ranging socioeconomic and environmental consequences. Jobs have been created from increasing agricultural productivity and more fragile marginal agricultural land is conserved to a greater extent. The encouragement of wheat growing in non-traditional areas, such as the Sudan, has increased family income, domestic household activities, education, social activities and gainful non-agricultural activities, and has reduced the number of farmers migrating to other professions (Ali, 1984).

In a review of the Green Revolution in Bangladesh, however, Alauddin and Tisdell (1991) reported that rising cereal production crowded out the growing of other crops. Production of pulses, fruit and spices has declined resulting in lower protein intake and a nutrient imbalance in the diet of some rural peoples. The proportion of pulses to cereals in diets in India decreased, on average, from 1 : 5 to 1 : 10 over the period of 1950–1977 (Roderuck and Fox, 1987). Protein in the 1 : 10 pulse : cereal diet is satisfactory for adults, but slower growth of children occurred at ratios greater than 1 : 4. Reduced lysine intake is a particular problem for growing children requiring a greater intake of, sometimes unavailable, alternative foods such as eggs, fish, milk and meat. This dietary impact of the Green Revolution has been called the 'protein gap' (Micke, 1983). It has encouraged a shift in plant breeding and agronomic efforts towards increasing protein concentration and quality in cereals.

Sustainability problems associated with the increasing adoption of modern varieties and higher fertilizer usage have been revealed in irrigated farming systems in Pakistan, which are thought to contribute to the unexpected stagnation of yields of wheat over the past two decades (Byerlee and Siddiq, 1994) despite substantial increases in fertilizer use. The negative factors detracting from yield improvement are thought to include soil structure, soil health problems and nutrient availability. Late sowing of wheat in the cotton–wheat and rice–wheat rotation in the Punjab region of Pakistan is also thought to be a contributory factor. Sodicity from poor quality tubewell water, an inefficiency of fertilizer use, increased weed problems (especially *Phalaris minor*) and disease losses could also be contributing to the yield gap between actual and predicted yields. Long-term studies in India on modern variety adoption have revealed a yield decline in rice, but not wheat (Byerlee, 1990). However, the wheat yields (at 4 tonne ha^{-1} or less) are often low considering the reported high levels of inputs and management. These low yields can sometimes be improved by application of micronutrients, higher levels of macronutrients and organic matter – suggesting a 'nutrient mining' approach in these farming systems (Byerlee, 1990). Closer monitoring of the 'resource base' has been advocated (Byerlee and Siddiq, 1994) with an increased emphasis on more 'sustainable practices' (Byerlee, 1990). These could include better rotations, more organic manure use, green manuring and better water quality. Improved extension and information services are encouraged for the better exploitation of future varieties and intensive husbandry practices to prevent further resource deterioration in South Asia (Pingali *et al.*, 1990).

MacKenzie (1982) argued that only those farmers who could afford to invest in fertilizer, appropriate agrochemicals and irrigation were likely to benefit from Green Revolution developments. This claim has recently been refuted in an analysis of modern rice variety introduction in seven Asian countries (David and Otsuka, 1994). The study revealed that overall income distribution in rural communities was not significantly poorer when both the direct and indirect effects as labour, land and product market adjustments are considered. There are, however, differences in different countries in these measures. The Green Revolution was associated with an increase in employment and significant increase in wage rates in the Punjab area of India during the late 1960s. However, this wage trend has since been reversed (Sidhu and Byerlee, 1992). In the 1970s and 1980s wheat consumers in India were the greatest beneficiaries.

Farmers who did not adopt new Green Revolution technology in marginal production environments are considered to have been disadvantaged when production increases in favoured environments depressed cereal prices. It is accepted that divergent rates of progress between favourable and marginal production environments exist (Morris *et al.*, 1991). Wheat area and yields in low rainfall countries have grown at only half the rate of well-watered countries, and this is also reported to be the case within individual countries, e.g. India. Rainfed wheat in India declined from 9 to 6 million ha, while irrigated wheat increased from 4 to 17 million ha between 1961 and 1986 (Morris *et al.*, 1991). In low rainfall environments the wheat yields remain low, and also tend to be highly variable.

Byerlee (1990) identified four phases of development involving improved seed and fertilizer usage in intensive production systems associated with the Green Revolution. The Pre-Green Revolution phase gave modest food production increases by increasing land utilization, in particular the irrigated area. In Phase 2 technological breakthrough resulting in the introduction of new high yielding varieties, more responsive to inputs which characterized the Green Revolution, dramatically increased productivity per unit of land area. Phase 3 was identified by an intensification of input use, in particular chemical fertilizers, substituting for increasingly scarce land. The present Phase 4 requires improved managerial and information skills to substitute for input use and increase input efficiency. Technical solutions and policy/institutional changes will be required, however, in support of Phase 4, to provide for more sustainable farming systems (Byerlee, 1990).

Increasing wheat production on more marginal land, such as the acidic savannah soil areas of South America, will probably be greatly facilitated by current breeding efforts. Varieties are being bred, in particular, for tolerance of heat, drought, salinity and waterlogging as well as more broad-based disease resistance (CIMMYT, 1995).

Wheat intensification and the environment

In richer countries with favourable climates there is concern about the impact of intensification on the environment, particularly when production of wheat is greater than utilization. Of particular concern is the distribution of nitrogen. Table 1.8 shows that in 1977, UK farmers harvested only 27 kg N ha^{-1} less in the grain than they applied as fertilizer. By 1992 this deficit had increased to 45 kg N ha^{-1}. This is not a direct measure of nitrogen loss as a small amount of nitrogen would have been incorporated in straw, and there are many other nitrogen sources apart from inorganic fertilizers. There is little doubt, however, that the shortfall between nitrogen input into the system, compared with nitrogen out, often increases under intensification. One source of loss is through volatilization of N as ammonia from the soil surface, which may be redeposited elsewhere. If this occurs on a natural or semi-natural ecosystem, the soil acidity (Powlson, 1987) and nitrogen content will be increased artificially and eutrophication of surface waters may ensue (Whitehead *et al.*, 1986). Another source of loss is through nitrate leaching. Nitrate in river and ground water supplies has increased in the UK since the 1940s (Tivy, 1990) and, as with ammonia volatilization, this particularly in the presence of phosphates can lead to eutrophication. There is also public anxiety over the potential health problems of methaemoglobin anaemia in infants and gastric cancer, although the actual risks of these are difficult to quantify. Most nitrate leaching from cereal fields in the UK does not derive directly from unused fertilizer but through the mineralization of organic nitrogen in autumn and winter when soil moisture content rises, temperatures are not too low, plant demand is low or non-existent and the land is often cultivated (Whitehead *et al.*, 1986; Powlson, 1987). Levels of 60–120 kg nitrate N ha^{-1} have been recorded following cereal crops growing on heavy soils in Eastern England (Jenkinson, 1986) and 50 kg ha^{-1} may commonly leach (Jenkinson, 1986) depending particularly on the amount of overwinter drainage (Powlson, 1987). In Denmark and Sweden there has been a close correlation between nitrogen application and leaching, particularly as applications rise above 100 kg N ha^{-1} (Joelsson and Petterson, 1984). Similar results have been reported for The Netherlands and Nebraska, USA (Kolenbrander, 1983).

Also of concern is the potential of agrochemicals to pollute and accumulate in the environment. Pollution incidents of water by crop protection chemicals can be caused through drift from crop spraying, direct contamination of drainage ditches, spillages and inappropriate storage and disposal (Anon., 1992b). The number of reported water polluting incidence averages about 50 per year in the UK (Anon., 1992b). The environmental impact is difficult to assess due to the wide range of active ingredients used, factors involved in chemical mobility and degradation, susceptibility of non-target organisms, and the complexity and variability of ecosystems. Also of importance is the more diffuse pollution, often arising as pesticides move from agricultural soils following good practice during application (Fawell, 1992). The environmental implications of this pollution are difficult to assess, although a repetition of well-documented problems arising from the widespread use of persistent chemicals such as the organochlorines (e.g. McEwen and Stephenson, 1979) becomes less likely as chemical registration procedures become more stringent in many countries. What is evident, however, is firstly the health threat perceived by the public (Slovic, 1984), particularly from contaminated drinking water (Fawell, 1992), and secondly the expense necessary for water authorities to comply with legislation for maximum admissible concentrations for pesticides in drinking water (Croll, 1987).

Agrochemicals also have both direct and indirect effects on fauna and flora. In the UK, for example, the use of herbicides has been implicated in the decline of at least five scarce species of arable plants, although other factors of intensive farming, such as high nitrogen rates and seed cleaning, have probably also contributed (Cooke and Burn, 1995). Some herbicides can also be toxic to earthworms and beneficial arthropods. Indirect effects of the intentional reduction in numbers of plants and plant species in a field can, however, have a greater impact on the fauna. A well-documented example in the UK is the decline of the grey partridge (*Perdix perdix*). Poor chick survival appears to have followed substantial reduction in food, which for the first six weeks of their life consists of invertebrates dependent on weed species. A decline in other farmland bird species such as the corn bunting (*Miliaria calandra*), linnet (*Carduelis cannabina*) and tree sparrow (*Passer montanus*) appear, at least in part, to be due to reduced availability of weed seeds. Allowing particular weeds to grow in headlands through reduced or selective herbicide management has increased populations of butterflies, true bugs (Hemiptera) and carnivorous ground beetles (Carabidae). The value of biodiverse headlands through specific species plantings and appropriate management for encouraging beneficial arthropods has recently been confirmed in integrated farming studies (Hart *et al.*, 1994). Increased weed seed abundance may also favour foraging by several mammals, including wood mice (*Apodemus sylvaticus*). Intensive use of some fungicides has been implicated in reduced earthworm populations, while the use of broad-spectrum insecticides can have detrimental effects on beneficial arthropods and other insectivores.

Another important impact of intensification upon the environment is its impact upon energy usage. The support energy for agriculture, i.e. the sum of all non-renewable energy inputs (White, 1981), received considerable attention following the 1973 energy crisis (Blaxter, 1975; White, 1979, 1981; Lewis and Tatchell, 1979) primarily due to the economic consequences of dependency on imported and limited sources of fossil fuels. In the 1990s the supply of support energy is still dominated by the burning of fossil fuels. In addition to economic consequences, there has been increasing awareness of the implications of the attendant production of greenhouse gases, especially carbon dioxide. Ozone depleting and acid rain producing gases may also be significant. The support energy requirements for agriculture are difficult to assess accurately, but approximate values of 75, 14, 8 and 250 MJ kg^{-1} are thought to be necessary in the production of N, P_2O_5, K_2O and crop protection chemicals, respectively (Gooding and Alliston, 1993). The use of these estimates suggests that support energy just for these inputs on wheat rose by 55% between 1977 and 1992 in the UK (Table 1.8). When all support energy requirements are included, i.e. when consideration is given to the fuel necessary to power and make farm machinery and crop handling equipment, values can approach 36 GJ ha^{-1} in the UK (Gooding and Alliston, 1993).

In specific regions of the world, intensification of production often involves increased water use and irrigation systems. Excessive exploitation not only reduces the water available for other essential uses, but can also lead to the raising of levels of salinity to the root zone. Additionally, inappropriate soil management leading to erosion can greatly reduce the sustainability of the system.

Extensive Production Systems

In many countries, therefore, there have been growing pressures to develop lower-input wheat production systems. Following the rapid increase in the use of nitrogen fertilizer and agrochemicals in the late 1970s and early 1980s, UK farmers have gradually reduced the levels of synthetic inputs (Table 1.8). A survey has revealed that 25% of UK respondents had reduced fertilizer and spray application over the previous five years (Fearne and Warren, 1993), while 10 and 23% of respondents said that they would further cut future fertilizer and spray applications, respectively, in response to lower crop values. In these circumstances farmers seek alternative ways of maintaining or improving financial returns and to modify their production systems accordingly (Agricultural Development and Advisory Service, 1994).

Consumer preference can also impact upon production systems. A willingness and ability to pay more for wheat which has been grown in a particular, less intensive way, can compensate the farmer for reductions in yield and ensure a market for the grain. This has contributed to the adoption of, for example, organic or biological systems by many specialist growers which do not permit soluble fertilizer use and most pesticide treatments (e.g. UKROFS, 1992). In the UK, wheat is the single largest organic commodity and is supported by strong demand for organic bread. The number of organic wheat growers increased dramatically during the 1980s (Yarham and Turner, 1992) encouraged by organic wheat often attracting twice the price of intensively grown wheat. In some studies of wheat production on the Great Plains of the USA, the premiums available to organic farmers together with the reduced cost of production have appeared to result in greater net returns than those possible with more conventional systems (Smolik and Dobbs, 1991). The authors also note that returns on systems not including commercial fertilizers or pesticides were less variable than the returns from more conventional systems. This is consistent with the observation that lower-input systems of production can be less vulnerable to economic loss under adverse growing conditions (Davies and Gooding, 1995), particularly in drought prone areas.

The third influence on farmers, causing a shift to less intensive strategies, is that through legislation and national schemes. Nitrogen applications might, for example, be limited particularly at certain times of the year for farmers in regions where nitrate movement into water supplies is particularly likely. Current schemes that affect fertilizer and farm manure use in the UK impact upon farmers under agreements in nitrate sensitive areas and for all farms in nitrate vulnerable zones. Garstang *et al.* (1994) concluded that in future all UK farmers will need to pay close attention to correct nitrogen use to meet pollution control requirements whilst maintaining profits.

Increasing interest in reducing synthetic inputs for cereal production, particularly in Europe, is therefore driven by intensified political and economic pressures, together with consumer and environmental concerns (Murphy, 1991; Jordan *et al.*, 1993; Holland, 1994). A number of experimental farming systems to investigate the long-term sustainability of integrated arable farming practices have been established in western Europe since the late 1970s (Holland, 1994).

In many other areas, where the climate and/or the pricing structures do not support intensification, wheat is grown in relatively extensive production systems as the norm. For example, low-input cropping systems are widely adopted for economic reasons in certain environments where land is abundant and rainfall is limited (Wagstaff, 1987). As stated earlier, wheat production in Australia is generally achieved with very low levels of nitrogen fertilizer. Although yields per hectare in Australia are commonly less than a third of those achieved in the European Union, production has been broadly profitable within a low-input–low-yield system despite falling real prices. Production costs have been cut by investing in larger machinery on larger farms with production extending into lower rainfall areas (Hall, 1981; Wagstaff, 1987).

Jordan *et al.* (1993) argue that to reduce agrochemical inputs, whilst maintaining profitability, quality production requires a systems approach in line with that defined by the International Organization of Biocontrol (El Titi *et al.*, 1993), i.e. 'An holistic pattern of land use, which integrates natural regulation processes into farming practices to achieve maximum replacement of off-farm inputs and to sustain income'. They also list a number of principles which are required to underpin these so-called integrated farming systems which are summarized here:

1. The production system and management requirements must be consistent with site ecology and thereby governed by geographical, climatic, geological and hydrological parameters of the area.
2. The pattern of land use must be influenced by regional constraints such as soil erosion and water conservation.
3. Establishment, maintenance and enhancement of natural habitats and species diversity of flora and fauna are desirable, and 3–5% of the total cropping area should be kept for non-agricultural vegetation.
4. Sustainability of soil fertility and of its non-agricultural functions is crucial while plant nutrients applied should not exceed nutrients harvested or otherwise exported.

5. Pest, disease and weed control should be considered only when levels exceed economic thresholds.

Integrated farming revolves around the choice of a balanced crop rotation which can potentially reduce pest, disease and weed problems, whilst hopefully maintaining soil structure and fertility. A rotation of at least four different crops is recommended, with a maximum of 50% cereals. Non-inversion tillage is encouraged to help preserve soil nitrogen, improve soil structure, and to encourage beneficial arthropods. The use of resistant varieties is also a key component, reducing the potential need for agrochemicals (Jordan and Hutcheon, 1994; Davies, 1995).

Rotations

Crop rotation is the regular and orderly change of different crops, in a time sequence, on a given location. Reducing dependence on inorganic fertilizers and chemicals increases the importance of rotation. Of particular impact in less intensive systems is the increased frequency of legume crops such as the pulses (beans, peas, lentils, etc.) and forage legumes (clovers, alfalfa (syn. lucerne), lotus, vetch, etc.) which can increase the availability of nitrogen to wheat grown in the same rotation. Hence, in organic systems, arable crops are often rotated with legume-rich grasslands which in the USA typically account for 30–50% of the farm area. In Europe, typical ratios of grazing to arable of 2 : 1 have been reported for organic farms (Wagstaff, 1987).

The rotation should also vary the cropping season and competitive environment so as not to favour the proliferation of one group of weeds. Traditionally, it has also been common to rotate *cleaning* crops with *dirty* crops. The cleaning crops being those which were grown on relatively widely spaced rows, such as turnips, sugar beet and potatoes, so as to allow easier mechanical inter-row weed control with cultivations. Dirty crops would include cereals in which it is more difficult to achieve weed control in this way. Rotations which have a high proportion of cereals, including wheat, can encourage grass weed problems. In the UK, for example, increased cereal growing, especially autumn sown cereals, has been associated with a higher incidence of wild oats (*Avena fatua* and *A. ludoviciana*), blackgrass (*Alopecurus myosuroides*), sterile brome (*Bromus sterilis*) and couch (*Agropyron repens*). Spring cropping is an effective method of reducing the occurrence of the autumn germinating blackgrass and cleavers. Rotation is also important for reducing the incidence of diseases, particularly those which survive in soil and crop debris. Of particular note is take-all (*Gaeumannomyces graminis* var. *tritici*) which can increase rapidly in successive crops of wheat in temperate climates and neutral to alkaline soils, but is reduced to minimal levels after a one year break from a susceptible crop.

Intercropping

One way of increasing the non-wheat component of wheat rotations is to intercrop the wheat with other crops such that they are growing in the same field at the same time. Total productivity of the land can be improved because: (i) the different crops may exploit different ecological niches allowing better utilization of nutrients and light; (ii) the alternation of susceptible and resistant plants in the field may reduce the spread of certain pests and diseases; (iii) the increased number of crops can increase the diversity of habitats for beneficial arthropods; and (iv) the risk of failure of both crops is less than the failure of one crop grown as a mono-culture. The inclusion of a legume can also increase the nitrogen input to the rotation and, if a pulse, may help alleviate the deficiency of protein in wheat based diets.

Variety choice

Using disease resistant varieties of wheat which are competitive against weeds is also of increased importance when fungicide and herbicide use is reduced. Traditional varieties with longer stem lengths may have a role when their possibly better ability to tolerate or compete with weeds may be valued (Balyan *et al.* 1991; Thompson *et al.*, 1992; Gooding *et al.*, 1993a; Cosser *et al.*, 1995) and the risk of lodging is reduced due to lower nitrogen availability. In one study, however, competition of wheat genotypes with annual ryegrass was not associated with crop height, rather competition was related to early root development (Reeves and Brooke, 1977). Taller varieties may, however, be less influenced by moisture stress than shorter ones (Thompson *et al.*, 1993b). In UK variety × nitrogen field experiments between 1982 and 1992 (Foulkes *et al.*, 1994), older varieties were often better at acquiring soil nitrogen, and more modern varieties, tended to require more fertilizer nitrogen than older varieties to reach their maximum yield. However, other characteristics of modern varieties such as the more efficient interception and use of light, and improved harvest indices, are important for extensive as well as intensive systems.

Soil tillage

Using modern machinery advances it is possible to establish wheat with reduced levels of soil disturbance. Reduced tillage can be an important method of conserving soil moisture and nitrogen; increasing soil biological activity and encouraging beneficial predators of key pests; reducing energy requirements; and reducing soil erosion. As a result some of the requirements for synthetic inputs are reduced (Radford *et al.*, 1992; Jordan *et al.*, 1993). Against this, however, relatively rigorous tillage systems, such as those relying on primary cultivations of complete soil inversion by ploughing, have been traditionally used to loosen soil, incorporate crop residues, and bury weed seeds of a number of important species below the depth from where they can successfully establish. The adoption of minimum soil cultivation and direct drilling (no till) systems has, for example, encouraged blackgrass and sterile brome infestations in the UK. Rotational ploughing to establish larger seeded crops (e.g. field beans) in cereal rotations is sometimes adopted as a cultural control in support of herbicide treatment in grass weed management strategies.

CHALLENGES FOR FUTURE PRODUCTION

Increasing Productivity

Despite the substantial increases in wheat production that have been achieved over the last 50 years, further increases are required to meet the demands of future population growth. An average projection is for world population to reach 6.2 billion by the year 2000 and about 8.3 billion by 2025, before possibly stabilizing at about 10 billion towards 2100 (Table 1.9). With constant per capita food consumption, therefore, yields need to be 57% higher in 2025 compared with 1990. Recent history

Table 1.9. Projected increases in world population by area (From Borlaug and Dowsell, 1993).

Region	Population (millions)			% Increase 1990–2025
	1990	2000	2025	
Low- and middle-income economies				
Sub-Saharan Africa	495	668	1229	148
East Asia and Pacific	1577	1818	2276	44
South Asia	1148	1377	1896	65
Europe	200	217	252	26
Middle East and N. Africa	256	341	615	140
Latin America and Caribbean	433	515	699	61
Sub-total	4146	4981	7032	70
Other economies (includes former USSR, Cuba, N. Korea)	321	345	355	11
High-income economies	816	859	915	12
World	5284	6185	8303	57

suggests that wheat production could match this rate of increase, at least over the next 10–15 years, as the average annual rate of increase in production over the 17 years previous to 1993 was 3.3%. This is greater than the population growth rates projected for the major wheat producing areas such as Europe, North and Central America, and Asia (Table 1.9). Whether or not the average rate of increase in production can be maintained, however, is difficult to determine. Although the average rate of increase in world production since the 1940s has been impressive, the rate of increase has varied. The maximum rate of increase occurred between 1976 and 1985 when wheat production increased at over 14 Mt year^{-1} (Fig. 1.24). In the ten years to 1996, however, the rate had slowed significantly to only 5 Mt year^{-1}, a rate slower than at any time since the mid-1960s. Rates of consumption increase were higher than that of yield in the years to the early 1990s causing a reduction in stocks which were falling at an average rate of 5 Mt year^{-1} in the ten years to 1996. Others have also noted a reduction in growth rate of wheat production throughout the

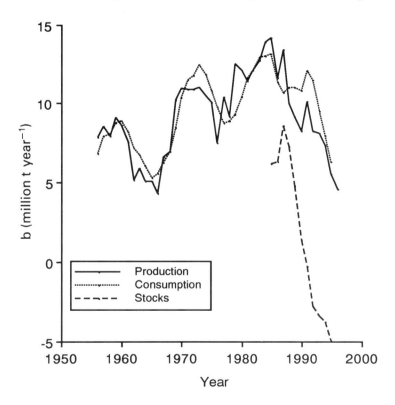

Fig. 1.24. Trends in world wheat production, consumption and stocks. Linear regressions used to estimate average rates of change (b) over the previous ten years.

world, particularly in developing countries (CIMMYT, 1993), and the United States Department of Agriculture has estimated or predicted falls in total world production from 1992 to 1995 relative to the peak yield achieved in 1991. Recent reductions have resulted from unfavourable weather conditions in major growing areas, as well as reduced production from the former USSR. Both these factors, however, may be transitory and further increases may be driven particularly by China which, since the fragmentation of the USSR, has become the world's leading single country for wheat production and where there are continuing strategies for increasing production in line with population growth. Virtually all Chinese farmers now use chemical fertilizers and by 1993 China was the world's largest producer, importer and consumer of nitrogen fertilizer reflecting a huge investment in chemical fertilizer facilities (Borlaug and Dowsell, 1993).

Past trends in wheat production do not, therefore, justify complacency. Despite the large increases in yield over the last 50 years, many peoples are still in deficit due to economic and political barriers to even distribution. In many of these areas (e.g. Africa) production is far lower than demand. If diets are to improve among the 1 billion hungry poor, food demand could increase by 100% between 1990 and 2025. To meet the projected food demands, Borlaug and Dowsell (1993) estimate that yield of cereals must be increased by 80% between 1990 and 2025. As most of the opportunities for opening new agricultural land have already been exploited, particularly in the densely populated regions of Asia and Europe, the vast majority of production increases must continue to derive from increased yields per unit area. It is estimated, therefore, that average wheat yields must rise to 4.4 t ha^{-1} by 2025 from the 2.4 t ha^{-1} achieved in 1990 (Borlaug and Dowsell, 1993). Achieving this level of production will have large implications for the environment. The increases in wheat production and yield in previous years have come about with heightened pressures and impact on the environment through increased support energy requirements, particularly in terms of synthetic nitrogen fertilizers and agrochemicals, and inappropriate soil management. Techniques need to be developed to minimize potentially harmful practices and to use synthetic chemicals and fuel as efficiently as possible. Further improvements will, however, almost inevitably depend on increased support energy use, although substantial gains through improved breeding techniques are likely to contribute.

Improving Quality

For chemical and other inputs to be used most efficiently it is essential that the wheat itself is closely suited to its intended end-use. This is true even in areas such as Africa where total production is in large deficit compared with demand. One of the reasons is that such areas have typically become accustomed to receiving imported wheat, frequently as 'food aid', from North America which is often of very high milling and bread-making quality. Local processors such as millers have adapted to utilizing high quality wheat and are less willing to accept locally grown wheats if these are of inferior character. Additionally, in many urban areas, the consumption of leavened bread is increasing because of the rising incomes of some consumers, and the interest in convenience foods and a diversified diet. Factors such as favourable government pricing policies providing subsidies for bread and the status symbol of leavened bread are also increasing demand. In some areas, such as Mexico until 1990, the price of wheat received by producers has been fixed regardless of quality, so farmers have previously had little incentive to exceed minimum quality standards. However, the liberation of markets, for example under the NAFTA and the General Agreement on Tariffs and Trade (GATT) has meant that locally produced wheat must increasingly compete with imported grain. Quality has become an important issue in this competition and variety choice, in particular, is being given more emphasis when marketing grain (Byerlee and Moya, 1993).

In areas where production has increased rapidly but population levels have remained relatively static, particularly in western Europe, farmers are increasingly under pressure to find outlets for their wheat and meet ever more stringent quality requirements for a range of markets. Understanding and meeting market specifications is, therefore, becoming increasingly important (Garstang *et al.*, 1994). Governments of some major exporting countries, such as Canada and Australia, actively maintain grain quality and promote exports by regulation of varieties and cleaning, and setting standards for other aspects such as protein content (Mercier and Hyberg, 1995). Quality differences are being actively exploited to promote sales of wheat on world markets. Even the world's largest exporter, the USA (40% share of wheat traded on the world market), which has relied heavily on financial incentives to expand its market share, is now under pressure to respond to the importance of quality as a source of competition (Webb *et al.*, 1995).

Farmers everywhere are consequently under increasing pressure to produce wheat to meet specific market and/or nutritional requirements, whilst utilizing inputs in the most economically efficient and environmentally sustainable manner. It is imperative to link production system factors, such as climate, soils, genetics, sowing methods, crop nutrition, water management, crop protection, harvesting and storing, directly with the requirements for end-use whilst assessing the environmental impact of the various strategies.

2

Grain Characters Influencing Utilization

Requirements for any particular wheat sample are governed by the market and intended utilization. The quality criteria for some markets are diametrically opposed to those of others. There is little possibility of growing 'quality wheat' with an all embracing combination of characters, only the ability to grow wheat in a way which makes it suitable for a particular end-use. The nature of grain quality, therefore, needs to be defined and criteria specified in relation to markets. The criteria used to grade wheat include physical, chemical and biological properties which influence the income of producers, cereal trading prices and end-user value (Mercier and Hyberg, 1995). It is the purpose of this section to review assessments made on grain which will define potential utilization and value.

GRAIN MOISTURE CONTENT

The moisture content of grain is of great importance for all users of wheat. Composition percentages, such as protein or starch content, are inversely related to moisture contents. Grain in the UK is often bought and sold on the basis of yield adjusted to a specific moisture content (mc) of often 14 or 15%. This value varies from country to country reflecting the common moisture contents of grain at harvest and/or moisture contents required for safe storage. Grain in Australia, for example, can be bought and sold on the basis of 11% mc (Brennan, 1990). Sliding scales regularly apply to grain with lower moisture contents than that specified to reflect the amount of dry matter in a given weight of grain. Excessively dry grain, however, often raises suspicion concerning the extent of drying and the possibility of excessive temperatures having been used which could have damaged the grain protein. Usually of more importance, however, is the presence of too much moisture. This may preclude sale as relatively wet grain is difficult to store safely as it will lead to increased respiration of grain, and pest and fungal problems. For example, between August 1985 and June 1986, 1328 UK grain loads were rejected at EU intervention stores on the basis of excessive moisture. This represented nearly 40% of all rejections or nearly 2% of all deliveries. Grain respiration equates to a direct dry matter loss from the wheat and also produces heat and water which increase the activity of pests and diseases. In severe cases the heat build up can result in charring of the wheat, and the increased moisture will cause caking and sprouting. Grains containing excessive moisture also soon loose their viability and therefore their value for seeding. Respiration in large, poorly aerated bins can lead to oxygen poor conditions, favouring the development of anaerobic bacteria. This can lead to the death of the embryos which become discoloured (Carter and Young, 1945). Such grain has a high acidity caused by the breakdown of fats in the embryo to fatty acids. This so-called *sick* grain has poor bread-making potential.

The moisture content required for safe storage depends greatly on temperature with grain being at risk from insect spoilage at temperatures above 23°C even when the moisture content is as low as 5%. Conversely, grain can be stored safely at moisture contents at and above 20% if the temperature is below 5°C. Under UK conditions, however, it is advised to store grain at no more than 15% moisture at temperatures at or below 15°C to give relatively low risk storage for 32 weeks under practical farm conditions, although there may still be a risk from mites. In the warmer conditions of Australia and the USA, moisture contents of 12 and 13% are more common. Large numbers of broken grain will make it more important to store at lower moisture contents and temperatures as damage exposes the endosperm to endogenous enzymes, pests and fungi, generating heat and moisture more readily than would occur with intact grains. In tropical conditions, such as South and Southeast Asia, seed needs to be stored throughout the hot and humid monsoon season at 20–25°C with a relative humidity of 75–100%. In these conditions it is practically impossible to store wheat without repeated sun drying and spraying agrochemicals (Clements, 1988).

Moisture content can be assessed in a number of ways. The most direct entails weighing samples before and after oven drying. The technique described by the International Standards Organization (ISO 712 1979) and adopted by the British Standards Institute (BS4317) involves drying a 5 g sample of milled grain for 2 h at 130–133°C. Alternatively, the UK livestock feed trade dry coarsely ground grain at 105°C for 4 h. Unfortunately, however, the two methods do not give the same results (Stenning and Channa, 1987). Other oven dry weights can be determined in a large number of ways where a reduction in temperature is compensated for by an increase in time. Drying at reduced temperatures may be particularly necessary where it is intended subsequently to measure amounts of relatively volatile compounds. More rapid and convenient measures are commercially available working on varied principles including electrical conductivity, electrical permittivity or the dielectric constant, gravimetric methods, near infrared reflectance (NIR) and nuclear magnetic resonance (NMR) (Stenning and Channa, 1987). Many of these techniques are used by farmers and users of grain. They are not, however, direct measures of moisture and results sometimes need careful interpretation.

SAMPLE PURITY

Purity has a direct relationship with the value of wheat samples for all end-uses and many markets will specify a total admissible level of admixture or impurity according to equation 2.1. Such levels commonly range between 2% for many milling and export markets (Dickie, 1987; Anon., 1995a), up to 12%, i.e. what was specified for acceptance by intervention for the European Union as common wheat in 1995.

$$\text{Impurities (\%)} = \left[1 - \left(\frac{\text{Total weight of sound grain of specified variety}}{\text{Total weight of sample}} \right) \right] \times 100 \quad [2.1]$$

A typical procedure involves passing 250 g of grain sample over grain sieves with slots 3.5 and 1.0 mm wide for 30 s each. Large grain remaining above the 3.5 mm sieve are returned to the sample but all other such material remaining above the 3.5 mm or passing through the 1.0 mm sieve are weighed as impurities. A subsample of the remaining grain (50–100 g) is then weighed and examined and broken grains, other cereals, sprouted grain, grains damaged by pests or frost, grains in which the germ is discoloured, weed seeds, ergot, damaged or decayed grains, stones, sand, husks and live and dead pests are identified, removed and weighed. The subsample is then passed over a sieve with slots 2 mm wide and shrivelled grain passing through this will be added to the impurities.

Contaminants of Grain Origin

For many end-uses the presence of certain impurities may be much more significant than just their contribution to total admixture as they will affect other quality characteristics. For example, impurities can include material such as shrivelled grain which will lead to poor flour extraction; grains with discoloured germs which will lead to tarnished flour; grains overheated during drying such that proteins are denatured; broken grain which can lead to poor storage properties; sprouted grain which will lead to poor baking quality; and grain of other varieties or cereal species not suited to end-use. These impurities will have a relatively large impact on wheat destined for milling compared with livestock feed, as nutritional quality is not necessarily impaired to the same extent. It should be remembered, however, that in areas where grain is not in surplus, impurities of grain origin that are screened out during cleaning by the end-user can have significant value. Such screenings can be added to ground, mixed livestock feeds or sold on a guaranteed chemical analysis basis without specifying the individual ingredients (Martin *et al.*, 1976). Brennan (1990), for example, calculated that a reduction in impurity from 2 to 1% would be worth 0.22% in extra financial value to Australian millers if the impurity could be sold for livestock feed. If the impurity was valueless, however, the same increase in purity would increase worth by 1%, in addition to reduced cleaning and handling costs.

Grain of contaminant varieties or other cereal species, however, are more difficult or impossible to separate economically. The problem is of particular importance when a variety or species not suited to the end-use is a contaminant. Tests have, therefore, been devised to examine variety purity on the basis of gliadin separation on electrophoretic gels (Cook, 1987a), a technique which has been adopted to investigate purity for seed certification and milling purposes. A more recent development which may have more application in the future is the use of computer image analysis techniques to identify varieties on grain dimensions and shape. A less complicated, but also less specific, method involves soaking grain in a solution of phenol which causes them to be stained according to one of three categories: black, intermediate staining, and light staining. Grains of one variety should all stain to the same extent.

Specific tests have also been devised to detect bread wheat admixture in durum wheat, and bread wheat admixture in spelt wheat. In these cases suspicion of foul play is raised as the contaminating species is cheaper to produce than the species desired for the end-use. Accidental contamination can, however, occur through poor volunteer control, and careless harvesting, storage and transport.

Weed Seeds

The presence of weed seeds receives particular attention for wheat being considered for seed certification. Of special importance are those weed seeds which are relatively similar in dimensions and/or ripen at the same time as the crop, so that they are harvested with the grain and difficult to separate from it. Weeds that fall into this category can achieve very widespread dispersal as wheat seed is moved from farm to farm and even country to country. A good example is common wild oat (*Avena fatua*) which is a noxious weed in many of the cooler, temperate wheat growing areas of the world. Once introduced, wild oat is difficult to eradicate because the seeds shatter readily and because many of the seeds are ploughed in, where they can lay dormant for many years, and then germinate and grow when they are turned up near the surface. The closely related winter wild oat (*A. ludoviciana*) and wild red oat (*A. sterilis*) can also cause similar problems in areas of, for example, the UK and USA, respectively. Wild oat contamination, therefore, will preclude the certification of samples intended for sale as seed (e.g. Table 2.1). Farmers should also be aware of the risks of using home-saved seed, which often contains far more weed seeds than certified seed. A survey of farm-saved seed in

Table 2.1. Quality for UK seed certification.

	Basic seed	Certified seed 1st generation	Certified seed 2nd generation
Higher voluntary standard			
Maximum no. of other seeds allowed in a 1 kg sample:			
All other species[a]	1	2	4
Other cereal spp.	0	1	3
Non-cereal spp.	1	1	2
Wild oats	0	0	0
Ergot pieces	0	1	1
Analytical purity (%,w w⁻¹)	99	99	99
Germination (%)	85	85	85
Moisture content (%)	17	17	17
Loose smut (%)	0.1	0.2	0.2
Minimum standard			
Maximum no. of other seeds allowed in a 500 g sample:			
All other species[b]	4	10	10
Other cereal spp.	1[c]	7	7
Non-cereal spp.	3	7	7
Wild oats	0[d]	0	0
Ergot pieces	1	3	3
Analytical purity (%,w w⁻¹)	99	98	98
Germination (%)	85	85	85
Moisture content (%)	17	17	17
Loose smut (%)	0.5	0.5	0.5

[a] Special standards apply to seeds of corn cockle, couch, sterile brome and wild radish.
[b] Special standards apply to seeds of corn cockle and wild radish.
[c] A second seed will not be counted as an impurity if a further 500 g is found to be clear.
[d] One seed will not be counted as an impurity if a further 500 g is found to be clear.

the UK between 1979 and 1981, for example, revealed 12% to be contaminated (Tonkin, 1987). Even low levels of contamination can be serious. One wild oat seed in 1 kg of wheat only represents a ratio of 1 : 20,000 in terms of wild oat : wheat grains. At average sowing rates in the UK, however, using this grain will still result in about 150 wild oats being sown per hectare.

Fungal Contaminants

For human and livestock consumption, grain needs to be free of ergots. Ergots are the sclerotia of *Claviceps purpurea* (Fig. 2.1) which attack rye and triticale as well as wheat, and occur in all major wheat growing areas of the world. The sclerotium is a hard purple–black mass which develops in the ear in place of a grain. When the sclerotia are shed or 'sown' onto the soil they overwinter and germinate in spring to produce perithecia

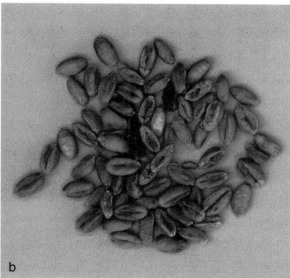

Fig. 2.1. Ergot in wheat. (a) In the ear, fourth spikelet from the base on the right. (b) In a grain sample.

which release ascospores. These are carried on the wind and penetrate the ovaries of susceptible florets. After a few days conidial spores are formed in honeydew which can cause secondary infections following splash, wind or insect dispersal. The infected ovaries enlarge and are converted into sclerotia. Wheat samples containing more than 1% ergot can cause serious illness, and sometimes death, due to the presence of poisonous alkaloids including ergotamine, ergosine and ergotoxine. The level of ergot contamination is, therefore, strictly controlled and 0.05% is the level that has been used for bread-making wheat offered to EU intervention (Scudamore, 1993). For standard seed certification in the UK there must be less than three ergot pieces in 500 g of seed or, for the higher voluntary standard (HVS), it must be fewer than three. As well as toxic effects, ergot contamination will also reduce flour extraction rate and cause flour discoloration, but effects on dough rheology and loaf volume are only slight.

Other mycotoxins, which can cause a variety of human and animal health problems at very low doses of parts per million or billion, can be produced by fungi infecting wheat in the field, particularly by *Fusarium* spp. causing ear blight or scab, such as *F. culmorum, F. avenaceum, F. graminearum* and *F. poae.* Infection occurs most readily at anthesis as it appears to be the anthers themselves which are the initial site of colonization (Parry *et al.*, 1995). Symptoms are usually first seen as small brownish spots at the base or middle of the glume. These then spread in all directions and a pinkish fungal growth may be seen along the edge of the glumes or at the base of the spikelet. Mycotoxins produced by *Fusarium* include zearalenone (known as giberella toxin) and deoxynivalenol (DON). These have been implicated in the development of alimentary toxic aleukia in humans, vaginal prolapse and vulva vaginitis in pigs, and stunted growth and poor feathering in poultry. The associated health risks should, therefore, preclude grain infected with *Fusarium* for use in baking and livestock feed.

As well as the release of mycotoxins *Fusarium*-infected grains shrink and become grey/brown with a floury discoloured interior (Parry *et al.*, 1995). Infection of the ear with races of *Fusarium graminearum* and *F. culmorum* causing blight or scab can, therefore, lead to reduced thousand grain weight, germination percentage and yield of nitrogen. Grain protein concentration can be increased but infection can alter the amino-acid composition in the grain and, therefore, protein quality (Szunics *et al.*, 1987). Bechtel *et al.* (1985) found that grain infection by *F. graminearum* destroyed starch granules, storage proteins and cell walls leading to reduced seed germination and vigour. In further studies *F. graminearum* decreased the proportions of albumin and glutenin proteins by 33 and 80%, respectively (Boyacioglu and Hettiarachchy, 1995). *Fusarium* infection will also influence seed crop quality as contaminated seed can pass infection to seedlings causing damping off or the development of foot rot.

Alternaria spp. also produce mycotoxins in culture (e.g. alternariol) but little is known about the occurrence of these in field grown cereals. This group of species have, however, been most commonly associated with blackpoint, i.e. brown and black discoloration of the embryo and crease regions of wheat grain. In the warmer cereal growing areas, the same symptoms can be associated with *Cochliobolus sativus* (conidial stage *Bipolaris sorokiniana*, syn. *Helminthosporium sativum*) (Zhang *et al.*, 1990), or in wet conditions with the bacterium *Pseudomonas syringae* pv. *atrofaciens* (Wiese, 1987). In recent UK experiments, isolates of *Alternaria alternata, A. infectoria, Stemphylium botryosum* and *Drechslera avenacea* all increased blackpoint severity when sprayed on to ears at GS 75. The same species were also re-isolated from grain at harvest (Table 2.2). Blackpointed grain can give rise to discoloured flour and hence rejection of grain loads for both bread and pasta production (Gooding *et al.*, 1993c). A survey of millers revealed that, on average, 4% of UK milling wheat a year is rejected or purchase price reduced on the basis of blackpoint severity. Conservative estimates, based on grain value reductions of between £5 and £20 a tonne as a result of blackpoint, are that these equate to losses to UK milling wheat farmers of between £1 million and £4 million a year (Ellis and Gooding, 1995). In years when blackpoint has been particularly severe, 15% of UK milling wheat grain loads have been rejected. Blackpointed grain have also been counted as impurities for acceptance into intervention by the EU (Anon., 1992c). Blackpoint associated with grain infection by *C. sativus* can reduce germination percentage but this does not appear to occur with *Alternaria* infection.

The bunt diseases of wheat also significantly reduce the quality of wheat. Common bunt (syn. stinking bunt) occurs throughout the world and is most commonly caused by *Tilletia tritici* (syn. *T. caries*), but also sometimes by *T. laevis* (*T. foetida*). Bunt spores contaminate wheat seed giving it a typical fishy odour and a dull, darkened appearance. The fungi

Table 2.2. The effect of spraying spore suspensions of different fungal species on ears of wheat at GS 75 on blackpoint severity and re-isolation from grain at harvest (from Ellis, 1996).

Treatment	Blackpoint severity (0–90)	Re-isolation frequency at harvest (%)			
		A. inf.	*A. alt.*	*S. bot.*	*D. ave.*
Water	12.4	2	0	2	0
Alternaria infectoria	17.2	62	0	0	0
Alternaria alternata	18.8	0	88	0	0
Stemphylium botryosum	17.8	0	4	82	0
Drechslera avenacea	27.1	0	4	0	84

persist as spores on the seed or in soil, and after germination can penetrate seedling tissues and grow intercellularly within the host until they inhabit the terminal meristematic tissues. Ovaries of infected flower primordia are displaced by spores developing within the pericarp. A *bunt ball* refers to the pericarp and the spores that it encases, often the same approximate shape and colour as a healthy grain (Fig. 2.2). The pericarp is, however, fragile and the bunt balls are easily burst at harvest, releasing spores to infect seed and soil. In severe cases clouds of black spores can be released at threshing which poses a fire risk due to their combustible nature. Where control measures have not been adopted bunt, in particular, can build up rapidly in farm-saved seed, increasing from undetectable levels to complete crop failure within two seasons (Garstang *et al.*, 1994).

Dwarf bunt (*T. controversa*) occurs in areas of North and South America, Europe and central Asia where winter wheat is covered in snow for prolonged periods. Effects on grain are similar to common bunt, but the plants are also severely stunted, and may grow to only 25–50% of normal height. Infection is mainly from soil borne spores which germinate at low temperatures (opt. 3–8°C), often under snow cover, to infect plants particularly at the two to three leaf stage (Wiese, 1987).

Karnal bunt (*T. indica*) occurs in northwest India and adjoining regions of Pakistan and Afghanistan. Teliospores on soil germinate to release wind

Fig. 2.2. Bunt balls (right) compared to wheat grain (left).

dispersed spores which attack spikelets and directly invade the glumes and ovary wall. The developing kernel is then partially replaced by spore producing pustules (sori). These are surrounded by the delicate pericarp which is disrupted at harvest to release spores and impart the same fishy odour, as with other bunt species (Wiese, 1987).

For seed certification purposes, grain samples need to be free of other fungal contaminants, for example, loose smut (*Ustilago nuda* syn. *U. tritici*) (Table 2.1). This disease occurs in all major wheat growing areas. The fungus resides as mycelium in the embryo of wheat grain and colonizes the seedling intracellularly, and later intercellularly as the plant matures, keeping pace with the growing point. The primordia of the ear are infected and the ovaries are replaced with developing spores. The smutted ears (Fig. 2.3) emerge at anthesis of the unaffected ears and the spores are blown on to the healthy flowers, and subsequently infect the developing embryo. Whilst in the embryo, the fungus does not alter the functionality of the endosperm so infected seed is acceptable for milling purposes. If infected seed are sown, however, nearly all of the contaminated grain will result in systematically infected plants. The level of infection can increase 20-fold in one season and crops grown from heavily contaminated home-saved seed have resulted in yield losses of up to 20% (Parry, 1990).

As well as being downgraded through infection by field fungi, the quality of grain can be seriously affected by fungi developing in storage, most usually because the grain is too damp. Common storage fungi include *Penicillium* spp. and *Aspergillus* spp. Both of these genera include species which can produce mycotoxins in culture which are also found in stored grain; ochratoxin A and citrinin by *P. verrucosum*, sterigmatocystin by *A. versicolor*, patulin by *A. clavatus*, and aflatoxins by *A. flavus* (Lacey, 1990; Scudamore, 1993). Of all the mycotoxins mentioned so far, the most

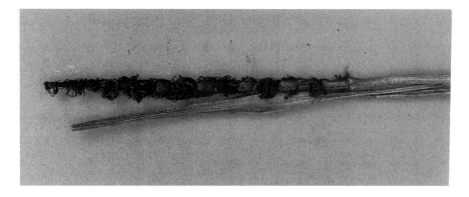

Fig. 2.3. A smutted ear.

common are those associated with *P. verrucosum*. For example, in analyses of 440 cereal samples for animal feed in England and Wales between 1981 and 1983, the most frequently detected mycotoxin was ochratoxin (12.9% of samples), followed by citrinin (11.2%), zearalenone (4.8%), DON (3.6%), sterigmatocystin (3.2%), aflatoxin (2.5%) and patulin (0.2%) (Moore, 1985).

Non-Millable Contaminants

Wheat sample contaminants include stones, soil, sand and even small pieces of metal. These require removal before milling to prevent food contamination and damage to mills.

Different levels of impurity in imported grain have become associated with different countries of origin. Wilson (1989), for example, found consistent differences in the amount of unmillable material in wheat from major exporting countries which might impact upon marketing and value. For example, in the past USA grain has rarely been cleaned on farm or at country elevators, because the loss from shrinkage and the cost of cleaning often exceeded the value of the screenings or reductions in price by the grain exporting agencies. Grain exported from Canada, however, has often been cleaned to defined levels and Webb *et al.* (1995) showed that there could be net gains to the US wheat industry if all US export wheat were to be cleaned to an impurity level between 0.35 and 0.40%.

SPECIFIC WEIGHT AND THOUSAND GRAIN WEIGHT

Specific weight is a measure of the bulk density of grain, i.e. the weight of grain that can be contained in a unit volume packed in a standard way. It is also known as the bushel-, test-, or hectolitre-weight of grain. It is a crude measure of grain shrivelling and badly shrivelled wheats may have specific weights as low as 38 kg hl^{-1} while very plump, well-filled grains can give specific weights of over 80 kg hl^{-1}. Because shrivelled grain contain proportionately more bran than endosperm compared with well-filled grain, it is used as a rough guide of expected flour yield by millers. Large, bold grain are also favoured for sowing and tend to have a high energy content as preferred for livestock feed. Specific weight also relates to transport and storage costs for a given weight of grain. This relevance to a wide range of markets renders specific weight one of the world's most frequently used measures of quality.

Many factors, however, contribute to specific weight not closely related to either the degree of grain shrivelling or flour yield and differences in specific weights above 74 kg hl^{-1} may not relate to either (Halverson and Zeleny, 1988). In a study of 359 wheat samples there was a correlation coefficient of only 0.20 between specific weight and flour extraction rate. Additional contributing factors include the average density of each individual grain, the shape of filled grain, and grain surface texture (Bayles, 1977). A low grain density may reflect air-filled spaces in the endosperm or separating layers of the pericarp. Frequent handling and moving of wheat may polish bran coats and increase specific weight (Halverson and Zeleny, 1988) while weathered grains with a rough surface can have reduced specific weight (Barnes, 1989). Grain specific weights also vary according to grain moisture content such that bulk density falls as moisture content increases above 12%. Unfortunately, however, the relationship between specific weight and moisture content differs between varieties (McLean, 1987) and specific weights at a given moisture content can alter depending on whether the grain was dried or wetted. There are, therefore, clear indications that the relationships between the specific weight of grain and its actual end-use value are questionable (McLean, 1987). Nonetheless, specific weight is firmly established as a specified quality criterion and values greater than 70 kg hl^{-1} are often a prerequisite for many wheat markets. For example, the minimum specific weight required for export from the UK is usually 76 kg hl^{-1} (Anon., 1995a) while wheat offered to intervention in 1994/95 received incremental reductions in price for specific weights from 76 kg hl^{-1} down to 72 kg hl^{-1} (Anon., 1994). Feed compounders have typically requested specific weights above 72 kg hl^{-1} (Lake, 1987). The actual value of increased specific weights will, however, vary with market. Nonetheless, as a guide for breeders, Brennan (1990) estimated that an extra 1 kg hl^{-1} was, on average, worth an extra 0.5% in financial terms.

Specific weights are usually measured using a vertical cylindrical mechanism known as a chondrometer (Fig. 2.4). The chondrometer consists of two chambers (A and B) separated by a removable slide with a weight placed on top. The lower chamber (B) is of a measured volume, usually 1 l, and slightly smaller than the upper chamber (A). Grain is first poured into A. When it is full the slide is removed to allow the grain to fall into B. The speed and evenness of falling, and therefore packing, is governed by a small air hole at the base of B. The slide is then carefully reinserted and the grain trapped above it is discarded. The grain in B can then be weighed accurately.

The weight of a thousand grains can also give an indication of flour yield on the basis that large, well-filled, dense grain will contain greater amounts of endosperm compared with bran. As with specific weight, however, correlations with flour yield are not particularly close, especially at high values and between varieties, and the measurement is not as convenient to undertake compared with specific weight. In durum wheat, however, kernel size, and therefore thousand grain weight, is considered

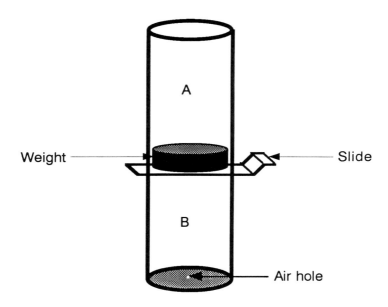

Fig. 2.4. Diagram of a chondrometer. See text for explanation.

the best index for potential semolina yield per unit weight of grain (Matsuo and Dexter, 1980).

ENDOSPERM TEXTURE AND WATER ABSORPTION

The endosperm texture of a particular wheat is often described as being either hard or soft and refers principally to its milling properties. Hardness relates to the amount of resistance encountered when milling the grain into flour particles – the harder the wheat, the greater the force required. Hard wheats have an endosperm which separates easily from the bran thus yielding a large amount of flour with low levels of bran contamination. The starch grains are embedded within thick walls of matrix protein, a large proportion of which is storage protein. When milled, hard wheats break down along the outlines of endosperm cell walls, or through the cells across the starch grains and proteins alike, to yield coarse, smooth sided granules which flow easily over surfaces and through sieves. Easy separation between bran and endosperm also allows a greater proportion of the total endosperm protein to be recovered in flour since protein levels are highest near to the bran (Ford, 1987). Conversely, in soft wheats it is more difficult to separate the endosperm from the bran and therefore flour extraction rates at a given level of bran contamination tend to be lower. The starch granules are only thinly covered in buffer soluble protein and produce a mass of fine, angular shaped particles of cell debris when milled. This is because there is little adhesion between protein, starch and cell walls, and fracturing through starch granules is as likely as fractures around the outside of granules. These particles do not flow easily and are thus more difficult to sieve and more likely to cause blockages in a mill compared with hard wheats. All other things being equal, therefore, millers would prefer to use hard wheats. However, endosperm texture can have an overriding influence on the amount of water that is absorbed by flours and this can determine which types are required for different markets. Damaged starch granules absorb much more water than those left undamaged and starch damage is easier to achieve when milling hard wheats because they respond to higher grinding pressures during roller milling (Blackman and Payne, 1987; Ford, 1987). It is difficult to obtain high starch damage in soft wheats without excessive pressures leading to overheating and bran contamination (Ford, 1987). Flours from soft wheats, therefore, absorb little water and are preferred for the manufacture of most types of biscuit where the lower water content reduces the required baking time, cost, and the risk of cracking during cooling after baking (Blackman and Payne, 1987). Soft wheats are also favoured for cake and pastry production. Flours from hard wheats absorb relatively large amounts of water which can lead to increased yields and longer shelf life of bread (Maga, 1975) while loaves from soft wheats tend to dry out more quickly. The use of hard wheat also has advantages for domestic flour because it packs more efficiently, flows easily from the bag and has a reduced sieving requirement before use (Ford, 1987). Very hard endosperm texture is required for pasta production so that the regular sized particles of semolina flour can be made without bran or fine flour contamination. Similarly, extreme hardness is required to make couscous which requires a very large particle size. Flours made from a mixture of hard and soft wheats can be suitable for noodles, chapatis and flat breads.

In wheat distilling there is a preference for soft endosperm texture. This is particularly so when grain whisky is produced following the cooking of whole grains, as soft wheat kernels are more efficiently disrupted by the cooking process than are hard wheats (Rifkin *et al.*, 1990). Hard wheats can sometimes be acceptable but only when they have a relatively low protein content.

Endosperm texture can be assessed in a number of ways including particle size indices, pearling index, time to grind, sound of grinding, starch damage, NIR, fineness, and crushing or slicing of individual kernels (Pomeranz, 1987; Halverson and Zeleny, 1988). In the UK, the proportion of flour passing through a 76 µm sieve has been used in the routine testing of varieties (Draper and Stewart, 1980). Soft wheats release

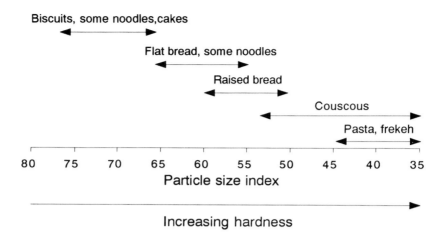

Fig. 2.5. The relationship between endosperm texture and suitability for different wheat products. Source: Canadian International Grain Institute.

a higher proportion of very small particles or *fines* than hard wheats. Higher numbers of fines are reflected in a higher particle size index (PSI). The broad relationships between PSI and suitability for different end-uses is shown in Fig. 2.5. If simple classification of 'hard' or 'soft' is required at grain intake into a mill, NIR has been shown to give adequate rapid determinations (Osborne and Fearn, 1983). Another test used widely at intake and in laboratory studies is that of the Stenvert hardness test whereby a sample of grain is ground in a hammer mill, and the time taken for a given volume to pass through a 2 mm grid is recorded; the longer the time, the harder the flour.

Water absorption can also be assessed in a number of ways. The Brabender Farinograph assesses the resistance of flour and water while they are being mixed together to form a dough. The level of maximum resistance can be adjusted to a predetermined optimum by altering the amount of water added. The ideal amount of water being a measure of water absorption (Kent and Evers, 1994).

VITREOUSNESS

The description of a grain as vitreous implies that the endosperm has a glassy, translucent appearance rather than a mealy or floury nature ('yellow berry'). It is used particularly for durum wheat where it correlates with semolina yields as yellow berry kernels pulverize more readily. Loose positive associations have been found between the number of vitreous grains in a sample and grain protein content. The incidence of yellow berry can increase when normally vitreous grained varieties are grown in low protein environments (E. Autrique, CIMMYT, Mexico, 1994, personal communication). It is possible for some varieties to have both vitreous and mealy areas of endosperm within the same grain. Vitreous areas of the endosperm of these grain have been found to be higher in protein than the mealy areas (McDermott and Pace, 1960).

FLOUR COLOUR

Wheat flour colour has several components. Firstly, flour tends to be slightly yellow due to the presence of carotenoids, and flavone compounds. A high β-carotene content in durum wheat is favoured for pasta, couscous and bulgur production to impart the characteristic yellow colour. Secondly, bran content will dramatically influence colour. Bran contamination will depend on the texture of the endosperm and the flour extraction rate. Flour extraction rates are, therefore, inextricably linked with flour colour, with the maximum rate depending on the maximum permissible level of bran contamination for a particular product. As well as bran contamination, however, flour colour will also be influenced by the inherent greyness of the endosperm. In the production of white bread, low bran contamination with low inherent greyness is desired. Unfortunately, high protein content is often associated with increased greyness (Barnes, 1989). This is thought to be because starch granules, which are highly reflective, are diluted by the protein.

Flour colour is regarded as one of the most important factors in assessing the value of flour for Japanese noodles where high levels of whiteness and brightness are preferred by consumers (Miskelly, 1984). In yellow or Chinese noodles, the noodles are partly coloured by the natural presence of flavones. For the production of chapatis white wheat varieties are preferred as red seed coats tend to increase discoloration.

There are several ways in which flour colour can be assessed. The Pekar test involves wetting a smoothed disc of flour. As the flour dries, tyrosinase activity deriving mainly from bran contamination, leads to the production of melanins which darken the disc of flour. Abrol *et al.* (1971a) found this reaction could identify suitable wheat varieties for chapati production.

The Kent Jones and Martin Colour Grader can be used to measure light reflectance of a flour and water paste to produce a flour colour grade (FCG) which is particularly strongly correlated with protein content. For assessing the yellow colour, particularly relevant for pasta production from durum, carotenoid pigments can be extracted with water saturated butanol and then quantified photometrically against a β-carotene standard.

PROTEIN QUALITY

Amino-Acid Composition

Amino-acid composition of wheat is rarely assessed by the end-user but it has a vital impact upon nutritional quality, and the welfare of dependent populations, which renders it worthy of mention in any review of quality. The quality of cereal protein can be considered relatively poor, being low in some essential amino-acids. The most serious lack is that of lysine (Altschul, 1965) which is only present at about 40% of the concentration considered by FAO to meet human requirements in pure wheat diets (Scrimshaw and Young, 1976). Increasing lysine to diets which meet near minimal requirements increases nitrogen retention markedly, in contrast to the effects of increasing tryptophan, cysteine or methionine (Clark, 1978). Unfortunately the amount of lysine is much less in white flour products because the endosperm protein only contains about 2% lysine, compared with non-endosperm protein which contains over 4%. Indeed, those storage proteins which impart the dough characteristics necessary for the production of leavened staples contain particularly low levels of essential amino-acids. As the protein concentration increases, e.g. from 7–15%, the relative proportion of these storage proteins increases such that the lysine content, expressed as a percentage of total protein, falls from 4–3% (Blackman and Payne, 1987). Lysine deficiency is of greatest importance in developing countries where a high proportion of the protein provision is directly from cereals. In the developed world, wheat protein makes a much smaller contribution to the total protein intake of the population, but even here Blackman and Payne (1987) argue that greater lysine content of wheat would reduce the present heavy reliance on soya bean and fish meal necessary for livestock production.

Dough Rheology

The rheological properties of doughs are often described in terms of resistance (elasticity) and extensibility (plasticity). A strong dough is one that gives a relatively high resistance as it is being extended, as well as a degree of elasticity, such that when released from moderate extension it will have a tendency to regain its former shape. A weak dough will extend readily with little resistance and have no elastic tendency. A tenacious dough is one that may have high initial resistance to extension but breaks after only a relatively small distance. These are often described as short, rather than strong, doughs and have little widespread use. Dough rheological characters are imparted by the amount, and more importantly the type, of storage proteins of the wheat grain, namely the gliadins and glutenins, which together form the vast majority of the insoluble protein content of the grain, i.e. the gluten. The extensibility of a dough is thought mainly to be due to the types and amount of gliadin present, while the high-molecular-weight glutenins play a major role in the resistance of a dough (Blackman and Payne 1987; Barnes, 1989). Weak flours, for example, can be converted to strong ones by increasing the proportion of the high-molecular-weight gluten protein present (MacRitchie, 1973).

Strong doughs are necessary for the production of leavened products such that gases produced during fermentation are captured within an extensive viscoelastic matrix, allowing the dough to rise with a crumb structure consisting of many gas filled pores. Strong doughs are therefore a prerequisite of nearly all leavened bread-making systems.

With regards to pasta production from durum wheat, cooking quality is mainly determined by the ability to absorb water while retaining firmness and shape and without becoming sticky, even after cooking (Sarrafi *et al.*, 1989). Both processing and cooking of pasta is, therefore, dependent on strong gluten and the sodium dodecyl sulphate (SDS) sedimentation test can be used to screen durum wheat gluten strength, as it can for the bread wheats (Dexter *et al.*, 1980; Kovacs *et al.*, 1995).

In biscuit production it is important that doughs can be rolled thinly with even thickness and should retain their shape after cutting. This requires a very extendible dough with no elastic properties.

Intermediate strength doughs can be used to produce steamed bread, chapatis and noodles but puffed breakfast cereals require a strong wheat to prevent disintegration when pressure is released after cooking. Flaked and shredded products are made from weak wheats similar to those for biscuits (Blackman and Payne, 1987; Lin *et al.*, 1990).

The strength of a dough can be measured directly or by a number of small scale tests. Direct measurement can be achieved by measuring the force required to extend the dough (resistance) whilst also measuring the distance achieved before the dough breaks (extension) (Fig. 2.6) (Draper and Stewart, 1980). This can be achieved with the Brabender Extensograph after the results of the Brabender Farinograph have been used to prepare a dough with the correct proportions of flour and water. Another relatively direct method is the Chopin Alveograph test. A disc of white flour dough is pressurized by a stream of air to force the production of a bubble in the dough. The maximum pressure to blow the bubble (P), the time for the bubble to burst (L), and the area under the pressure × time curve (ω) are used to describe the dough properties (Fig. 2.7). A low $P : L$ ratio is indicative of a weak, extensible dough suitable for biscuit making (e.g. below 0.6 or 0.5 depending on market) while a high ω figure is indicative of strong gluten suitable for bread production (e.g. above 150 or 200) (Fenwick, 1993; Anon., 1995a). Alternatively the Pelshenke test is used in many countries and by plant breeders to assess dough strength.

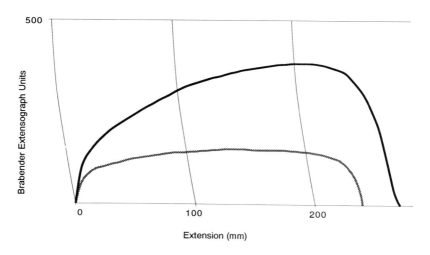

Fig. 2.6. Extensograph of a relatively strong, elastic dough with potential for bread-making (solid line) and a weak, extendible dough with potential for biscuit making (shaded line).

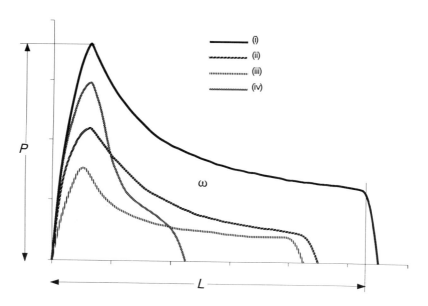

Fig. 2.7. The measurements taken from a Chopin Alveograph and typical relative curves from doughs exhibiting (i) a $P/L < 0.8$ and $\omega > 200$ suitable for pan bread production; (ii) a $P/L < 0.6$ possibly suitable for blending with stronger samples for bread production; (iii) a $P/L < 0.5$ and $\omega < 110$, suitable for biscuit production; and (iv) a $P/L > 1$ unsuitable for most markets other than livestock feed (after Cooksley, 1995).

Balls of whole wheat dough are made with a yeast suspension and immersed in water. Very weak doughs disintegrate quickly (< 30 min) while very strong doughs may remain intact for over 400 min. The Zeleny test is used extensively within the EU for assessing dough strength and relies on suspending white flour from specially milled wheat in a weakly acidified solution (lactic acid), sometimes with isopropanol, in a graduated cylinder. The height of sediment (mainly swollen gluten and occluded starch) is recorded after 5 min of standing. Very weak and strong wheats can vary from 3 to 70 mm, respectively, but a more common range is between 20 mm (weak) and 50 mm (strong). For grain offered into intervention in 1994/95 a Zeleny of at least 20 mm was required. A similar test used in the UK is that of the SDS sedimentation volume test. This involves the addition of SDS to the acid solution and a longer settling time (20 min) compared with the Zeleny test, but gives a better prediction of UK bread quality. It also has the advantage over the Zeleny test in that it is sufficient to use wholemeal wheat samples, rather than white flour, so milling requirements are less complicated (Axford et al., 1979; Halverson and Zeleny, 1988). The acidified SDS solution dissolves much of the protein but not the high-molecular-weight proteins, particularly the glutenins, which remain in the swollen layer of sediment (Fullington et al., 1987; Barnes, 1989). The SDS and Zeleny sedimentation tests are, however, similar in principle and regressions have been calculated between them – the closest of which (Fenwick, 1993) is shown in equation 2.2:

$$\text{Zeleny} = (0.81 \times \text{SDS}) - 9.5 \quad \{\text{SE} = 7.1\} \qquad [2.2]$$

Although Zeleny and SDS sedimentation volume tests have proved useful for wheat breeders and buyers in selecting and screening for breadmaking quality, they have on occasion failed to predict baking performance. Tests are, therefore, still being developed and one of the most promising involves gel protein (Pritchard, 1993). This is the insoluble gel layer that forms when flour is extracted at 10°C with 1.5% SDS solution followed by high speed centrifugation. The amount of gel protein is similar to the SDS sedimentation and Zeleny tests in relating to the amount of functional protein present. It is, however, possible to study the rheology of the gel protein itself to give additional information about probable dough characteristics.

A test used widely within the EU which is loosely related to dough rheology is that of the machinability test. This involves a trained operator deciding whether a ball of dough produced in a mixing machine sticks to the apparatus or not, and results are expressed just on a pass or fail basis (Fenwick, 1993). In 1994/95 this was only used for intervention wheat if the Zeleny value was below 30.

The quality of gluten can be dramatically reduced if grain has been exposed to excessive heat. Heat damaged gluten can be detected in the gluten turbidity test. Excessive heat reduces the solubility of glutenins normally soluble in acetic acid. Dilute acetic acid extracts are precipitated by addition of alkaline ethanol and the precipitate is quantified spectrophotometrically (Kent and Evers, 1994). Alternatively, damage to the gluten during drying or storage can be assessed at mill intake with the gluten washing test. Grain is ground and sieved before being placed in gluten washing apparatus in which the starch is washed away to leave a ball of gluten. This is assessed for colour and rheological properties by hand, and its size can also give an indication of protein quantity.

GRAIN NITROGEN AND CRUDE PROTEIN CONCENTRATION

The last section emphasized the importance of gluten and associated proteins in the functionality of doughs made from wheat grain flour. The amount of gluten in grain is partly determined by the amount of nitrogen available for the synthesis of protein and, therefore, the nitrogen concentration in grain has long been used as a quality criteria, particularly for bread and pasta production. It is often assumed that there is a constant relationship between the amount of nitrogen present and the amount of protein such that crude protein content of wheat is calculated as the nitrogen content multiplied by 5.7 (Draper and Stewart, 1980). Crude protein contents calculated in this manner can vary between 6 and 25%. More recently it has been recognized that 5.83 is a more accurate multiplier for assessing protein content of wholemeal flour (Kent and Evers, 1994). It should, however, be remembered that not all nitrogen is present as protein. Many molecules, including free amino-acids, amines, nitrates, nitrogenous glycosides, sphingolipids, glycolipids, B vitamins and nucleic acids also contain nitrogen. Conditions that change the ratio of these compounds to true protein could contribute to crude protein being a less reliable predictor of the amount of gluten present. In spite of this, most markets use the crude protein content calculation as the basis of threshold levels. Confusion can arise because crude protein content requirements for different markets can be quoted on the basis of different moisture contents. For example, UK millers often quote crude protein content requirements on the basis of 14% moisture, while EU specifications are on a dry matter basis. In the USA, the Federal Grain Inspection Service reports protein content on a 12% moisture content basis (Hoffman et al., 1988).

For a given species or variety of wheat there is often a positive relationship between grain nitrogen content, loaf quality (Finney et al., 1957) and sedimentation tests (Gooding et al., 1993b). For those markets which require a strong gluten, relatively high concentrations of protein are, therefore, needed. For pasta products and hearth breads values over

13% are commonly required while pan bread requirements, including the Chorleywood bread-making process, are usually between 11 and 13% (14% mc).

Because of the important role of protein in the functional and nutritional quality of leavened products, support and/or guaranteed minimum prices offered by national and international agencies are often dependent on protein concentration. For example, the Canadian federal government offered a premium of C\$33 t^{-1} for Western Red Spring Wheat reaching 12.5% protein compared with 11.6% in 1994 (Anon., 1995c). In 1994 the Australian Wheat Board instigated financial incentives to increase the protein content of its main wheat grade, Australian Standard White (ASW). This was to halt a trend in reduced average protein contents from 10.4–9.6% since 1982, which if continued would have seriously limited the competitiveness of Australian wheat in milling markets (Reuter and Dyson, 1990; Anon., 1995c). In 1994/95, therefore, growers received an additional A\$0.50 ton^{-1} for each 0.1% in protein above 10%, and penalties of A\$1.00 for each 0.1% below 10%. Bonuses of A\$0.50 t^{-1} for each 0.1% in protein above 11.5% for Australian Prime Hard wheat were continued. A campaign was launched to increase the average protein concentration of Australian wheat by 0.5–10% by the year 2000. In Europe, wheat offered to intervention agencies of the EU in 1994/95 received a 1% reduction in price for every 0.43% below 9.9% protein content.

Those products needing a weak, extensible dough such as biscuits, cakes and pies usually require a protein content of less than 11% and levels down to 8% can be tolerated for some products (Orth and Shellenberger, 1988). Steamed bread, chapatis and noodles often have requirements of 9–11%, 10–11% and 8–10%, respectively (Blackman and Payne, 1987; Lin *et al.*, 1990). For many of these products, white wheats are often preferred. In some white wheat growing areas, particularly in the USA, the climate in certain years leads to grain protein contents above the desired range, especially if nitrogen fertilizer has been used (Rasmussen and Rohde, 1991). This can restrict the marketing and value of this type of wheat.

As for grain whisky, the distiller requires starch to convert into alcohol. Higher protein reflects lower starch in grain so very low levels of grain protein concentration are favoured (Taylor *et al.*, 1993).

Grain nitrogen concentration can be determined directly with the Kjeldahl method, developed in 1883 by Johan G.C.T. Kjeldahl. Nitrogen in a weighed sample is transformed into ammonium sulphate by digestion with sulphuric acid. The solution is then alkalized and the amount of resulting ammonia produced determined by titration. This method is still commonly used for variety testing and breeding work, as well as in wheat experimentation (Draper *et al.*, 1979). It does, however, have disadvantages, which mostly derive from the digestion process. This reduces speed of throughput; carries a risk of exposure to concentrated sulphuric acid; and produces corrosive and toxic fumes at high temperatures. Quicker, less hazardous, but also less direct methods have therefore been developed for assessing grain at intake to mills such as those using NIR (Osborne and Fearn, 1983) and, more recently, oxidative combustion (Wilson, 1990).

α-AMYLASE ACTIVITY, HAGBERG FALLING NUMBER AND SPROUTING

The Hagberg falling number is a measure of the viscosity of a mixture of water and ground wheat mixed in a tube and placed in a water bath at 100°C. α-Amylase breaks down starch to produce a mixture of glucose and maltose and, therefore, reduces viscosity. The Hagberg falling number is the time in seconds required for stirring (60 s) plus the time taken for the stirrer to fall through the flour suspension while it is being liquefied by the enzyme. Hagberg falling number is, therefore, an indirect measure of gelatinized starch (55–65°C) after it has been degraded by α-amylase during a 30–40 s period before the reaction mixture exceeds the temperature when the enzyme is denatured (75–80°C) (Vaidyanathan, 1987).

Hagberg falling numbers can be assessed readily at the point of trade using automated equipment. The procedures are designed to minimize errors (Perten, 1964) and have been adopted for official purposes by the International Association for Cereal Chemistry (ICC Standard No. 107), and as a British Standard method (BS 4317 pt. 9). It is, however, possible to obtain different Hagberg falling numbers for the same level of α-amylase activity (Olered, 1967). This is primarily due to differences in the amount of damaged starch, but overall, the Hagberg falling number is much more sensitive to α-amylase activity than any other grain component.

Some α-amylase activity is required in bread production to release sugars and therefore aid fermentation. However, excessive activity results in a sticky crumb of poor resilience and texture, and a darkened crust as a result of the sugars caramelizing. The crumb is also turned brown as the sugars combine with some amino-acids by the Maillard reaction. Notably, lysine is sequestrated so nutritional quality is also impaired (Kent and Evers, 1994). A more serious problem for producers of cut loaves arises because the sticky dough attaches itself to slicers. The loaves then deform as they pass through the slicer and slice thickness becomes irregular. The slices stick to each other and finally the loaves break up in the slicer bringing production to a halt (Chamberlain *et al.*, 1982).

Grain for producing bread commonly give Hagberg falling numbers of at least 300, although samples with lower levels might be accepted by millers if they can be blended. However, the relationship between Hagberg falling number and α-amylase activity is not linear and relatively large amounts of wheat with high numbers are needed to be added to wheat with low numbers to achieve a satisfactory level. This non-linear relationship is described by the liquification number (Fig. 2.8). The common falling number thresholds set by UK millers for the Chorleywood breadmaking process range between 220 and 250. Noodles and pasta also have a requirement for high Hagberg falling numbers (250–300) but other products such as biscuits and chapatis can have a much less stringent specification (e.g. 120 minimum) (Orth and Moss, 1987). For some uses such as a thickener for soups, flour is heated, which destroys the enzyme (Blackman and Payne, 1987) making the quantity of α-amylase present largely irrelevant.

MINERAL, FIBRE AND VITAMIN CONTENT

In addition to protein, there are several minerals in wheat grain that have dietary implications for human and livestock populations. An indication

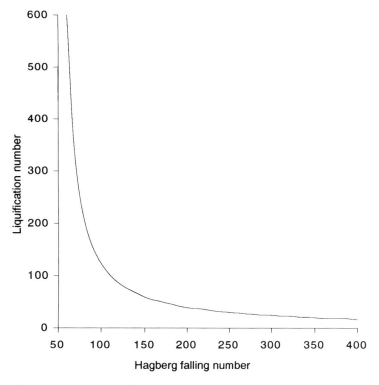

Fig. 2.8. The relationship between Hagberg falling number (HFN) and liquification number (LN) (LN = 6000/(HFN – 50)) (after Perten, 1964).

Table 2.3. Mineral composition (mg kg⁻¹ DM) of winter wheat surveyed in the UK.

	1982 (McGrath, 1985)	1992 (Chaudri et al., 1993)
Phosphorus	3800	3400
Potassium	5000	4900
Sulphur	1700	
Calcium	449	420
Magnesium	1096	1000
Sodium	38.6	
Iron	41.4	36
Aluminium	5.64	
Titanium	0.62	
Zinc	28.6	24
Copper	5.12	4.5
Chromium	0.44	
Manganese	27.9	23
No. of samples in survey	238	400

of amounts present in UK grain is shown in Table 2.3. There appears to be a general reduction in the mineral content of wheat surveyed in 1992 compared with 1982. This is possibly due to dilution with carbohydrate as grain yields increased over this period by about 1 t ha⁻¹. Sander *et al.* (1987) consider that the amount of phosphorus and potassium contained within cereal grain will always be adequate for people and animals, largely independent of the type of soil within which the crop was grown. In contrast, levels of calcium will invariably be deficient, and other sources of this element need to be found when the diet is cereal based. Concentrations of metal ions in wheat can have considerable consequence for human health. Cereals, for example, contribute about 40% of the daily intake of iron in UK diets and 10% of this is derived from fortification, mainly through statutory additions to flour (McGrath, 1985). Zinc deficiency has been reported in certain areas of the Middle East where diets have been dominated by wheat (Sander *et al.*, 1987) and decreased contents of the mineral in the grain have intensified the problem (Hambridge, 1981). Improvement in wheat zinc content could, therefore, be of particular value (Ascher *et al.*, 1994). One of the most important factors determining metal ion availability is the type and degree of processing. The milling of grain to extract white flour can remove up to 50% of the iron, zinc and copper, which increases the fortification necessary for nutritional adequacy (Pederson and Eggum, 1983).

Selenium is an essential element for humans and livestock, but not for wheat. Selenium contained within wheat grain appears to relate to the amount of selenium contained within the blood of humans following consumption (Sander *et al.,* 1987). Low levels of selenium in wheat of areas in the Heilongjiang and Yunnan Provinces of China have been associated with the incidence of Keeshan disease. This is a congestive heart problem, particularly prevalent among young children, which was virtually eliminated by selenium supplementation of the diet. High levels of selenium can be toxic but wheat grains containing potentially damaging levels of selenium are very rare, even when the wheat has been grown on selenium-rich soils, for example, in North America (Sander *et al.,* 1987).

Unfortunately, not all the minerals contained within wheat products are nutritionally available. Therefore, not only is the mineral content important, but also its bioavailability. Increased dietary fibre, for example, reduces the amount of mineral absorbed from wheat sources (Johnson and Mattern, 1987; Alldrick, 1991). This presents a conflict with diets in many developed countries which are considered to be deficient in fibre. For example, a number of symptoms and diseases including constipation, irritable bowel syndrome, colonic diverticulosis, gall stones, coronary heart disease, diabetes mellitus, and obesity are sometimes relieved by high-fibre diets (Alldrick, 1991). As wheat can be a major source of dietary fibre, a number of wheat products have fulfilled the market demand for more healthy foods. As dietary fibre particularly derives from the bran layer, products such as bran flakes or certain brown or wholemeal breads have a high bran content. Consumers in the developed world often receive enough minerals from various non-cereal sources which may thus negate the importance of wheat fibre reducing mineral bioavailability. Dietary fibre itself is complex in composition and difficult to determine precisely. Increased public awareness of its importance in nutrition, however, has led to the development of a number of techniques for fibre assessment. Bell (1985) evaluated a sample each of white, brown and wholemeal bread with seven different techniques. The means of these methods would suggest dietary fibre contents of about 3, 6 and 10% dry matter (DM), respectively.

In addition to fibre, phytate, the phosphoric acid ester of inositol, of which 87% is found in the aleurone layer, forms insoluble complexes with some minerals thus reducing their bioavailability. Calcium may exacerbate the effect of phytate in reducing the bioavailability of zinc. The presence of certain minerals also reduces the uptake of others, hence calcium and magnesium antagonize the uptake of each other as do copper and zinc, and manganese and iron.

Wheat is an important source of several vitamins (Table 2.4) including vitamins E, thiamine, niacin and B_6. The concentrations of these can vary between sites, seasons and varieties (Sander *et al.,* 1987), but they are also

Table 2.4. Vitamin content ($10^3 \mu$g kg^{-1}) of wheat from different sources.

	cv. Chris	Mean of 406 cvs
Thiamine	9.9	4.6
Riboflavin	3.1	1.3
Niacin	48.3	55
Biotin	0.056	
Folacin	0.56	
Pantothenic acid	9.1	
Vitamin B$_6$	4.7	

greatly modified by processing as they are concentrated within the germ and/or bran.

Digestibility and Energy Value for Livestock

Many of the quality criteria already mentioned have implications for livestock systems. Increased protein content will, for example, reduce the amount of protein supplementation required from other sources. Lake (1987) considered that the UK compound feed industry would prefer minimum protein contents, specific weights and moisture contents of 10.5% (fresh weight), 72 kg hl^{-1} and 16%, respectively. Even these relatively modest requirements, however, were not being consistently attained. Latterly it has been recognized that different wheat samples can produce very different growth rates in livestock, particularly in poultry, which are not easily related to variations in the quality characteristics already discussed.

The primary role of wheat grain for livestock farmers is as an energy source. For ruminants, this is often stated in terms of metabolizable energy (ME), which is the amount of energy absorbed from a feed during digestion minus the energy lost in the production of methane gas and urine. The ME, determined *in vivo* for ruminants can vary significantly. In the UK for example, the ME of 25 winter wheat samples had a standard deviation of 0.62 and varied between 12.3 and 14.7 MJ kg^{-1} DM (Ministry of Agriculture, Fisheries and Food, 1992). Energy of cereal grains in general, and wheat grain fractions in particular, is inversely related to the amount of lignin and cellulose present. This can be estimated in determination of acid detergent fibre (ADF), i.e. the residue left after refluxing with 0.5 M sulphuric acid and cetyltrimethylammonium bromide (Fig. 2.9).

For poultry the most frequently used measure of energy content of wheat is apparent metabolizable energy (AME), i.e. the amount of gross

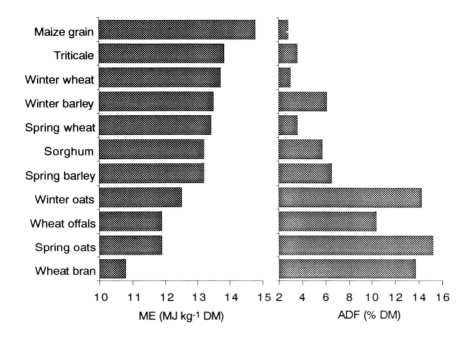

Fig. 2.9. The metabolizable energy (ME) and acid detergent fibre (ADF) of cereal grains used as energy feeds for ruminants in the UK (plotted from data in Ministry of Agriculture, Fisheries and Food, 1992).

energy consumed in the feed less that voided in the excreta (Wiseman *et al.*, 1993). As with ME for ruminants, AME for poultry can vary considerably between samples with values ranging from 10.4 to 16.6 over different surveys (Wiseman, 1990). The major energy yielding component of wheat is starch. This is the principle form of carbohydrate in the grain, located in granules of varying size throughout the endosperm. Concentrations typically fall in the range of 60–75% DM (Wiseman, 1990). By comparison all other polysaccharides only contribute between 10 and 14% DM, including pentosans in the cell walls (6–7% DM) and β-glucans (< 1% DM) (Wiseman, 1990). Despite their small contribution to total DM, increased contents of the non-starch carbohydrates may, however, sometimes be responsible for reduced AME values for wheat. Digestibility of the pentosan complex in the cell walls is itself low but, possibly more importantly, pentosans may increase the viscosity of the feed while it is being digested; increase water holding capacity thereby cause swelling and reduce feed intake; and increase nutrient binding which would reduce absorption generally (Wiseman, 1990).

GERMINATION AND SEEDLING VIGOUR

The germination rate and seedling vigour are particularly important for seed grain. The germination rate describes the percentage number of grain germinating in a normal way under standardized conditions with adequate levels of moisture and temperature. Abnormal germination will cause a seed to fail the test and can be due to any number of morphological irregularities such as the absence of shoot and/or roots, excessively small shoot and/or roots, the first leaf breaking out of the side of the coleoptile, or an abnormally short first leaf (Anon., 1986). Many tests are in operation but the official test of untreated seed in England, Wales and Northern Ireland involves replicated samples being placed into, and then covered by, moist sand. These are then kept in darkness under controlled temperatures. The number of normal seedlings are counted and removed after 4 days, and again after 7 days. More rapid guidance tests are available which rely on viable germs being stained red after immersion in tetrazolium. For example, seeds can be cut in half longitudinally and soaked in a solution of tetrazolium in a vacuum at 55°C for ten minutes, after which a red colouring should develop on viable embryos

Germination tests, although measuring viability under good conditions, do not always relate directly to establishment under field conditions. Seedling vigour, therefore, refers to the ability of grain to overcome stress in the field during germination and establishment. There are no standard methods for assessing wheat seed vigour and different grain attributes may have varying degrees of importance, depending on the particular stress or stresses imposed during establishment. Large seeds will have more energy and protein reserves than small seeds and may, therefore, be better able to survive deeper sowing. Seed weight also sometimes relates to subsequent young seedling weights, so grain with high thousand grain weights are desirable. Closer correlations with seedling weights, however, have been found with total protein weight per seed (Ayers *et al.*, 1976), rather than total seed weight, suggesting that total nitrogen reserves might be more important than energy reserves. Phosphorus content of the seed has also been related to subsequent increases in seedling emergence rate, leaf areas and root lengths (De Marco, 1990). This suggests that phosphorus contained within the seed could partially compensate for low phosphorus availability in the soil during early seedling growth. When manganese availability in the soil has been low, the amount of this element in the seed has also been positively associated with wheat dry matter production (Ascher *et al.*, 1994).

Irrespective of weight, grain of poor germination percentage often has even less vigour, which emphasizes the need to begin with grain with very high levels of viability when establishing wheat.

3

Genotypic Effects on Grain Quality

Choice of wheat variety by a farmer is nearly always the most important factor under his/her control determining the market to which the grain will be suited. Bread, biscuit, cake, pasta, noodle and chapati products, and the distilling industries can have very distinct preferences for wheat variety. Practical choice within a country or state may be limited to one or two suitable varieties for a particular market to the exclusion of other genotypes. Specific varieties with certain characteristics may, therefore, be the only ones considered for receiving financial premiums.

In many countries varieties are grouped into classes relating to potential end-use. For example, in the UK the National Association of British and Irish Millers (NABIM) has produced lists giving a clear indication of how millers view wheat varieties, to assist farmers in making marketing and planting decisions. Group 1 varieties are favoured for bread-making, which includes hard red milling winter and spring types with appropriate protein quality. At time of writing this group included cvs Hereward, Mercia, Spark, Avalon (all winter varieties), and Axona, Alexandria and Canon (all spring varieties). Group 2 includes newer varieties or wheats with more limited bread-making potential. This is a large group which, for various reasons, does not have as wide an appeal to millers of bread-making flour. Varieties in this group may require special attention due to unusual dough rheological properties, e.g. cvs Pastiche, Torfrida and Fresco. Also included are some soft wheats, e.g. Shiraz. This group is less likely to attract the same price as varieties in Group 1. Group 3 varieties are soft wheats suitable for the production of biscuit, cake and other flours and currently include cvs Riband, Galahad, Consort, Admiral, Apollo, Beaver and Encore. Group 4 includes other varieties which have no special value to flour millers. These hard and soft wheats are often sold for livestock feed.

In Germany, wheat varieties are classified into A, B and C quality groups. 'A' signifies the best baking and mixing quality which can be used for the improvement of lower quality wheat. 'B' wheat possesses good baking quality but has no mixing quality. 'C' wheat has little or no baking quality and is mainly used for industrial purposes or as feed. These grades are further divided to give a more precise indication of baking performance. 'A' wheat is graded from 6 to best quality varieties at 9 (e.g. Rektor and Bussard); 'B' wheat from 3 to 5 (e.g. cv. Orestis); and 'C' wheat classified as either 1 or 2.

Australia categorizes its wheat into four main classes constituting prime hard, hard, standard white and general purpose. This is based mainly on the variety and protein characteristics. In addition, other variety related characteristics such as the thousand grain weight and ash content can markedly influence price premiums (Ahmadi-Esfahani and Stanmore, 1994).

Many of the breeding advances since the 1960s have resulted in yield increases which are not always matched by quality improvements. For example, in both India and Pakistan there has been a price premium of at least 20% for quality grain of old tall varieties over more modern short genotypes (Byerlee and Moya, 1993). More latterly, however, combining grain yield with quality has become an important objective of the world's wheat breeders. For the developing world, CIMMYT now employs sophisticated screens for quality, particularly based on dough rheology, to ensure that all material released has a high end-use potential. The increased interest in quality has led to a greater knowledge of its genetic basis and inheritability for use in breeding programmes.

GRAIN SPECIFIC WEIGHT

Varieties of wheat can vary significantly with respect to specific weight, and this feature can impact upon the marketability of grain for a large number of outlets. This is despite the observation that varietal differences in specific weight do not always relate to differences in grain plumpness, thousand grain weight, or flour yield. Pushman and Bingham (1975), for example, found that varietal differences in specific weight of ten UK winter wheat varieties were related to individual grain density, rather than thousand grain weight and flour extraction rate. Differences in grain density were associated with grain protein content. This is despite the densities of wheat starch (1.51) being greater than wheat gluten (1.29) (Ghaderi and Everson, 1971). Increased protein may, however, lead to greater grain density because of the packing of protein into the spaces between starch granules (Pushman and Bingham, 1975). The association between specific weight and protein content is still evident in modern UK varieties. For example, there is a significant correlation ($r = 0.55$, $P = 0.01$)

between the specific weight and protein concentration of the 20 varieties in the recommended lists for 1995 (Anon., 1995b). Twenty-five varieties grown in the absence of nitrogen fertilizer and agrochemicals showed an even greater degree of correlation ($r = 0.64$) (Thompson, 1995). In contrast there is no such association with thousand grain weight, indeed, the variety listed as having the lowest specific weight, Haven, had the highest score for thousand grain weight (Anon., 1995b).

In addition to grain protein content, variation in grain morphology can also explain some of the varietal differences in specific weight. Pushman and Bingham (1975) suggested that the poor packing efficiency of Maris Huntsman was due to morphological features of the floret and developing grain which gave rise to transverse folds in the ventral surface of the mature grain. Much of the improvement in grain specific weight of UK wheats since the mid-1970s has, therefore, been through the replacement of this once widely used variety. Counter to this trend, however, is the observation that dwarfing genes can reduce specific weight (Allan, 1986; Cosser *et al.*, 1996). This effect appears to be related to the much reduced grain sizes (Gale and Youssefian, 1985) resulting from the greater numbers of grain produced, but may also at least partly be due to reduced grain protein contents (Table 3.1).

MILLING PROPERTIES AND WATER ABSORPTION

Endosperm texture is largely under genetic control with varieties either being classed as soft or hard, a character which is simply inherited and

Table 3.1. The effect of the *Rht* dwarfing genes in isogenic lines of Maris Widgeon winter wheat on grain yield and quality (from Cosser *et al.*, 1996).

Variable	No *Rht*	*Rht1*	*Rht2*	*Rht1+2*	SED (26 d.f.)
Ear population (No. m^{-2})	218	248	243	267	12.0
Grains (No. m^{-2})	9900	13,300	11,300	14,000	950
Thousand grain wt. (g DM)	38.1	33.2	36.4	30.4	2.0
Yield (tonne ha^{-1})	2.96	3.59	3.25	3.30	0.195
Specific weight (kg hl^{-1})	74.5	73.6	73.8	70.3	0.58
Crude protein (% DM)	11.2	10.6	10.6	10.7	0.14
SDS sediment vol. (ml)	80.8	77.3	79.2	77.4	1.08
Hagberg falling number	239	280	294	311	10.6
N yield in grain (kg ha^{-1})	68.7	79.0	71.5	74.5	4.96

SED, standard error of difference; df, degrees of freedom.

Table 3.2. The endosperm texture, flour yield, flour colour and water absorption of eight wheats grown at three sites in the UK in 1970 (adapted from Dyke and Stewart, 1992; Stewart and Dyke, 1993).

Cultivar	Endosperm texture	Flour yield (%)	Grade colour figure	Water absorption (% @ 14% moisture)
Cappelle-Desprez	Soft	72.0	3.1	52.3
Champlein	Soft	71.8	3.2	50.9
Joss Cambier	Soft	71.6	2.9	49.7
Maris Widgeon	Hard	76.1	2.4	54.5
Maris Beacon	Soft	71.3	3.2	49.8
Maris Ranger	Soft	70.2	3.7	49.1
West Desprez	Soft	72.8	3.0	53.6
Cama	Hard	72.4	3.3	58.3

easily selected in breeding programmes. As explained in Chapter 1, hard varieties often give higher flour yields than soft varieties. In a selection of 27 wheat variety samples (Evers *et al.*, 1990), for example, hard varieties gave flour extraction rates averaging 78.4% (SD = 1.40) which compared with 75.6% (SD = 2.07) in soft varieties ($P = 0.0006$). The importance of variety on associated flour properties is also demonstrated in Table 3.2 with the hard wheat cv. Maris Widgeon giving a high yield of light coloured flour and a high water absorption compared with soft wheats. The other hard wheat in the experiment, however, only gave a moderate flour yield and colour suggesting that the relationship between hardness and the other variables can be modified.

Experiments exploring the relationships between the extractability of different proteins and endosperm texture has implicated a protein known as friabilin. It has been suggested that in soft wheats, which possess this protein in large quantities, friabilin prevents strong binding between starch and protein. Immunolabelling studies, however, reveal that friabilin is present in both soft and hard wheats from very early in grain development (four days after anthesis), and can be found within the starch granules. Brennan *et al.* (1993) surmise, therefore, that friabilin could be involved with starch granule formation and that differences in the structure of starch granules resulting in either hard or soft milling properties might result from different locations of this protein.

NITROGEN AND CRUDE PROTEIN CONCENTRATION

The major influences on grain crude protein (CP) concentration are environmental, particularly climate and nitrogen availability. The genetic

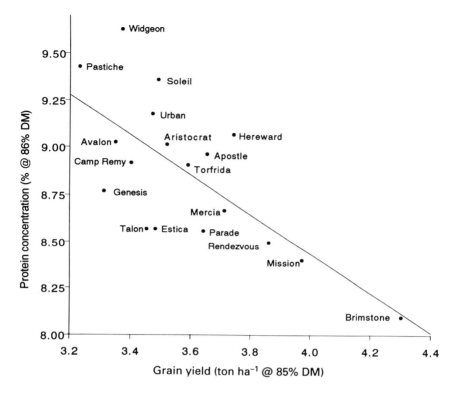

Fig. 3.1. The relationship between grain crude protein concentration and yield of 18 UK bread-making varieties of winter wheat grown without inorganic nitrogen (after Thompson, 1995).

impact, though, is also significant (Stoddard and Marshall, 1990) and varieties often rank in similar orders in different environments, i.e. the variety × environment interaction is relatively small compared with yield. The increases in grain yield achieved through improved harvest indices have, however, often outstripped improvements in the uptake and partitioning of nitrogen to grain. This has contributed to the often reported negative relationship between the yield of varieties and their grain nitrogen concentration (Fig. 3.1). Increases in yield through variety choice have, therefore, tended to depress CP concentration. High protein concentration in some varieties has not been associated with greater protein, but by reduced carbohydrate accumulation (Singhal *et al.*, 1989). Recent analyses (Gooding *et al.*, 1997) suggest that, at a given nitrogen fertilizer application rate, the varieties used in 1993 in the UK reduced grain CP content by 0.8% dry matter (DM) compared with 19 years earlier (Fig. 3.2). The size of this effect is supported by field experiments comparing old and new varieties directly (Gooding *et al.*, 1993b) and can be partly

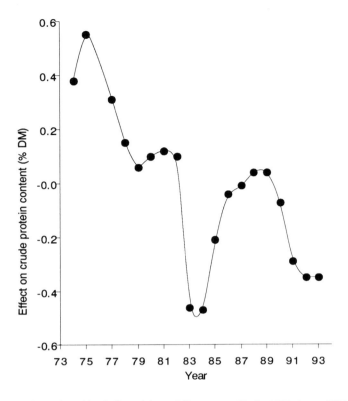

Fig. 3.2. The estimated combined effect of the varieties surveyed in the UK between 1974 and 1993 on the average national crude protein content (after Gooding *et al.*, 1997).

attributed to the adoption of varieties containing dwarfing genes (Gale and Youssefian, 1985; Cosser *et al.*, 1996; Table 3.1). Modern, shorter stemmed varieties have, however, a greater yield response to nitrogen and increased nitrogen fertilizer applications since 1974 have contributed to increased CP. Overall, therefore, the introduction of modern varieties has led to improved CP even though on average they may give lower grain protein concentrations at a given nitrogen application rate compared with some of their predecessors (J.I. Ortiz-Monasterio, CIMMYT, Mexico, personal communication). Even at the same levels of nitrogen availability, plants containing dwarfing genes can have increased offtakes of nitrogen in the grain (Table 3.1).

There is a debate concerning whether the negative relationship between the protein and yield of varieties has a strong causal basis which will be difficult to overcome. An alternative view is that it is, at least partly, an artifact of the breeders concentrating more on methods which increase carbohydrate accumulation rather than nitrogen accumulation, such that nitrogen concentration falls. From a biochemical view, Penning de Vries *et al.* (1974) calculated that 1 g of glucose produced by photosynthesis can be used by the crop to produce 0.83 g of carbohydrate or 0.4 g of protein when nitrate is the N source. Therefore, increasing protein content will use more photosynthate and decrease available photosynthate for starch synthesis and yield. However, regressions of variety grain yield against protein content are not always particularly close, ranging between –0.25 and –0.77 as reviewed by Beninati and Busch (1992). Even relationships towards the top of this range, for example that shown in Fig. 3.1 ($r = -0.71$), still demonstrate that varieties with the same yield can have substantial differences in protein concentration. Maris Widgeon, for instance had more than an extra 1% protein concentration compared with cvs Talon and Estica which gave comparable yields. Nitrogen accumulation in the grain can be increased by exploiting important genotypic differences for both nitrogen uptake and the degree of nitrogen partitioning to the grain (Beninati and Busch, 1992). With more emphasis given to protein concentration in breeding programmes it seems possible, therefore, to combine high yielding varieties with high protein contents. For example, Cox *et al.* (1986) and Beninati and Busch (1992) report good evidence for the existence of major genes conferring increased protein concentration without adverse effects on yield. Ali (1995) detected mutant lines which had higher protein contents, but similar yields to their mother line. Loffler and Busch (1982) were also able to increase both yield and protein concentration simultaneously. Soft wheat varieties developed in North Carolina, USA, are also reported to have higher protein content than others by 2–4%, which was not associated with yield depression (Johnson and Mattern, 1987).

An alternative approach is the growing of mixtures of high yielding varieties with high protein varieties. Some preliminary investigations have suggested that it is possible for such a mixture to yield as much as the high yielding variety with an improved grain protein content. These benefits, however, have not always been realized (Thompson *et al.*, 1992). More needs to be known about the interactions between contrasting varieties grown in field conditions, especially with respect to grain quality. The growing of variety mixtures has been promoted mainly to reduce foliar disease build up, and has focused to date mainly in the UK on mixtures of feed wheat varieties.

PROTEIN QUALITY

Lysine Content

Early work showed that the proportion of lysine in protein was negatively associated with the protein concentration of a variety (Fig. 3.3), between

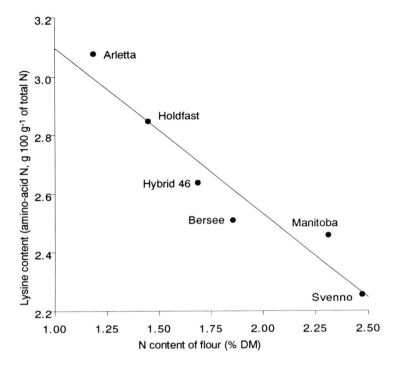

Fig. 3.3. The relationship between lysine content of protein and nitrogen concentration of white flour in six wheat varieties (after McDermott and Pace, 1960).

the endosperms of grain of the same variety, and even between regions of the endosperm of the same grain (McDermott and Pace, 1960). The ability of wheat breeders to improve lysine content whilst maintaining or increasing protein content is, therefore, a problem. Additionally, the total genetic variation in lysine content amongst a large, worldwide wheat collection (12,613 samples) was only found to be 0.5%, which is only one third of the amount required to bring lysine into balance with other essential amino-acids (Blackman and Payne, 1987).

Dough Rheology

Leavened bread production is largely limited to the genomes coding for the necessary proteins to generate an elastic dough suitable for the capture of gas in bubbles during fermentation, thus allowing the dough to rise. This property varies greatly, however, even within bread wheats (*T. aestivum*), such that only certain varieties are considered suitable.

These varieties are those which can give a strong dough when their flour is mixed with water.

The significance of the type of gluten in defining its dough strength and baking quality was demonstrated experimentally by Booth and Melvin (1979). They extracted different protein fractions from a good and poor bread-making variety and interchanged them by substitution. The role of gluten composition and its genetic basis has, therefore, led to it receiving a great deal of attention by molecular biologists and plant breeders in many parts of the world with the aim of improving the bread-making potential of wheats. At gluten subfraction level, varieties which have large quantities of gliadin relative to glutenin tend to have viscous, extensible doughs often suitable for biscuit-making. Those having a relatively low gliadin : glutenin ratio have better elasticity and strength more suitable for baking (Khatkar *et al.*, 1995).

Within the glutenin subfractions, the high molecular weight (HMW) subunits appear to be particularly important because allelic variation is closely linked with variation in bread-making quality and dough resistance. These subunits have been found to be composed of 580–730 amino-acid residues (Tatham *et al.*, 1987) and are easily observed and separated on sodium dodecyl sulphate polyacrylamide gels following electrophoresis (SDS–PAGE) (Fig. 3.4). They are controlled by genes at loci on the long arms of chromosomes 1A, 1B and 1D and are signified by *Glu-A1*, *Glu-B1* and *Glu-D1*, respectively. Varieties of bread wheat contain between three and five major HMW-glutenin subunits. Two of these are coded by genes at *Glu-D1*, one or two by *Glu-B1*, and one or none by *Glu-A1* (Payne, 1987a). Each locus appears to contain only two genes which encode for the two potential subunits, and are designated x and y on the basis of their electrophoretic mobility and cysteine content. The levels of cysteine are reportedly about 50% lower in x types than in y types. The 1A encoded subunit is of the x type. The would-be 1Ay subunit appears to be non-functional due to the presence of a stop codon in the middle of the coding sequence.

The different HMW-glutenin subunits detected by SDS–PAGE have been assigned numbers between 1 and 22, and their appearances alone or in pairs have been given recommended allele designations as shown in Table 3.3. A score has been devised by which each allele is rated for effect on bread-making quality and dough resistance. Total variety scores can then be calculated by adding the scores for the *Glu-1* alleles (Table 3.3). This approach has shown that 60% of the variation in bread-making quality of UK varieties could be accounted for by variation in the *Glu-1* quality scores (Payne *et al.*, 1987). This was further improved to 67% by adjusting scores for the presence of a translocated chromosome which consists of the long arm of *1B* and the short arm of *1R* from rye, which is known to be associated with poor bread-making quality and present in several UK varieties (Payne *et al.*, 1987). Similar studies were conducted

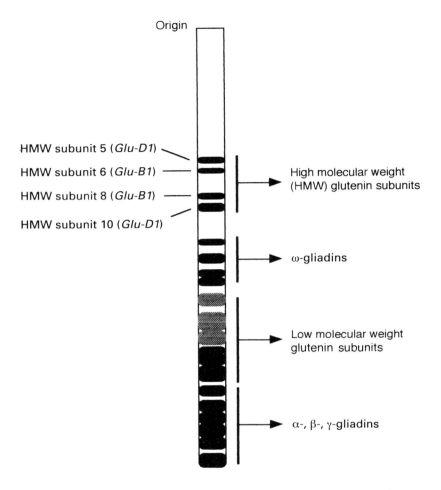

Fig. 3.4. A diagrammatic representation of an SDS–PAGE separation of gluten proteins of cv. Mercia, showing the HMW subunit designations and gene loci.

on Spanish wheat varieties, where *Glu-1* quality score was strongly correlated with dough strength, Zeleny sedimentation volume and the 'ω' measurement of the Chopin Alveograph (Payne *et al.*, 1988). Concurrently, the same *Glu-1* quality score was found to explain 59–69% of variation in a baking strength index of 70 Canadian grown wheat varieties (Lukow *et al.*, 1989). Further evidence for the strong influence of HMW-glutenin subunits on the bread-making quality of wheat varieties comes from back-crossing studies (Moonen and Zeven, 1985), random selections from wheat crosses (Lagudah *et al.*, 1988), and from studies on the bread-making

Table 3.3. Allele and subunit designations, bread-making quality scores and compositions for selected varieties of HMW-subunit glutenins (after Payne *et al.*, 1987; Lukow *et al.*, 1989).

Locus	Allele	Subunit(s)	Quality score	Dough strength	Example UK varieties			
					Avalon	Brock	Mercia	Widgeon
Glu-A1	a	1	3	Strong	✔			✔
	b	2*	3	Strong				
	c	null	1	Weak		✔	✔	
Glu-B1	a	7	1	Weak		✔		✔
	b	7+8	3	Strong				
	c	7+9	2	Strong				
	d	6+8	1	Weak	✔		✔	
	e	20	1	Weak				
	f	13+16	3	Strong				
	g	13+19						
	h	14+15		Strong				
	i	17+18	3	Strong				
	j	21						
	k	22						
Glu-D1	a	2+12	2	Medium	✔			✔
	b	3+12	2	Medium				
	c	4+12	1	Weak		✔		
	d	5+10	4	Strong			✔	
	e	2+10						
	f	2.2+12						
Total variety score					6	3	6	6
Bread-making score					B	D	B	A
Biscuit-making score					D	A	D	D

qualities of homozygous wheat lines lacking particular HMW-glutenin subunits (Lawrence *et al.*, 1988).

Table 3.3 shows the subunit association between bread-making quality and dough strength. Varieties with high *Glu-1* scores, therefore, are generally unsuitable for biscuit-making. However, it can also be demonstrated that UK varieties, even those rated A for bread-making quality, do not approach the maximum *Glu-1* quality score of 10. This highlights the historical fact that varieties grown in the UK have generally produced weak doughs and have been selected principally for the high yields attainable in a maritime climate, which are more suited for the biscuit-making and livestock feed industries. Given that dough strength has been a major limiting factor on bread-making quality in the UK it is, perhaps, not surprising that the alleles conferring dough resistance have such a major impact upon overall bread-making quality. This may not be the case in other areas of the world. For example, Payne (1987b) found that the average *Glu-1* quality scores for variety collections from Argentina and Australia were 9.5 and 8.0 out of 10, respectively, which compared with only 5.2 in the UK. In these countries, therefore, differences in bread-making quality are unlikely to be due to *Glu-1* allelic variation. Even in the UK there have been difficulties with the adoption of relatively new varieties, such as Fresco and Torfrida, which were selected partly for their HMW-glutenin subunits. These have proved to have a dough that is too strong for the Chorleywood process, which has been widely adopted to cope with the generally comparatively weak UK varieties. It was suggested that these varieties could still prove useful in the production of products which require particularly strong doughs such as wholemeal bread, puff pastry products and hamburger buns.

Despite the large amount of evidence implicating the *Glu-1* encoded subunits in dough strength, the precise causal mechanisms have not been defined. It may be that the qualitative differences in the structure of the different subunits are not as important as the associations between certain subunits and the amount and size of the glutenin macropolymer present (Weegels *et al.*, 1996).

As well as the major effect of the HMW-glutenin subunits mentioned above, other proteins present in the gluten must also have additional influences. For example, low molecular weight (LMW) glutenin subunits and some gliadin proteins have also been found to correlate with bread-making quality tests (Wrigley *et al.*, 1981; Pogna *et al.*, 1982; Gupta *et al.*, 1989). More recent studies have focused on these particular proteins because HMW-glutenin subunits fail to explain all the variation in dough and loaf characteristics. For example, cv. Darius only has a *Glu-1* quality score of 4, but it is one of the highest graded French varieties for bread-making (Branlard and Dardevet, 1994). Crossing experiments suggested that it was the absence of three ω-gliadins that was responsible for increased dough resistance and strength, and reduced extensibility of Darius. The ω-gliadins are coded for on the short arms of the same chromosomes as the HMW-glutenin subunits and the loci are given similar notation, namely *Gli-A1*, *Gli-B1* and *Gli-D1*. These loci are complex, containing 9–15 genes, encoding for LMW-glutenin subunits and γ-gliadins as well as ω-gliadins (Payne, 1987a). The null allele for the ω-gliadins was located at *Gli-D1*. Recent studies (Weegels *et al.*, 1994) have involved extracting specific gliadin fractions, in conditions designed to minimize changes in protein functionality, and exploring the effects of adding these fractions to flours of different varieties. Fractions containing hydrophobic gliadins were beneficial to loaf quality, while hydrophilic fractions also containing

LMW-glutenin subunits had smaller effects. Glycolipids extracted with gliadins could be detrimental to loaf volume. The authors conclude that it is a delicate balance of lipid, as well as protein, that governs bread-making quality of a wheat variety. This supports the results of McCormack *et al.* (1991) who found that loose, but repeatable associations could be detected between lipid composition and baking quality of different varieties grown at six levels of nitrogen fertilizer. In addition, variations in starch properties between varieties may also impact upon baking quality. Soulaka and Morrison (1985), for example, found that differences in loaf volumes were also closely related to the gelatinization temperatures of large (as defined by sedimentation) starch granules.

Because gluten properties play a large part in determining pasta quality, associations between storage proteins and the gluten strength of durum wheat have also been researched. A gliadin band labelled 45 has been associated with gluten strength and gliadin band 42 with gluten weakness (Kosmolak *et al.*, 1980). Payne *et al.* (1984), however, showed that genes coding for gliadin bands 45 and 42 were closely linked to genes coding for glutenin subunits LMW-2 and LMW-1, respectively. They concluded, therefore, that associations of gluten strength with the presence of the gliadin bands were more likely caused by the LMW-glutenin subunits.

Although leavened bread production with durum wheats is very limited, partly due to the lack of the D genome, variation in bread-making quality does exist within the species and some varieties can approach the baking potential of good bread-making varieties of *T. aestivum* (Boggini *et al.*, 1995). Pena *et al.* (1994) examined the flour protein, SDS-sedimentation volumes, Chopin alveographs and loaf volumes of 26 durum wheats and found that differences in bread-making quality-related characteristics among durum wheats were dependent on both LMW and HMW glutenin subunit composition. They found that the best varieties for quality were those containing both the *Glu-1B* 7 + 8 HMW and the LMW-2 subunits. They concluded, therefore, that these would be most suitable in breeding programmes involving hybridization with bread wheats. This is consistent with the recognized value of 7 + 8 for bread-making quality in the bread wheats (Table 3.3), and the association of LMW-2 with gluten strength in durum wheat (Payne *et al.*, 1984). Latterly, Boggini *et al.* (1995) have confirmed a strong association between LMW-2 and dough strength. They suggest that the best dough rheological and baking properties were obtained when LMW-2 was combined with HMW subunits 2+, 1 or 1* encoded at the *Glu-A1* locus. They also report varieties with novel LMW and HMW subunits which gave higher loaf volumes.

α-AMYLASE ACTIVITY AND HAGBERG FALLING NUMBER

There appear to be three major processes by which varieties differ to alter the production of α-amylase and therefore Hagberg falling number. The first is the production of α-amylase in sound, ungerminated grain, the so-called *prematurity* α-amylase (Cornford and Black, 1985; Flintham and Gale, 1988). The activity can be equivalent to that in grains undergoing germination as evidenced by obvious shoot and root growth. Varieties which are particularly susceptible to producing damaging levels of α-amylase before germination have been recorded in Europe, the former USSR, South Africa and Australia. This production appears to be increased in these varieties by delayed grain drying due to cool and/or moist conditions and could be due to a single recessive gene. Cornford and Black (1985) noted that many high amylase varieties shared cv. Professor Marechal in their parentages.

Prematurity α-amylase levels can be modified by the presence of the dwarfing genes *Rht1*, *Rht2* and *Rht3*. These genes render wheat less sensitive to gibberellin activity, a hormone implicated in the initiation of α-amylase production by the aleurone layer. *Rht3* is the most effective at reducing prematurity α-amylase activity but causes excessive dwarfing for wide scale incorporation into modern genotypes when compared with *Rht1* and *Rht2*. Recently, some evidence of a systematic, positive relationship between grain size and prematurity α-amylase activity has been reported (Evers *et al.*, 1995). It is suggested that grains above a certain threshold size have anatomical abnormalities that can lead to failure of the enzyme control mechanisms operating in smaller grains. Varieties that had the largest grain sizes were frequently placed in the worst category for excessive α-amylase activity. Cornford and Black (1985) also cite possible breakdowns in control mechanisms to explain high α-amylase activities in certain wheat varieties, in explaining changes in the sensitivity of the aleurone to gibberellin of immature wheat grain.

The second process which impacts upon α-amylase activity is the delay in sprouting of mature grain even in conducive environmental conditions, i.e. seed dormancy. The earliest genetic link noted between α-amylase activity and variety characteristics was the link between seed coat colour and dormancy, i.e. red wheats are less likely to germinate before harvest than white wheats under similar environmental conditions (Nilsson-Ehle, 1914). The genes for red coat are dominant and occur on the long arms of group 3 chromosomes, designated *R/r* loci on chromosomes 3A, 3B and 3D. Evidence suggests that there is a causal link between red seed coat and increased dormancy, rather than just close genetic linkage (Flintham and Gale, 1988). This implies that either the red pigment, or precursors in its biochemical pathway, cause a temporary inhibition of germination. This relationship tends to favour the use of red wheats in climates where rainfall is relatively high close to harvest time, such as the UK. The use of red varieties, however, does preclude the production of wheat for markets where white varieties are desired, e.g. for the principal wheat foods of India, Pakistan and other Asian and African countries including chapatis and noodles.

Traditionally, therefore, white wheat production has been limited to areas receiving very low levels of rainfall prior to and during the harvest period. There are, though, breeding programmes designed to increase pre-harvest sprouting resistance in white wheat. Programmes using artificial rain and germination tests have allowed the selection of white wheat lines with equivalent sprouting tolerance to red wheats (McCaig and DePauw, 1992). Separate dormancy genes have been found in white wheats (Bhatt and Derera, 1980) and DNA markers have been identified which are associated with resistance to pre-harvest sprouting in white wheats (Anderson *et al.*, 1993). More recent research has investigated associations between physical characteristics of waxes covering the embryo and dormancy (Evers and Flintham, 1994).

The third way in which varieties can have altered α-amylase activity is through differences in the rate of α-amylase production once germination has started. As with prematurity α-amylase, this can be reduced by the *Rht* alleles. Table 3.1 shows, for example, that *Rht1* can increase Hagberg falling number by 40 s when levels are close to threshold values for bread production. The production of α-amylase may also be influenced by morphological characters. For example, the presence of awns increases the rate of imbibition of rainwater leading to increases in pre-harvest germination (King and Richards, 1984). Varieties which are predisposed to lodging will also tend to have lower Hagberg falling numbers due to the more moist microclimate in flattened crops.

There are, therefore, several ways in which varieties can differ with respect to Hagberg falling number. The strong genetic component is modified by the climate, and can sometimes interact with nitrogen availability (Gooding *et al.*, 1986b) but generally varieties rank in a similar way, irrespective of husbandry. Thompson (1995), for example, found that the Hagberg falling numbers of 10 UK bread-making wheat varieties grown at one site in an extensive, organic system were correlated ($r = 0.76$) with their average performance in nationally surveyed intensive systems. The large differences that can occur between varieties grown on the same farm in the same season, and the non-linear relationship between α-amylase activity and Hagberg falling number, increase the potential damage of cross-contamination during harvest, handling and storage of varieties destined for bread-making with varieties giving low Hagberg falling numbers. Vaidyanathan (1987), for example, found that a single grain of cv. Fenman with a falling number of *c.* 75 in 19 grain of cv. Avalon with a falling number of 440 was enough to reduce the combined falling number

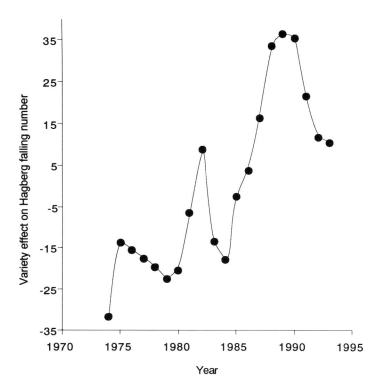

Fig. 3.5. The estimated combined effect of the varieties surveyed in the UK between 1974 and 1993 on the average national Hagberg falling number (after Gooding *et al.*, 1997).

to 250. Care must also be taken when choosing varieties to be grown as a blend together in the same field.

The increased knowledge concerning the genetics of α-amylase production, together with a greater appreciation by plant breeders and variety evaluators of its damaging effects for many markets, has led to long-term improvements in the Hagberg falling number. In the UK, changes in variety usage appears to account for an increase of over 30 s in national falling numbers from 1974 to 1993 (Fig. 3.5).

MINERAL CONTENTS

Surveys of UK grain by McGrath (1985) and Chaudri *et al.* (1993) found statistically significant differences between winter wheat varieties for phosphorus, potassium, calcium, magnesium, copper, zinc, manganese and iron. The magnitude of difference was not, however, particularly large.

Chaudri *et al.* (1993) argued that they were not sufficient to be agriculturally important in the UK, and may partly have arisen through negative associations with grain weights. The fact that genuine differences between genotypes do exist, however, provides a basis for variety selection to help rectify nutritional inadequacies in animal and human populations which are heavily reliant on wheat grown on deficient soils. New mineral enriched varieties of wheat are currently being developed to be grown in areas deficient in particular in iron and zinc (R.D. Graham, Adelaide, Australia).

FUNGAL CONTAMINANTS

Varieties of wheat differ significantly with respect to all the major fungal contaminants of grain. Resistance to pathogens attacking the flower, namely ergot, loose smut and bunt can be achieved in certain varieties that have a large proportion of florets that remain closed, or only open for short periods during anthesis. Physiologically based disease resistance has been observed against loose smut in some varieties. Both loose smut and bunt races exist, though, that could challenge the basis of this resistance. A possibly more stable form of field resistance against loose smut has been suggested by Jones *et al.* (1995). Systemic infection by the fungus alters gibberellin-induced responses causing infected plants to grow taller and head earlier than healthy plants. This facilitates the movement and dispersal of spores from infected heads to healthy flowers when they are in anthesis. In varieties containing dwarfing genes such as *Rht1* and *Rht2*, which confer reduced sensitivity to gibberellin, loose smut infected plants generally grow shorter than healthy plants and head later. Smut spores may, therefore, be less able to transfer from below the healthy canopy to uninfected ears above. Meanwhile, the healthy ears are likely to have progressed beyond the susceptible stage (anthesis) before spores are released. These observations may explain the near extinction of loose smut of wheat in the British Isles where it was once a very common problem.

Varietal differences in blackpoint can be significant and consistent over sites, seasons and production systems (Gooding *et al.*, 1993c; Ellis *et al.*, 1994, 1996). The susceptibility of varieties is sometimes associated with grain size (Thompson *et al.*, 1993a; Ellis *et al.*, 1996), leading to the suggestion that small seeded varieties may be resistant if their glumes remain closed and do not permit the passage of spores to the grain. Recent experiments (S.A. Ellis, Royal Agricultural College, UK, unpublished data) have revealed, however, that larger grained varieties may have a longer period of grain drying, possibly allowing greater amounts of subepidermal fungi, including *Alternaria* spp., to develop. These studies have also shown that blackpoint severity on different varieties is more closely correlated with grain size and moisture content during grain development, rather than at harvest. Variety × time interactions during grain maturation may therefore explain why grain characteristics at harvest are only related to variety susceptibility in some seasons.

ENERGY FOR NON-RUMINANTS

Compared with the amount of knowledge concerning the relationship between varietal characteristics and quality for various human foods, the factors relevant to livestock feed are poorly understood. Wiseman *et al.* (1993), however, found that varieties of wheat differed significantly with respect to apparent metabolizable energy (AME) when fed to young poultry. In a trial with eight varieties, each grown at two sites, variety means ranged from 9.01 to 13.67 MJ kg^{-1} DM. Variety effects, however, interacted significantly with site and with the rate of wheat inclusion in the diet. There was a close association between the AME and the concentration of digestible starch present in the wheat. There was also a negative relationship between the AME and the intake of insoluble non-starch polysaccharide (NSP) present. This might be because the NSP reduces the digestibility of starch *in situ* but this has yet to be confirmed. AME of varieties was not correlated with endosperm texture (Wiseman *et al.*, 1993), thousand grain weight, specific weight or Hagberg falling number (Wiseman and McNab, 1995). Further work confirmed that the AME of varieties interacted strongly with site, but not with nitrogen treatment (Wiseman and McNab, 1995). Varieties also differed with respect to digestible energy contents for pigs but, as with poultry, effects were not consistent over sites.

In studies on pigs the level of digestible nitrogen from rations based on wheat is considered particularly important, and this varies between varieties (Wiseman, 1993). These experiments revealed that the higher the nitrogen content, the greater the digestibility of the nitrogen. As discussed earlier, however, wheats with higher nitrogen content have greater amounts of protein fractions containing reduced concentrations of essential amino-acids. It was concluded that the lower concentration of essential amino-acids with high grain nitrogen content might be offset by a higher overall nitrogen digestibility. Lysine and methionine are important dietary requirements for non-ruminants (Wiseman, 1993) and differences in amino-acid composition and digestibility between wheat varieties is of particular significance.

GENETIC ENGINEERING OF WHEAT

The knowledge of the molecular basis for many of the quality attributes of wheat grain has contributed to the interest in delivering genes conferring desired traits through genetic engineering rather than traditional plant breeding methods. The successful transformation of wheat, however, has lagged behind that of many other crops. Early transformation of crops, for example tobacco in the early 1980s, was mediated through the use of a bacteria known as *Agrobacterium tumefaciens*. This organism is responsible for inducing tumours, known as grown gall, in susceptible plant species. It incorporates some of its DNA into the host plant genome as part of its natural disease cycle and, therefore, is a convenient vehicle by which genes can be directly passed into plants. Unfortunately, the host range of *A. tumefaciens* does not extend to grasses and cereals and it was not until 1992 that fertile transgenic wheat plants were reliably produced (Vasil *et al.*, 1992). The gene introduced into the transformed plants conferred resistance to the broad-spectrum herbicide phosphinothricin (PPT). The delivery of the gene was achieved by coating either tungsten or gold particles with the desired DNA (Vasil *et al.*, 1991) and then using these to bombard cells derived from immature wheat embryos. Cells exhibiting resistance to the herbicide were selected and regenerated to produce adult plants with micropropagation techniques (Redway *et al.*, 1990).

Since this initial report a number of others have successfully produced transformed wheat using the particle bombardment technique (Weeks *et al.*, 1993; Becker *et al.*, 1994). Shewry *et al.* (1995) have applied the technique to commercial bread-making and feed wheat varieties rather than varieties predisposed to higher success rates for transformation and regeneration. They are also developing systems to produce plants transformed with HMW-glutenin subunits and claim that 'genetic engineering will have a major impact on the end use properties and utilization of wheat [before 2005]'.

4

The Crop Environment and Grain Quality: Weather and Soils

Wheat is essentially a temperate crop, which is increasingly spreading to warmer, non-traditional regions. The majority of the important wheat growing areas such as the prairies of North America, the Pampas in Argentina and the Steppes of Central Asia are typified by environments which would otherwise support a grassland type of native vegetation. The same does not apply, however, for major wheat areas in eastern China and western and southern Europe.

In comparison with other cereals, wheat has much broader adaptation to a wide range of environments, as outlined in Chapter 1, and it is grown in many different circumstances in various parts of the world. For many years, wheats have been developed to be better suited to cooler climates than the crop's original habitat in the Middle East. With modern inputs, these semi-dwarf, more disease resistant types are highly productive in such temperate areas. Since 1982, however, considerable efforts have been made to broaden the genetic variability for the more marginal and warmer climates (Curtis, 1988). As a result, tropical wheat production has developed in two basic environments. Firstly, in situations of high temperature, dry short seasons with few disease problems; and secondly, with high temperature, humid short seasons with disease challenges. Within the tropics, of course, temperate regions exist at higher altitudes – such as the Chiang Mai area of northern Thailand where winter wheat can be grown in a long dry cool season from October to April. In these conditions, few serious disease problems develop (Saunders, 1988). In contrast, in the wetter and higher temperature conditions of the Cagayan Valley of Luzon, in the Philippines, heavy disease pressures have sometimes resulted in complete crop loss. Dry conditions during the growing season may be mitigated by residual soil moisture from earlier rains, and soil storage of water is particularly significant in some regions, such as upland areas in parts of India.

As well as encouraging disease, excessive rainfall can cause additional problems, including increased lodging, delayed harvesting, difficult threshing, disadvantaged soil preparation and drilling, and increased leaching of nitrates from the soil (Martin *et al.,* 1976). Rainfall during the latter stages of grain maturation can delay harvesting and stimulate pre-harvest sprouting of the grain. Despite the harmful effects of excessive rainfall, however, adequate soil moisture is a prerequisite for germination, nutrient uptake and efficient photosynthesis. This is emphasized by the high yields achieved in some regions of northwestern Europe (Fig. 1.4). These areas receive relatively high levels of rainfall, distributed evenly throughout the year, without the excessive temperatures which lead to heavy disease pressures and profuse photorespiration.

Klapp (1967) felt that the most favourable climate for wheat would be a mild winter followed by a warm summer with high radiation, without excessive cooling summer rains. Optimum temperature for growth is usually within the narrow range of 20–25°C (min. 2°C, max. 30°C). Ideal annual rainfall can vary widely, between 250 and 1000 mm depending on seasonality and stage of growth (Stoskopf, 1992). Air temperatures and precipitation considerations alone, however, may be highly misleading. Fischer and Byerlee (1991) point out that in warmer climates the crop canopy may be up to 10°C cooler than air temperatures as a result of evapotranspiration, which in turn is influenced by the vapour pressure deficit in the crop environment. Root temperatures in the soil are also considered very significant. Fischer (1985) stresses the value of humidity differences, as well as the rate of temperature change; diurnal temperature ranges; solar radiation and water supply. These factors have been used in classifying the wheat growing areas, mentioned previously, in non-traditional production regions (Fischer and Byerlee, 1991).

The diversity of wheat adaptation can be illustrated by reference to particular production areas. In the hot rainfed areas of central and southern India wheat production has relied upon the drought tolerant variety C306 which was first released in 1967. In this instance, drought tolerance is attributed, in part, to a deeper root system and longer vegetative period (Aggarwal and Sinha, 1987). High yields have also been obtained at low rainfall sites and without irrigation in Bangladesh, from varieties grown in deep silty soils which retain high moisture status when the crops are produced with appropriate agronomy. The heat tolerance of particular varieties when grown in irrigated, hot, dry, disease-free, regions can also result in respectable yields in unfavourable circumstances. Good examples include cv. Debeira in the Sudan and cv. Fang 60 in Thailand (Fischer and

Byerlee, 1991). Reactions to high temperature have been found to interact with radiation level and high humidity in phytotron studies in Australia. With the temperatures explored (30–25°C versus 18–13°C), grain number was depressed by higher temperatures during booting, but responses varied substantially between wheats from different climates (Table 4.1). High temperatures during grain growth, which accelerate canopy senescence and reduce grain growth, may be mediated by the temperature influences on roots (Kuroyanagi and Paulsen, 1988). Effects may also depend on soil mechanical impedance in the root zone.

Wheat thrives on well-drained, fertile, medium to heavy textured soils, particularly silt and clay loams with a high nutrient status. Soils of high water capacity are also favoured to reduce reliance on summer rains. Conversely, very sandy or inadequately drained soils are poorly suited to wheat production. Effects of soil type, however, interact closely with weather and cropping system. Excessive soil moisture can result in soil pores being filled with water, impeding gaseous diffusion and resulting in anaerobic conditions which in turn can severely restrict root function, growth and possibly plant survival. On fine-textured soils (e.g. clays) water will tend to be held more tightly and rainfall will limit occasions when tractors and cultivation equipment can operate without damaging soil structure. On coarse-textured soils (e.g. sands) increased rainfall can wash mobile nutrients such as calcium and nitrate nitrogen through the soil profile, below the root zone. Heavy rainfall in tropical areas induces laterization through excessive leaching of soluble salts, or alternatively in contrasting circumstances the excessive accumulation of surface salts through evaporation of soil moisture. In both cases this can limit plant growth but, as with weather factors, the influence of soil conditions depends on the variety grown. As part of the strategy to extend the production of wheat into the warmer areas, selection of varieties for tolerance to various soil mediated stresses has, and is, being achieved. Production in the warmer areas can take place on a wide range of soil types, including

Table 4.1. Influences of temperature increases (from 18–13°C to 30–25°C) at different stages on different yield components of wheat (after Dawson and Wardlaw, 1989; Wardlaw *et al.*, 1989).

Variety type	Impact of high temperatures during booting on grain number (%)	Impact of high temperatures during grain filling on grain weight (%)
Temperate European wheats	−47	−46
Middle Eastern wheats	−22	−32
Tropical Philippine wheats	−14	−30

alfisols, oxisols, ultisols, vertisols and entisols. Of these, the oxisols and ultisols are highly leached, have a low pH, and a high soluble aluminium concentration. These soils and conditions are typical of much of tropical America, Southeast Asia and lowland tropical Africa. Alfisols are more fertile and less acid. Phosphorus deficiency is, however, a common problem on these soils. Calcium can also be low but this affects wheat less than other crops. Vertisols have a high content of swelling clays which, in the dry season, cause deep cracks to form. They tend to be very sticky in the wet season and very hard in the dry season and therefore tillage is troublesome in areas where primitive implements are still in use. The soils can, however, be very productive with appropriate mechanization, fert-ilization and irrigation and are used in, for example, subtropical regions of Australia (north-east) for wheat production. Entisols lack highly devel-oped soil horizons, having formed relatively recently over new geological formations (e.g. lava, land recently exposed, and stabilized sand dunes), soils degraded by erosion, rocks which resist weathering or are toxic to plants, or alluvial soils recently deposited by rivers and streams. Entisols are used extensively for wheat production in subtropical India.

Acid conditions can be tolerated to an extent by wheats derived from crossing old Brazilian wheats and semi-dwarf Mexican wheats in South American growing areas (Wall *et al.*, 1991). Micronutrient deficiencies are common on these soils. Zinc, for example, is limiting in many acid wheat soils of Africa and Latin America; copper is limiting in soils in Kenya and Brazil, and boron deficiency is widespread. Soil organic matter breakdown is encouraged by cultivations and higher temperatures, and reduced or-ganic matter content contributes to capping problems, compaction, water-logging and erosion. Poor performance of wheat in some rice–wheat systems is thought to be due, in part, to poor root channels as a result of the slower breakdown of rice root debris. These nutrient and structural limitations adversely affect both yield and quality.

Wind can exacerbate drought conditions or reduce grain moisture before harvest through a drying effect. Crop husbandry techniques such as crop spraying can be delayed, or the efficiency reduced, by excessive wind at optimum spray timings. Wind after stem elongation can also cause the crop to lodge with detrimental consequences for both yield and quality (Weibel and Pendleton, 1964), in particular when weak strawed varieties are grown in high fertility situations.

As discussed in Chapter 1, varieties have specific requirements for vernalization and photoperiod. These criteria, together with other variety characteristics, will often interact with market standards to define which class of variety is grown. This is demonstrated by the distribution of wheat classes grown in the USA (Fig. 4.1). Spring varieties, for example, are grown in areas where the harsh winters preclude successful winter wheat production. The major area of white wheat production, Washington State,

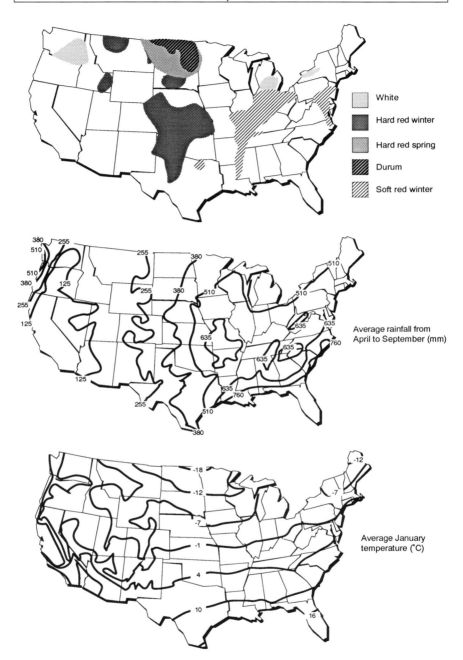

Fig. 4.1. The distribution of wheat classes in the USA in relation to summer rainfall and January temperatures.

is in an area with very low summer rainfall. This reduces the chance of pre-harvest sprouting which can otherwise be a problem for these varieties. The major areas of hard and soft winter red varieties can also be partly distinguished on the basis of summer rainfall. The hard wheats are favoured for leavened bread production. Bread production, however, also requires high protein contents in the grain. As discussed later, this is more readily achieved with lower summer rainfalls compared with the soft wheats destined for biscuit production which require lower protein contents. Once these broad regional distinctions are defined there is the possibility of positive feedback, increasing the association between a particular wheat class and its principal regions of production. For example, if a hard red wheat is grown in an environment which helps it achieve the market requirements to which it is suited, plant breeders will also endeavour to produce hard wheats which will be more productive in that type of environment.

As well as associations between climate and wheat class grown, large changes in both yield and quality will inevitably occur on a yearly and area basis due to the weather and soil properties. The importance of year on year variations in weather on wheat quality is exemplified in Fig. 4.2, which shows changes in UK wheat quality in national surveys. Dramatic falls in specific weights in 1985, 1987 and 1992, Hagberg falling numbers in 1977, 1985, 1987 and 1993, and very high protein contents in 1976 are much larger than can be explained by changes in husbandry and variety choice and were undoubtedly due to the weather conditions experienced. These influences have large implications for utilization, and the worth of the wheat, particularly in regions where wheats are normally near to threshold levels required for high value markets. In the UK, the proportion of home-grown wheat used by UK millers for bread-making generally increased between 1974 and 1994 (Fig. 4.2). This was due, in part, to the adoption of the Chorleywood bread-making process and the imposition of import tariffs. The trend, however, has frequently been interrupted, particularly in years when Hagberg falling numbers have been excessively low (Fig. 4.2). In 1977 and again in 1987 the poor availability of home-grown wheat of sufficient quality resulted in the price of UK bread-making wheat increasing dramatically compared with the value of wheat destined for livestock production, which has to date been less sensitive to quality characteristics.

An ability to understand the influences of the climate on wheat quality has many applications. If wheat quality could be predicted early enough before harvest, the information could influence the way farmers grow, harvest and store the crop. Alternatively, prediction of grain quality for the prospective wheat harvest would be of interest to millers when deciding how much grain, and at what price, should be bought on a futures market or on a deferred delivery basis (Catania, 1993).

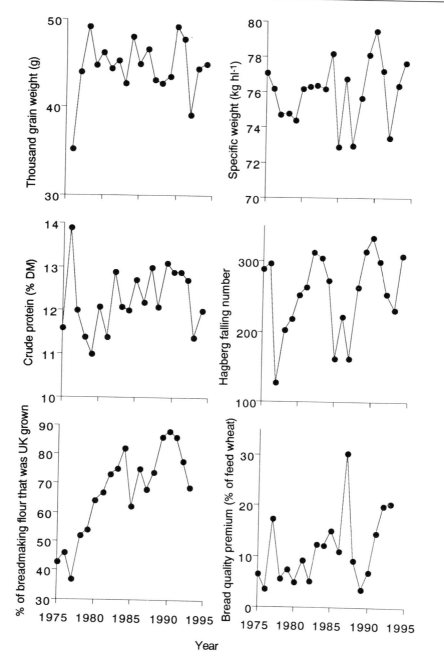

Fig. 4.2. The quality, contribution to bread-making grists, and price premiums paid for bread-making quality wheat in the UK between 1975 and 1994. Source: Home-Grown Cereals Authority, London.

GRAIN NITROGEN AND PROTEIN CONCENTRATION

The most intensively studied climatic effect on grain quality is grain protein content and its importance has been recognized for more than a hundred years (Sander *et al.*, 1987). The significance of climate, relative to husbandry, on grain nitrogen concentration is emphasized in a study by Rixhon and Vandam (1987). They reviewed yearly results between 1968 and 1985 of grain nitrogen concentration from the same site. Average annual results ranged from 1.88% in 1974 to 2.61% in 1971. The greatest range achieved within a season by using contrasting cultivation techniques and varieties was, however, only from 1.93% to 2.32% in 1984. In addition, it has long been recognized that the high nitrogen concentration of the hard wheats in the USA is due to the environment in which they are grown. When grown in soft wheat regions the grains of hard wheats are starchy and only marginally, if at all, higher in grain nitrogen contents than soft wheats (Fifield, 1945).

In many instances, grain nitrogen concentration appears to increase with increasing temperatures and reduced rainfall. Such influences are revealed by regression techniques relating annual crude protein concentration in surveyed wheat samples in England between 1974 and 1993 (Anon., 1974–1994) with national rainfall, temperature and nitrogen fertilizer data. Following the removal of the variety effects shown in Fig. 3.2, variations in yearly protein contents in different regions can be described by equation 4.1. This shows that crude protein (CP) content (% dry matter (DM)) was positively associated with summer temperature (mean daily temperature 25 June to 15 July, °C; Student *t*-value = 12.89, 105 d.f.), nitrogen fertilizer (kg N ha^{-1}; t = 9.54), and summer rainfall (28 May to 8 July, mm; t = 3.81), and negatively associated with spring rainfall (5 March to 27 May, mm; t = 8.95).

$$CP = 5.73 + (0.339 \times \text{summer temp.}) + (0.0122 \times N)$$
$$- (0.00774 \times \text{spring rain}) + (0.00625 \times \text{summer rain}) \qquad [4.1]$$

The relationship between the observed values of national protein content and those predicted by equation 4.1 is shown in Fig. 4.3.

Similar associations have been found in many countries and are reviewed as follows:

1. Other UK studies have found a positive association between temperature from 17 June to 14 July and grain nitrogen concentration in grain from one site between 1957 and 1973 (Benzian and Lane, 1986). Farrand (1972) found grain protein to increase with June rainfall, the major period covered by the summer rainfall coefficient in equation 4.1. It may be that June rainfall is encouraging mineralization at a time when nitrogen availability has a large impact on grain nitrogen concentration.

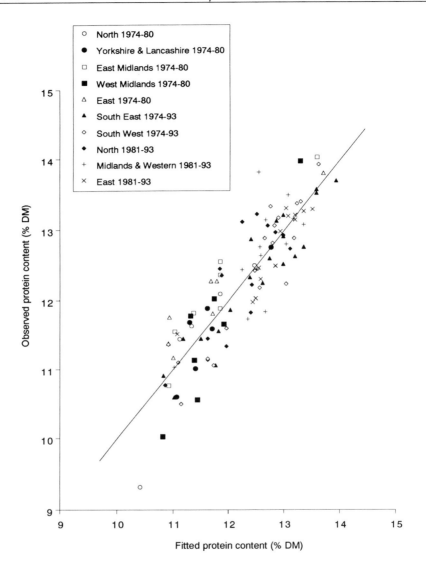

Fig. 4.3. The relationship between values of crude protein predicted from equation 4.1 and the crude protein as surveyed in regions of England over a 20-year period. The correlation coefficient (*r*) is 0.90 (after Smith and Gooding, 1996a, 1996b).

2. In Germany, Schipper (1991) found temperature during grain filling to have an important positive influence on protein concentration.

3. In Australian variety trials extending over 27 years, grain nitrogen concentration was positively associated with the number of hours above 35°C during grain filling (Blumenthal *et al.*, 1991).

4. In China, grain nitrogen concentration in durum wheat was positively associated with average daily temperatures during grain filling and negatively correlated with growth period (He *et al.*, 1990).

5. In the Pacific northwest of the USA, higher temperatures during grain filling generally increased grain nitrogen concentration of soft white winter wheat, but the same climatic variables did not consistently have the same effects in all locations and years (Rao *et al.*, 1993). Also in the USA, protein content of winter wheat grain in the semi-arid central Great Plains of the USA (e.g. Nebraska and Colorado) was negatively associated with rainfall between 40 and 55 days before maturity, and positively associated with maximum air temperature for a five-day period 15–20 days before maturity (Smika and Greb, 1973). The relationship with temperature was curvilinear, protein content increasing up to temperatures of 32°C, followed by reductions in hotter conditions.

6. In Canada, above average protein concentrations in west Canadian hard red spring wheat have been associated with below average midsummer rainfall and above average July temperatures (Hopkins, 1968).

In smaller scale field trials, nitrogen concentration in the grain has been negatively associated with high levels of natural precipitation or increased moisture deficits (Hera *et al.*, 1986; Ivanic and Vnuk, 1988; Campbell *et al.*, 1991).

These national and field experiment results are consistent with many studies in more controlled environmental conditions showing that an increase in air temperature during grain growth increases nitrogen concentration (Sosulski *et al.*, 1966; Partridge and Shaykewich, 1972; Sofield *et al.*, 1977; Spiertz, 1978; Kolderup, 1979; Campbell *et al.*, 1981; Vos, 1981; Kalinin, 1986, 1988; Ahmad *et al.*, 1989; Milosev, 1989; Triboi *et al.*, 1990). Increased moisture availability has also decreased grain nitrogen concentration in protected environments (Sosulski *et al.*, 1966; Campbell *et al.*, 1981; Kalinin, 1988) but Benzian and Lane (1986) found no significant relationship between rainfall and nitrogen concentration.

Mechanisms to explain the effects of temperature and moisture have been postulated. Increasing temperature can reduce carbohydrate accumulation more than nitrogen accumulation as the rate of senescence is increased thereby reducing photosynthesis and grain growth. Spiertz *et al.* (1971) found that the duration of green leaf area of the flag leaves on temperate spring wheats lengthened as temperature after anthesis was reduced (Table 4.2). Petr (1991) reports, from central European experiences, that high temperatures during grain formation give smaller grain and lower yields. The size and number of starch granules per endosperm is also reduced in hot conditions (Tester *et al.*, 1995). Concurrently, the rate of relocation of nitrogen from the vegetative parts of the plant to the grain is increased by higher temperatures (Neales *et al.*, 1963). The rate of

Table 4.2. Temperature influences on the size of the green area of the flag leaf as a percentage of total flag leaf area following anthesis (Spiertz et al., 1971).

Temperature (°C)	Four weeks after anthesis	Six weeks after anthesis
10	85	42
15	69	36
20	31	0
25	2	0

protein synthesis in the grain is also promoted by warmth more than the rate of starch synthesis such that protein concentration increases (Spiertz, 1977). High soil temperatures also favour the mineralization and uptake of nitrogen (see Chapter 6). Smika and Greb (1973) cite this to explain a very close relationship (correlation coefficient (r) = 0.95) between wheat grain protein on the Great Plains and average daily soil temperature from spring until the soft dough stage.

The above review reveals that relationships with temperature are more frequently reported than those with rainfall. The mechanisms underlying the effect of rainfall and soil moisture content are complex. High levels of water availability often increase total nitrogen yield in the grain, at the same time as reducing nitrogen concentration. Water can increase nitrogen availability to the crop because it increases root growth; the mass flow of water and therefore nitrogen towards the plant; mineralization of nitrogen from soil organic matter; movement of nitrogen fertilizers into the root zone; and the availability of other nutrients which can have large influences on root growth and distribution (Sander *et al.*, 1987). In severe drought conditions, however, the grain tends to be shrivelled with little starch accumulation and therefore a relatively high content of bran and correspondingly high concentration of nitrogen. Rainfall prior to grain filling has been thought to encourage dilution of early nitrogen reserves by vegetative proliferation in arid environments, and leaching and other forms of soil nitrogen loss in humid environments. Powlson *et al.* (1992), for example, found that loss of nitrogen applied to wheat in the spring was closely related to the accumulated rainfall in the following three weeks. Later rainfall, during grain filling, can cause a relative nitrogen dilution by extending leaf life and maintaining photosynthesis and carbohydrate translocation (Taylor and Gilmour, 1971).

Influences of climatic factors other than temperature and moisture are poorly defined. Campbell *et al.* (1969) and Bremner (1972) found that reduced light intensity lowers the grain nitrogen and grain weight to the same extent such that nitrogen concentration remains the same. In contrast, Benzian and Lane (1986) reported a negative relationship between irradiation in July and August (late grain filling) and grain nitrogen concentration such that an increase of 1.0 MJ m^{-2} day^{-1} gave a reduction of 0.03% N. In pot experiments Spiertz (1977) also found reductions in grain nitrogen concentrations with increases in light intensity at four different temperatures, whilst Sofield *et al.* (1977) reported greater nitrogen accumulation at reduced illumination during grain development. These results are consistent with shading experiments showing grain yields to decrease and protein content to increase as shading is made more severe (Pendleton and Weibel, 1965).

Other than soil moisture influences, often inferred from precipitation data, there has not been much work concerning the effects of other soil properties on grain nitrogen concentration. Increasing salinity is often more detrimental to grain yield than nitrogen accumulation, resulting in increased protein concentration (Francois *et al.*, 1986; Devitt *et al.*, 1987). Soil textural differences are difficult to interpret because they are difficult to examine within one field experimental site. A large survey of grain protein contents, between 1935 and 1959 for 161 survey districts in Canada, did reveal certain soil categories to give differences in protein content not predicted by models including just rainfall and temperature data (Hopkins, 1968). These categories included interzones between brown (natural vegetation of prairie grassland) and black (parkland), and between grey (northern forest) and black soil types. In contrast to the effects of climate, soil textures giving rise to higher grain yields can simultaneously be associated with high grain nitrogen contents (Carr *et al.*, 1992). This is often indicative of fertile conditions where soil nitrogen availability is maintained relatively late into the growing season.

PROTEIN, DOUGH AND LOAF QUALITY

Despite the observation that higher temperatures and soil moisture deficits increase grain nitrogen concentration, there is evidence that these factors, particularly when severe, can alter the functionality of the nitrogen. Increases in grain nitrogen concentration would normally be expected to increase the proportion of vitreous grains, sedimentation volumes, gluten concentration, dough strength and loaf quality. Improvements in some of these measures following warm dry weather during grain growth have been reported (Pelikan and Belan, 1987). The relationships between nitrogen concentration and protein quality criteria may, however, alter depending on weather patterns. For example, Schipper (1991) found that increases in grain nitrogen content resulting from high temperatures during grain development were associated with greater dough resistance but lower extensibility than that of grains with similar nitrogen content produced at lower temperatures but with higher nitrogen application rates. Ivanic and Vnuk (1988) found that although nitrogen concentration reduced in high precipitation years, the proportion of grain nitrogen present as true protein increased. A study of wheat varieties from eight sites in Nebraska, USA, in each of two years (Graybosch *et al.*, 1995) found that sodium dodecyl sulphate (SDS) sedimentation volume was negatively associated with hours of temperatures above 32°C and hours of relative humidity below 40% during grain filling. This inferred changes in protein quality because similar relationships with protein content were not detected. In support of this suggestion, it was also observed that high temperatures and low relative humidities were associated with reduced glutenin and gliadin contents, respectively. This could explain the earlier observations of Finney and Fryer (1958) who found a negative relationship between air temperatures above 32°C during the last 15 days before harvest and loaf volume. Work by Blumenthal *et al.* (1993) in Australia suggested that periods of high temperature stress favoured gliadin production at the expense of glutenin production thereby effecting a reduction in protein quality. Excessively high temperatures and low relative humidities during grain filling, despite increasing grain nitrogen concentration, have therefore reduced dough strength, gluten quality and loaf rating in some studies (Halverson and Zeleny, 1988; Blumenthal *et al.*, 1991).

Increasing salinity can also have important influences on the qualitative aspects of the nitrogen present. For example, Devitt *et al.* (1987) found contrasting influences on the concentrations of different free amino-acids in the grain and Francois *et al.* (1986) found that increasing salinity increased the quality of durum wheat but not of bread wheat despite increases in grain nitrogen concentration. In contrast, Rhoades *et al.* (1988), and Kelman and Qualset (1993) found that the use of brackish irrigation water in Californian valleys increased the loaf and SDS sedimentation volumes of wheat. Hunshal *et al.* (1989) found that gluten content was significantly increased under saline conditions as were scores for flavour and the texture of chapatis.

HAGBERG FALLING NUMBER AND α-AMYLASE ACTIVITY

α-Amylase activity in grain is associated with the process of germination. Factors encouraging germination, therefore, will tend to result in increased enzyme activity and a corresponding drop in the Hagberg falling number. Situations which give rise to damp conditions around the maturing ear can lead to pre-harvest sprouting and hence lower Hagberg falling numbers.

These conditions can develop in a standing crop of wheat with high levels of rainfall close to harvest. Additionally, if rain and wind combine to lodge the crop, a humid microclimate encouraging germination within the mat of laid straw and ears is almost inevitable.

As well as a reduction in Hagberg falling number due to wet conditions close to harvest, it is also recognized that prolonged ripening exhibited as delayed drying of the grain under cool and moist conditions can have a similar effect in the absence of visible sprouting (Flintham and Gale, 1988; Hough, 1990; Astbury and Kettlewell, 1992). This effect may be particularly pronounced in certain varieties (Gold *et al.*, 1990) but Hagberg falling number in the UK can be related to the temperature in June and July which coincides with the period of grain growth as well as rainfall in August which coincides with harvesting and possible stimulation of pre-harvest sprouting. Regression analysis of Hagberg falling numbers (HFN) surveyed in regions of England between 1974 and 1993, following removal of variety effects, reveals a negative association with August rainfall (mm; $t = 3.5$, 100 d.f.), and positive associations with June temperature (mean daily temperature, $°C$; $t = 6.13$), August temperature (mean daily temperature, $°C$; $t = 6.44$) and nitrogen fertilizer (kg N ha^{-1}; $t = 3.61$), as shown in equation 4.2.

$$\text{HFN} = -300 - (\text{August rain} \times 0.494) + (\text{June temp.} \times 15.8) + (\text{August temp.} \times 19.3) + (\text{nitrogen} \times 0.318) \quad [4.2]$$

The relationship between the observed values of Hagberg falling number and those predicted by equation 4.2 is shown in Fig. 4.4.

In earlier years (1959–1970), rainfall in July, August and September were all positively associated with increased α-amylase activity, which would be expected to reduce Hagberg falling number (Farrand, 1972). These findings are consistent with those of Karvonen *et al.* (1991), who reviewed Scandinavian variety trials between 1968 and 1988. They found that if relative humidity was greater than 80%, and average maximum daily temperature was less than 13°C during grain filling, Hagberg falling number was below 120. Conversely, if relative humidity fell below 70% and average maximum daily temperature was above 16°C and radiation exceeded 12 MJ m^{-2} day^{-1}, Hagberg falling numbers were above 230. Additionally, Kettlewell (1993) found that Hagberg falling number was positively related to the potential evapotranspiration rate calculated one week before and after the grain moisture content was at 40%.

SPECIFIC WEIGHT

Specific weight is a complex character assessing as it does both the density of individual grains and their packing properties. Low values can

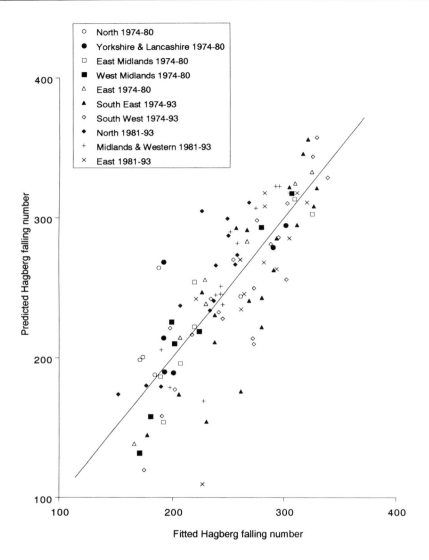

Fig. 4.4. The relationship between values of Hagberg falling number predicted from equation 4.2 and the Hagberg falling number as surveyed in regions of England over a 20-year period. The correlation coefficient (*r*) is 0.83 (after Smith and Gooding, 1996a, 1996b).

indicate very poor grain filling (Bayles, 1977) and, therefore, climatic influences leading to grain shrivelling can impair specific weight through reduced packing efficiency. Such conditions can include severe drought during grain filling, weather conditions conducive to rapid disease spread, and/or lodging (Weibel and Pendleton, 1964). Adams *et al.* (1988) listed a number of potential contributory factors to unusually low specific weights including drought stress and high temperatures. They observed that clayey soils usually produced their highest specific weights when rainfall was at or below average. Repeated wetting of grain close to harvest can lead to a reduction in specific weight (Braken and Bailey, 1928). Favourable conditions during grain site formation may lead to greater numbers being set than can adequately be filled later on in the season.

Weather conditions conducive to high grain nitrogen concentration may have a positive effect on specific weight due to increased density of individual grains. Such conditions, however, often do not lead to optimal grain filling thus compounding the complexity of weather influences on specific weight.

EAR AND GRAIN INFECTION

Interactions of wheat with potential pathogens are mediated by the weather, influencing both the macro- and microclimate of the crop. Infection, subsequent disease development, and the resultant impact on yield and grain quality will be governed in particular by the prevailing temperature and humidity.

In the maritime climate of the UK, cool dull wet weather after anthesis can result in a complex of so-called ripening diseases which can cause necrosis of leaf tissues and infections of the ear (Yarham, 1980). The impact on yield and quality will be affected by weather interactions on grain filling and ripening as much as disease development. Wet weather in May and June can encourage *Septoria nodorum* and subsequent glume blotch on ears. Weak pathogens including *Ascochyta graminearum*, *Didymella exitialis* and *Cladosporium herbarum* can invade ageing leaves, accelerate senescence, and contribute to contaminating sooty mould development on ears. Other fungi can be stimulated by the flowering process (Yarham, 1980). Pollen grains on the ear can provide the substrate for weakly pathogenic fungi such as *Botrytis cinerea* and *Fusarium poae*, allowing them to invade glume tissues under humid conditions.

Strange *et al.* (1978) demonstrated that invasion of glumes by *Fusarium graminearum* is stimulated by choline and betain diffusing from anthers under humid conditions. *Fusarium* head blight of wheat is widespread in temperate production regions. It is particularly favoured in areas having mild winters and humid springs such as occurs in the southern cone of South America, southern Africa, Australia, eastern and western Europe and North America (Luzzardi, 1985). Losses of 10–40% of yield and toxin build up in ears, giving up to 1 million tonnes loss of grain, have been reported from areas of China (Zhong, 1988). Sutton (1982) reports that disease initiation and subsequent epidemics are greatest at temperatures above 15°C and continuous moisture over three days during flowering. Parry *et al.* (1995) suggest that optimum conditions for *Fusarium* ear infection occur when temperatures above 15°C coincide with wet periods (100% relative humidity) of at least 24 h during anthesis. Over the season they conclude that epidemics are likely when there have been warm dry soil conditions early in the season to promote the development of *Fusarium* foot rot and the production of inoculum on stem-bases. Intense rainfall during anthesis can then effectively disperse *Fusarium* inoculum to the ears.

Ergot dispersal and infection is heavily dependent on the weather. The sclerotia that are distributed on the soil appear to be stimulated to germinate by undergoing a chilling period of between 5 and 60 days at temperatures of 0–10°C. Subsequent development, however, requires higher temperatures depending on the location of the ergot population. In the UK, cereal ergots appear able to germinate between mid-May and the end of August, although at any one site the duration of germination will be less. The most important factor appears to be the time of onset of warmer weather. Ascospore liberation begins about seven days after germination and is particularly intense at 77–86% relative humidity. The release of ascospores does not often coincide with the flowering period of wheat but secondary infection of wheat may occur after the ascospores have caused primary disease on grasses such as blackgrass (*Alopecurus myosuroides*). Subsequent infection then depends on cool wet weather at anthesis which favours the production and dispersal of fungal spores and extends the period of host susceptibility. Having an extended period over which florets are open due to cool wet weather also lengthens the time available for loose smut to invade the ovary, thus influencing the value of the crop for seed purposes.

Weather conditions also play an important role in the development of blackpoint. Potential causal organisms such as *Alternaria* and *Helminthosporium* spp. require a grain moisture content of at least 30–33% to grow (Christensen and Kaufmann, 1965), so high rainfall and/or heavy, extended dew periods causing delayed drying of the grain would be expected to increase disease severity. Hanson and Christensen (1953) observed that blackpoint was most severe in wetter seasons and Svetov (1990) recorded particularly high levels of blackpoint in years with greater rainfall at the milky-ripe growth stage. Languasco *et al.* (1993) found wet conditions during flowering to extend the duration of a water film over the developing ear and hence spore germination and colonization of tissues by *Alternaria* spp. They developed a model to forecast blackpoint severity on the basis of rainfall and temperature after ear emergence.

Spores of common bunt (*T. tritici*) in soil or on seed germinate and produce infectious hyphae in response to moisture and cool temperatures. In an analysis of climatic data and results of Swedish field experiments performed during 1940–1988, Johnsson (1992) showed that the temperature during the first 11 days after sowing was strongly correlated with the frequency of winter wheat spikes infected. Infection was most frequent when the mean temperature during this period was 6–7°C. Wiese (1987) quotes a wider range of 5–15°C. In the UK most infection has been through seed borne spores but high levels of infection due to soil borne inoculum may be partly due to unusually dry weather, favouring the longer term survival of spores in the soil (Yarham, 1993b).

The incidence of karnal bunt (*T. indica*) is often associated with damp weather during the flowering period (Sharma *et al.*, 1994) and Mavi *et al.* (1992) developed a regression model relating severe infection to the number of rainy days and intensity of evening relative humidity during specific weeks. Loose smut (*U. tritici*) infection is also favoured by wet weather during the flowering period, in association with temperatures below 23°C (Verma *et al.*, 1986; Wiese, 1987).

IRRIGATION

The single most limiting factor to wheat productivity is soil moisture. The relative importance of irrigation, however, depends greatly upon natural precipitation, crop yield potential, the cost of irrigation and the availability of water. As already stated in Chapter 1, much of the increase in productivity of wheat in the developing world was associated with irrigated systems. Much of this irrigation is achieved by temporary flooding, with water guided through surface channels or furrows at key developmental stages of the wheat (Fischer *et al.*, 1977). This method requires almost level or contoured land and it can be the least efficient in terms of water utilization and land conservation. The machinery necessary, however, is not complicated and water can be applied to large areas relatively quickly. Alternative irrigation systems include a large number of types involving rotary sprinklers and rain guns. These can arguably apply water to the crop at more precise rates but often incur larger capital and running costs.

In countries where summer rainfall is often more plentiful, such as the UK, wheat is not seen as a crop that regularly justifies irrigation. It is generally less dependent on high soil moisture availability than more valuable crops such as vining peas, potatoes and sugar beet. In these areas, however, irrigation may still be justified on coarse-textured or sandy soils in drier seasons and areas, particularly if the farm has already obtained mobile irrigation equipment in order to produce other crop species in the rotation. Wibberley (1989) reviews a number of UK trials and notes that responses to irrigation are most likely on sandy soils, where nitrogen supplies are good, ear diseases developing subsequent to irrigation are controlled, and lodging risk is reduced by appropriate choice of varieties and use of growth regulators. He suggests best yield effects are achieved if the soil moisture deficit is prevented from falling below 25–50 mm.

In more arid areas, irrigation is often needed shortly after seeding in dry soils to initiate germination, and/or after three to four weeks to help the development of nodal roots (Bhardwaj *et al.*, 1975). Following crop establishment, however, work in many contrasting areas suggests that yield responses to irrigation are greatest when they alleviate drought during flowering. In Mexico field conditions, however, Fischer *et al.* (1977) found that the period of sensitivity extended from 25 days before flowering to 20 days after flowering. This concurs with UK experiments when best results were obtained from irrigation during late stem extension, especially just before ear emergence (Wibberley, 1989). Yield improvements by irrigation at this time are mostly due to increases in grain numbers rather than grain weight. Significant increases in thousand grain weight do occur but these contribute less to grain yield effects except when irrigation is delayed until grain filling (Fischer *et al.*, 1977).

Many of the influences of irrigation are similar to the effects of natural precipitation. For example, grain protein concentration often declines with increasing irrigation (Barber and Jessop, 1987; Nakhtore and Kewat, 1989), although total yield of protein is raised. However, compared with natural precipitation, irrigation treatments can be better controlled with regards to amount and timing. Early field experiments in Australia showed that sprinkler irrigation interacted strongly with nitrogen application timings for both yield and protein concentrations (Wood and Fox, 1965). When no nitrogen was applied, or when nitrogen was applied at flowering, irrigation increased both yield and protein percentage simultaneously. When nitrogen was applied at sowing, or split between sowing and flowering, irrigation increased yields but decreased protein concentration.

SUMMER FALLOW

Where irrigation is not financially viable or practical, and where rainfall is insufficient to support a crop every year, land can be left uncropped as a fallow. The principal reason for this is moisture conservation but fallowing can also improve the available nutrient status of the soil, and help suppress some weeds. Fallowing has been a common practice in many of the major wheat growing areas (see agroecological systems 1 and 2 in Chapter 1) and can be particularly profitable where annual rainfall is less than 380 mm. In such situations, between 15 and 30% of rainfall falling in the fallow period can be stored within the soil (Martin *et al.*, 1976). Fallowing is most effective when most of the rain falls in the winter and where summer evaporation is low, but is ineffective on coarse sandy soils or in high rainfall areas. For successful summer fallowing, Martin *et al.* (1976) list the following principles that must be observed:

1. The surface of the soil must be kept sufficiently rough to absorb rains and prevent wind erosion.
2. Weed growth must be suppressed to conserve soil moisture.
3. The operations must be accomplished at low cost.

So-called 'stale seedbed' techniques can be further exploited for moisture retention by utilizing no-till (direct) drilling systems subsequently to the benefit of germination and crop establishment.

Relay cropping can also be exploited in warmer circumstances to limit soil and seedbed deterioration prior to wheat establishment. Moisture in the soil can be exploited by broadcasting wheat seed into standing rice 15–21 days before harvest. Strong establishment can ensue which can remove irrigation requirement, and be more beneficial than a no-till situation in rice–wheat systems in Bangladesh.

5

Crop Establishment

Many factors associated with sowing a wheat crop have a large impact on plant establishment. The degree to which the plants are then able to compensate determines the subsequent effects on yield. Final grain quality can also be influenced by sowing factors, but the reported effects are often small (Kiyomoto, 1986) particularly when considering small variations in practices around those normally adopted to optimize yield. Plant population and yield relate in a parabolic fashion in wheat, with a reported yield plateau between 200 and 400 plants m^{-2} in the UK. Wibberley (1989) reports little difference in yield from populations between 100 and 400 plants m^{-2}, but recommends a target of 250 plants m^{-2} for UK conditions. The elements which can have an influence include seedbed preparation, sowing date, seeding rate, sowing method, drilling depth, and row widths and patterns.

SEEDBED PREPARATION

The soil tillage that is employed to produce a seedbed has many objectives and can take many forms depending on soil type, timing and time available, rainfall, equipment and power available, soil erosion risk, and the control of weeds, pests and diseases (Wibberley, 1989). The first, and sometimes main requirement, will be to bury and incorporate trash of the previous crop. This hinders the development of certain established weed, pest and disease problems through interrupting life cycles. Small seeded weed species with low levels of dormancy and inadequate reserves to establish from depth will also be suppressed. Any surface applied organic manures or fertilizers can also be incorporated.

In many wheat growing areas of the world these functions are achieved by mouldboard ploughs, which were traditionally exploited for weed control and trash burial. Modern variants differ greatly in size, shape and attachments to obtain different furrow widths, depths, and stubble distribution. The general principle, however, is that the plough shears off a furrow slice by forcing a triple wedge through the soil which is then inverted and partially broken up. The surface is rendered suitable for the use of other implements to produce a finer tilth, for example with discs followed by tines and harrows. The tilth produced should be fine enough to give intimate contact between the seed and soil to allow sufficient water imbibition for germination, and to assist in the accurate placement of seed at optimum depth.

The tilth should be deep enough to allow roots to penetrate fully. Soils are aerated by disturbance and relieved of surface compaction, thus contributing to an adequate oxygen : carbon dioxide ratio in the soil atmosphere. As well as aiding root growth, such aeration and soil movement may also result in a flush of nutrients, particularly nitrate, due to increased microbial activity. Better root growth and nutrient release contribute to increased yields with increasing plough depths, as demonstrated by early work in the USA (Fig. 5.1). Optimum fineness of tilth, however, is difficult to quantify. Seedbeds with soil aggregates 2–3 mm in diameter resulted in earlier emergence and higher wheat yields than when aggregates exceeded 4 mm on a loam soil (Braunack and Dexter, 1990). On clay soils, seedbed aggregates of 1–2 mm with a dry bulk density of 1.2–1.3 g cm^{-3} have been optimal (Jaggi *et al.*, 1972). When establishment is protracted, for example in late sown winter crops, there are benefits in having a much more cloddy seedbed to prevent the formation of a hard soil cap. Cloddy seed beds also help reduce erosion when there is a high risk of soil blowing. Aggregates less than 1 mm tend to restrict aeration and may reduce emergence in wet conditions, but have given earlier emergence in drier seasons. When soil is flood irrigated, results can be radically different due to complete collapse of the aggregates.

Despite the proven benefits of ploughing there are also a number of disadvantages associated with inverting the soil from a depth of 20–30 cm, together with the subsequent multiple use of disc, tine and harrow operations to prepare a seedbed. In the first place, plough based systems are relatively costly in terms of both finance and energy usage for machinery, fuel, and labour requirement. Secondly, much of the nitrate made available is lost as leachate as the seedling roots and young crop demand are too small to utilize the available nutrients fully. Thirdly, the natural biology and physical structure of the soil is disrupted, with soil damage and compaction likely on wet and/or heavy soils. Lastly, repeated movement of the soil tends to dry it out.

Because of these factors many systems have been developed for re-duced, and minimal, tillage. Many of the implements designed to replace

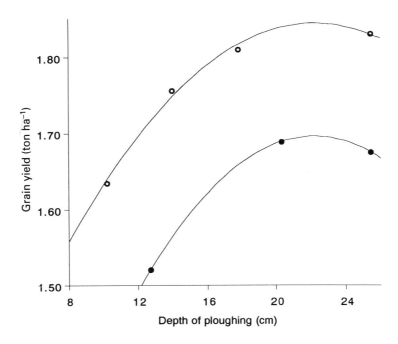

Fig. 5.1. The relationship between plough depth and grain yield in Nebraska (○) and Utah (●) (from data presented by Martin *et al.*, 1976).

ploughing for cereal establishment are reviewed by Davies (1988) and are only summarized here. Early implements were heavy, such as chisel ploughs and large fixed tine cultivators which increased work rate but frequently left uneven, cloddy surfaces except on sandy soils. Subsequent development attempted to avoid this by cultivating soil from the top down to produce a fine tilth at the surface and loosened soil below, but without large-scale movement and inversion which would bring large clods to the top. This might be achieved, for example, with banks of tines arranged at increasing depths on the same implement, or with different implements applied in different passes. Davies (1988) suggests that compaction at lower levels is eventually relieved such that only the upper layers need working, allowing a true shallow cultivation system to be maintained. As well as tines, discs can achieve shallow cultivation although they can cause localized smearing and compaction in moist soils. They may also have difficulty penetrating hard dry soils. Ploughs with the capability of working at 10–15 cm depth can also achieve a shallow cultivation as well as providing burial of surface trash. Davies (1988) considers them to be applicable to medium textured soils, but can be unsuited to hard clays with penetration problems.

The benefits of reduced cultivation techniques for lowering establishment costs were clearly demonstrated by Patterson (1975). When averaged over three years and two soil types (clay loam and silty loam) establishing winter wheat with two passes of a chisel plough (13 cm), followed by two passes of a disc harrow and then drilling was only 86% of the cost of ploughing (20 cm) followed by a disc harrow and drill. Costs were further reduced to 53% when considering a shallow plough (10 cm) followed by a combined cultivator and drill system.

The most extreme form of reduced or minimal tillage is direct drilling. Direct drills deliver seed into a slot, formed either by a disc or tine, in otherwise uncultivated land. This is often referred to as a no-tillage system (Sprague, 1986). Wilkinson (1975) summarized the site and soil conditions that increased the chance of direct drilling being successful in the UK. These include a high degree of soil uniformity, self-mulching tilth properties, resistance to compaction, high porosity, a high rate of water infiltration and profile drainage, level site, and high biological activity.

The benefits of reduced tillage systems for labour and fuel energy requirements are revealed in Table 5.1. Adoption also makes it possible to grow wheat on land previously in grassland where it was too clayey or too shallow to cultivate with conventional ploughing. Other benefits include an increased level of organic matter in the surface layers (Griffith *et al.*, 1986). Repeated use of direct drilling creates a gradient from high concentrations of organic matter near the surface. In contrast, long-term use of ploughing results in a fairly uniform distribution in organic matter throughout the ploughed layer. In some cases total amounts of organic matter can be greater with direct drilling, due to more rapid decomposition from ploughing. This is because the greater aeration, increased contact between soil and decomposing material, and increased exposure of fresh organic surfaces encourage microbial breakdown. Organic matter

Table 5.1. Labour, fuel energy and area capability assessments for cereal cultivation systems in the UK (from Davies, 1988).

Cultivation system	Depth of primary tillage (cm)	Labour requirement (h ha^{-1})	Fuel energy (kWh ha^{-1})	Area that can be covered by one man in a season (ha)
Plough, cultivator, drill	23	4.0	89	88
Shallow plough, combined cultivator drill	13	2.0	52	178
Heavy duty spring tine × 3, drill	5.0–7.5	1.9	38	180
Sprayer, direct drill	0	1.0	11	353

breakdown is faster in warmer wheat growing areas, such as parts of South America, Africa and Southeast Asia, where reduced tillage techniques are seen as having major benefits over traditional practices for organic matter conservation (Lal, 1991; Wall *et al.*, 1991). In cooler areas, such as the UK, however, some field experiments have sometimes shown that the differences between ploughed and direct drilled land for microbial activity and organic matter have either been small or non-existent (Powlson and Jenkinson, 1981).

Shallow cultivations, and direct drilling in particular, can result in greater amounts of crop residue being left on the soil surface. This helps reduce soil erosion and water runoff (Griffith *et al.*, 1986). Water evaporation during cultivation is also minimized. Direct drilling leaves the soil more consolidated or compacted as indicated by a higher bulk density. This would normally be detrimental to root growth and water movement within the soil. This is not, however, commonly observed partly because the lack of ploughing leads to an increase in earthworm populations and activity which creates natural channels and aeration. Earthworms also bring significant quantities of subsoil to the surface and take down decaying organic matter (Edwards, 1975). If the soil experiences wetting and drying cycles, vertical planes of weakness may also develop and be preserved. Further channels may be formed along the roots of a previous crop (Russell *et al.*, 1975). Compaction that can arise from reduced cultivation systems tends to be less severe and nearer the surface than the plough pans and compaction developing from inversion tillage. It is, therefore, often easier and less costly to correct.

Continued use of reduced cultivation systems does, however, lead to new problems, not previously important when traditional ploughing was the norm. Generally, reducing the use of non-inversion tillage has increased reliance on herbicides. Wicks (1986), for example, considers that the traditional reason for ploughing on the Great Plains of North America was weed control and that the availability of herbicides has now rendered it possible to establish wheat with reduced tillage techniques. Some weed species that have increased in severity, however, are not easily controlled. In the UK, for example, reduced tillage has been associated with an increase in grass weeds, particularly sterile brome (*Bromus sterilis* L.) and blackgrass (*Alopecurus myosuroides*). Both of these species are small seeded and do not remain viable in the soil for many years. Conventional ploughing gave suppression in the past. Reduced cultivation practices, particularly on heavy land, have led to them becoming troublesome on many farms. Active ingredients for selective control of sterile brome are rare and therefore expensive, while herbicide resistance and poor control of blackgrass is increasingly being reported. Similar problems with direct drilling have been encountered in other countries. In Australia, direct drilling has been associated with a build up of giant brome grass (*Bromus diandrus*) (Heenan *et al.*, 1994). Effective grass weed control is considered essential for adoption of minimum tillage systems now in the light of experience. To counteract the long-term build up of small seeded grass and perennial weeds in reduced tillage systems it may be more appropriate to use rotational ploughing, i.e. ploughing each field at least once every five or six years. This is the approach utilized in the EU-funded Less-Intensive Farming and the Environment (LIFE) project (Jordan *et al.*, 1993). Most arable crops in this system are established with Dutzi equipment which cultivates and sows seed in a one pass, non-soil inversion operation. Field beans established by broadcasting followed by ploughing are undertaken one year in six.

As well as influences on weeds, direct drilling can also lead to shifts in pest and disease pressures. In the UK, direct drilling led to increases in slug and wireworm attacks, but damage by stem boring fly larvae was greater after ploughing (Edwards, 1975). In warmer areas certain aphids and stem borers have been suppressed by non-tillage systems (Dubin *et al.*, 1991). Yarham (1975) considered that non-ploughing techniques increased the risks from diseases that overwintered on plant debris on the soil surface such as *Septoria* spp. and eyespot (*Pseudocercosporella herpotrichoides*). In warmer areas, zero tillage increases inoculum density and severity of tan spot (*Drechslera triticirepentis*) and spot blotch (*Bipolaris sorokiniana*), creating the need for rotations to reduce the survival of the pathogens on wheat residues (Dubin *et al.*, 1991). As a result of greater crop protection pressures there are requirements for greater agrochemical use in no-till systems.

Minimal tillage reduces the amount of nitrate released for seedling growth. In the UK, 25 kg N ha^{-1} fertilizer to the seedbed is recommended in autumn when direct drilling. In terms of support energy usage, therefore, the benefits of direct drilling for reducing fuel energy can be almost completely lost when the energy required for extra nitrogen and herbicide manufacture is taken into consideration (Table 5.2).

With such a large range of influences on soil, weed, pest and disease factors it is not surprising that the benefits of reduced cultivation over ploughing or vice versa are not consistent, and depend on particular factors limiting growth. In a review of reduced cultivation and direct drilling experiments in the UK between 1957 and 1974, Davies and Cannell (1975) reported lower yields of winter wheat than after ploughing. After 1970 variations amongst the cultivation techniques were much less. They attributed the relative improvement in shallow and no-tillage techniques to an increase in the availability and use of herbicides, and also to better machinery and management. In a continuation of this work, Cannell *et al.* (1980) found that direct drilling gave inferior yields to ploughing in wet years, but the position was reversed in comparatively dry years. This better performance was attributed to increased water availability and

Table 5.2. Support energy requirements (MJ ha^{-1} year^{-1}) for different cultivation systems used for establishing wheat in the UK (estimated by White, 1981).

	Conventional establishment (plough based, 150 kg N ha^{-1})	Direct drilling system (175 kg N ha^{-1}, twice the amount of herbicide)
Nitrogen	10,950	12,775
Phosphate (50 kg P$_2$O$_5$ ha^{-1})	700	700
Potassium (50 kg K$_2$O ha^{-1})	400	400
Seed	719	719
Herbicides	139	278
Fuel for cultivations	1,846	212
Fuel for harvesting	625	625
Cultivation equipment	701	627
Tractor	489	166
Combine harvester	1,590	1,590
Drying plant	550	550
Grain drying	2,436	2,436
Total	21,145	21,708

deeper rooting in spring in direct drilled plots. In a number of experiments, the responses to nitrogen are greater with direct drilling and shallow cultivation. At low nitrogen fertilizer application rates, yields are often higher in ploughed systems, but at higher rates, yields from reduced tillage systems approach those achieved following ploughing and may, in some cases, be greater (Dowdell and Crees, 1980; Vaidyanathan and Davies, 1980).

For quality influences different cultivation methods have given variable effects. Cannell *et al.* (1980) found that both direct drilling and shallow cultivation techniques led to reduced protein concentration in the grain compared with ploughing. This was associated with reduced total nitrogen yields. In contrast, Gorbacheva *et al.* (1989) found that ploughing increased yields but had no effect on grain protein concentration compared with shallow cultivation. In Canada, the protein concentration and yield have been increased simultaneously in no-tillage systems (Aulakh and Rennie, 1986). Direct drilling in Australia after legume crops has sometimes led to reduced grain protein contents compared with crops drilled conventionally following cultivation (Heenan *et al.*, 1994).

SOWING DATE

The sowing date is largely governed by climate and the requirements of a rotation. Figure 5.2 indicates the diversity of sowing and harvest dates for wheat production in selected countries. In general, the higher the latitude, the cooler the summer temperatures, the longer the cropping period, and the earlier the drilling. At high latitudes wheat is in the

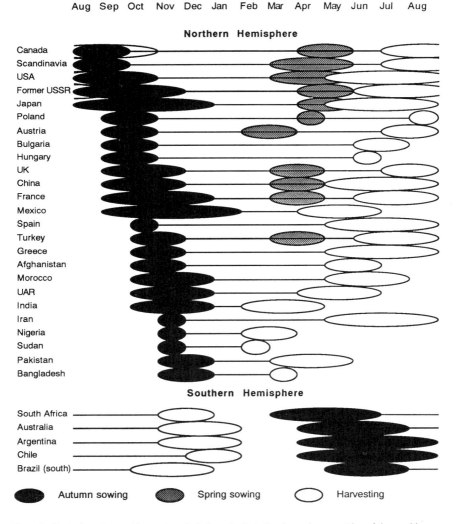

Fig. 5.2. Typical sowing and harvest periods for selected wheat growing countries of the world (adapted and updated from Martin *et al.*, 1976).

ground for a substantial period and thus in a position to trap more light energy through photosynthesis and accumulate more dry matter. Root development can also be extensive, improving anchorage against frost heave and winter kill in severe conditions, and reducing the impact of summer droughts. In these areas sowing date in the autumn is often determined by the date of harvest of the previous crop. To ensure good crop establishment in the autumn, at a time when weather and soil conditions may be rapidly deteriorating, speed and timeliness is very important. This is particularly true where large areas are to be drilled, leading to the greater popularity of more minimal cultivations. Early sowing means, however, that there is less time for trash disposal which may carry disease or pests or damage young seedlings directly through the release of toxins. There is greater risk of disease carryover from one crop to the next, and an extended period in the autumn suitable for disease and pest outbreaks. Weed control is more difficult as more weed germination occurs after the crop is drilled (Cosser *et al.*, 1995). Improvements in potential yield through early sowing may, therefore, be obscured when weed, pest and disease control is compromised, and adds to growing costs through high crop protection inputs. Early sown wheat may also undergo excessive tillering leading to taller weaker straw susceptible to winter damage and lodging (Fielder, 1988).

In hotter climates the cropping period is restricted to the cooler winter months. In extremes, for example in the Sudan, sowing date is critical to make sure that the reproductive phase coincides with the coldest part of the season (Ageeb, 1994), and the total growing season is restricted to just 90–100 days. In Southeast Asia, the sowing of wheat is often within rice rotations such that wheat can be planted in the cooler, drier season between November and March, when land would otherwise be fallow from rice (Hobbs *et al.*, 1994). Early sowing, however, exposes the seedlings to excessive temperatures leading to reduced tiller numbers, vigour, and grain sites and an increased risk of stem base and root rot diseases. Conversely, late-planted crops experience excessive rainfall from anthesis to maturity, resulting in poor grain set, increased foliar diseases, and a higher percentage of blackpointed and sprouted grain (Razzaque and Hossain, 1991).

The precise influences of early or late sowing, therefore, depend on the factors limiting the cropping period in the first place. Regarding grain quality, there are many reports in the literature of the effect of early and delayed sowing in a large number of locations. It is, however, difficult to quantify because of factors specific to location. For example, in Saudi Arabia, delaying sowing from November to December reduced thousand grain weight and water absorption but also reduced blackpoint severity and sprouting damage. These latter beneficial effects were ascribed to earlier sowing increasing the chance of rains coinciding with grain maturation (Mustafa *et al.*, 1987). In several experiments, crude protein concentration has been increased by sowing at dates which have given reduced yields. In Australia this has occurred by sowing in April rather than June (Batten and Khan, 1987) and in the Ukraine by sowing in late, rather than early, October (Netis, 1987). Delayed sowing has also increased crude protein contents in South Africa (Manley and Joubert, 1989), Finland (Kivisaari, 1985), Iran (Khodabandeh, 1985), Poland (Fatyga, 1991), China (Li *et al.*, 1991) and in the UK (Hayward, 1990a), often in association with reduced yields, thousand grain weights, and/ or specific weights (Kivisaari, 1985; Hayward, 1990a; Fatyga, 1991). Irrespective of increases in protein content, sowing excessively late can sometimes reduce Pelshenke values (Mahajan and Nayeem, 1990) and test baking (Khodabandeh, 1985) suggesting a possible reduction in protein quality. Mazurek and Kus (1991) found that late sown wheat produced grain with reduced grain viability and seedling vigour. In other cases, however, a wide range of sowing dates has had very little effect on quality in general, and protein concentration in particular, despite very large effects on the rate of early plant establishment (Campbell *et al.*, 1991). Effect of sowing date on Hagberg falling number has only been assessed in a minority of cases but these have shown that even dramatic changes in sowing date has had either little or no impact (Hayward, 1990a,b). Where early sowing causes weakness of the straw and greater lodging, an associated detrimental effect on Hagberg falling number might be expected.

INTERACTIONS WITH VARIETY

Some experiments have shown the effect of sowing date to interact with variety with respect to yield, thousand grain weight, specific weight, protein content and Hagberg falling number (Fielder, 1988; Cutforth *et al.*, 1990; Spink *et al.*, 1993). Over a range of sites and seasons in the UK, Fielder (1988) demonstrated that some winter wheat varieties were particularly suited to early (mid-September) rather than normal (late September) or late (mid-October) sowing. Varieties not suited included those with poor standing power as earlier sowing generally increased straw length and, therefore, increased the risk of lodging. Earlier sowing of these varieties might still be beneficial, however, if they could be kept standing by plant growth regulator treatments. Early sowing can also increase the severity of eyespot, *Septoria tritici*, and sometimes take-all. Varieties which were particularly susceptible or intolerant of these diseases, therefore, often performed less well when drilled early. Sometimes, variety interactions on yield were also reflected in thousand grain weights and specific weights but effects on quality were not as consistent as those on yield. The significant interactions that are repeatable may, however, be exploited such that certain varieties might be chosen for particular sowing dates as governed by certain cropping sequences (Spink *et al.*, 1993). Drilling dates are also important in relation to vernalization requirements of winter wheats, and the latest safe sowing dates are sometimes recommended for particular varieties (Anon., 1995b).

More striking variety × sow date interactions have been observed by drilling wheats out of season, for example by sowing spring wheats in autumn and vice versa. The main biological difference between winter and spring varieties is requirement of a vernalization period of the former. The late autumn/early winter sowing of spring wheats is a relatively common practice in the UK where many winters are relatively mild and when previous crops may not be harvested until October or November (Hayward, 1990b). This practice attempts to combine the relatively high protein quality traditionally associated with the spring wheat varieties with the higher yields attributable to autumn sowing. This strategy has often been successful but has resulted in large yield reductions at sites and in years with particularly severe winters and/or when the spring wheat has been sown too early in autumn. In a demonstration of the benefits of autumn sowing spring wheats, Hayward (1990b) drilled both winter and spring wheat varieties at four dates from November to April. The spring wheats had comparable yields and protein content to the winter wheats when sown in November. As drilling date was delayed, grain yield and specific weight declined, and the protein content increased. This was most dramatic with the winter varieties, presumably due in part to inadequate vernalization. In Chile, spring wheats have also given heavier yields and specific weights when sown in autumn but, in line with the UK results, gave the highest values for protein concentration, sodium dodecyl sulphate (SDS) sedimentation volume and loaf quality when sown in the spring (Hevia *et al.*, 1988).

Interactions with Seed and Soil Borne Pathogens

Wheat seed is prone to infection by a wide range of pathogens, which may develop from the seed or soil. Influences on plant establishment are highly variable and can depend on levels of inoculum, variety and soil conditions. The influence of seed treatments and sowing date on emergence from seeds infected with *Fusarium* was demonstrated by Wainwright (1995). He showed that plant numbers per 10 m row were reduced from 246 to 117 by sowing on 3 December rather than 26 October. These figures were increased to 587 and 757, respectively, when the seed was treated with triadimenol plus fuberidazole fungicide.

SEEDING RATE

The amount of seed sown per unit area is governed by several factors. The aim is to establish a target population of plants. The weight of grain needed to do this will depend on the mean grain weight and the proportion of seeds expected to produce adult plants (equation 5.1):

$$\text{Seed rate (kg ha}^{-1}) = \frac{\text{target population (plants m}^{-2}) \times \text{thousand grain weight (g)}}{\text{expected establishment (\%)}} \quad [5.1]$$

Seed rates should, therefore, be increased as the expected establishment rate declines. Higher seed rates are required for poorer seedbeds, and for grain of reduced germination capacity and vigour. Seed rates should also increase with mean grain weight, although many small grains may also reflect poor seedling vigour.

Establishing higher numbers of plants increases the amount of light interception and crop competitivity, particularly in the early growth stages of the crop. Several studies have shown negative relationships between sowing rate and subsequent weed infestations (Marwat *et al.*, 1989). There are, however, a number of attendant risks. More densely sown crops often have an increased susceptibility to a number of diseases such as powdery mildew and *Septoria*, possibly due to the increased humidity within the canopy and duration of leaf wetness, respectively (Tompkins, 1992). Interplant competition may also lead to weaker, taller stems. Stapper and Fischer (1990), for example, found lodging to increase as sowing rate was increased from 50 to 200 kg ha^{-1}. When moisture resources are scarce, drought problems may be exacerbated as the increased leaf area leads to greater evapotranspiration. Recent studies have shown, however, that the relationship between water use and sowing rate is complex. Under conditions of high drought stress, Tompkins *et al.* (1991) demonstrated that increasing seed rate from 35–140 kg ha^{-1} increased total water use by the crop but reduced water loss by evaporation. The effect of the contrasting seed rates on water use for transpiration depended on growth stage. Before anthesis the water use was highest for the higher seed rate but after anthesis this position was reversed. Sowing 140 kg ha^{-1} resulted in 21% higher grain yields and a 9 kg grain cm^{-1} H$_2$O higher water use efficiency. This, together with earlier evidence (Martin *et al.*, 1976), counters the idea that wheat must be sown thinly under semi-arid conditions to avoid subsequent crop failure.

Increasing seed rate also has financial implications. Most commentators stress the relative cheapness of seed, faced with potentially higher costs of poor establishment in adverse circumstances from false seed economies (Martin *et al.*, 1976; Wibberley, 1989).

Lowering seed rates places greater reliance on the plants ability to compensate, particularly by increasing the number of tillers per plant, but also through increasing the number of grains per ear (Fig. 5.3). Reduced plant numbers may also lead to increased thousand grain weights (Ellen, 1990) but this is a less common compensatory mechanism to reduced seed rates compared with an increase in grain number per unit area (Fig. 5.3; Cromack and Clark, 1987). Plants with a greater chance to tiller can, therefore, be sown at reduced seed rates. Early sown winter crops can be sown at lower seed rates than later sowings. Wibberley (1989) suggests that 160, 185 and 210 kg ha^{-1} of seed be sown for UK winter wheat crops drilled in September, October and November, respectively. This is supported with experimental evidence showing the response of yield to seeding rate being greater when sowing after September (Fig. 5.3). In similar conditions, spring sown wheats, which usually have a reduced capacity for tillering, are suggested to be planted at 200–225 kg ha^{-1}. The response of yield to increasing seed rates is, however, relatively flat and in many countries average seed rates will be lower than those recommended for use in the UK. Most areas in the USA and Canada, for example, sow less than 100 kg seed ha^{-1}.

Despite this, excessive reliance on compensatory growth can lead to several problems. The importance of the survival of individual plants is increased. Pest attacks in the seedling stage which kill the growing points of only a relatively few plants could leave large gaps, resulting in poor utilization of light and nutrients. Prodigious tillering resulting from reduced seed rates may also be the cause of variable and delayed maturation (Tompkins, 1992; Thompson *et al.*, 1993b), making the crop uneven and difficult to manage and harvest.

In the majority of experiments, seed rate has had no significant effect on quality (Campbell *et al.*, 1991; Cromack and Clark, 1987; Mazurek and Kus, 1991; Jedel and Salmon, 1994). In organic farming experiments, however, Samuel and East (1990) found that low seed rates could result in low protein values when weed pressures were high. At sites where weeds were not so much of a problem protein concentration declined as seed rates and yields increased. In contrast, increasing seed rates in semi-arid Canadian conditions has increased both grain protein content and yields in association with improved water use efficiency (Tompkins *et al.*, 1991). The protein and gluten contents of rainfed wheat in northern Iraq have also increased at higher seed rates (Kalid and Wali, 1988). Samuel and East (1990) found that Hagberg falling number decreased and specific weights increased as the seed rate was raised, but the effects were only slight. For example, an increase in seed rate from 200 to 600 m^{-2} only resulted in a reduction in Hagberg falling number from 323 to 306, and an increase in specific weight from 75.2 to 76.5 kg hl^{-1}. In other similar experiments looking at less extreme rates of seeding (200–400 m^{-2}) there have been no significant effects on specific weight, protein concentration,

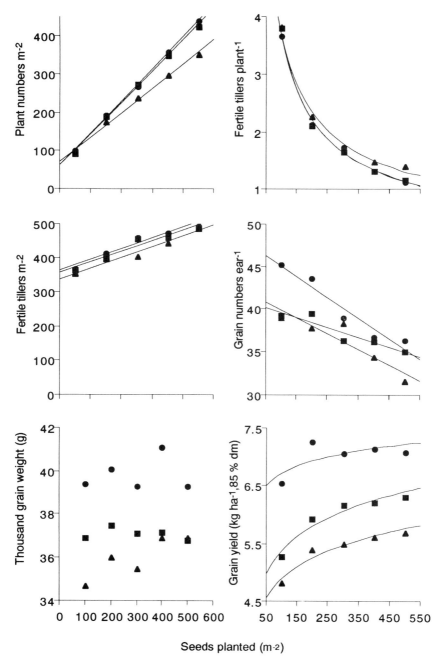

Fig. 5.3. The effect of seed rate on yield and yield components of winter wheat grown on a shallow calcareous soil in the UK (plotted from data presented by Cromack and Clark, 1987) when sown in September (●), October (■) and November (▲).

Hagberg falling number, SDS sedimentation volume or blackpoint (Thompson *et al.*, 1993b).

SOWING DEPTH

The optimum sowing depth of wheat is, like that of most crops, a compromise and commonly ranges from about 2.5 cm to 10 cm. Wibberley (1989) suggests that 3 cm is ideal for cereals for most temperate soil conditions. Sowing deeply reduces the risk of the seed being eaten, particularly by birds. The seed also experiences less fluctuation in temperature and soil moisture. Surface frosts are thus less likely to damage the seedling roots, but more importantly, the seed has a greater chance of remaining in contact with soil moisture. When rains are infrequent during crop establishment and irrigation is not available there can, therefore, be benefits to sowing more deeply. In the UK, spring wheats should usually be planted deeper due to drier conditions during establishment compared with what is normally encountered by autumn sown crops. Light sandy soils which are drought prone and less able to retain moisture in the surface layers can also render deeper sowing more suitable.

Deep sown crops, however, commonly take longer to emerge and delayed development may extend at least until anthesis (da Silva, 1991). Commonly there is a reduction in the total emergence and an increase in the numbers of abnormal seedlings as sowing depth is increased (Fig. 5.4). Seedlings are often weaker and less able to tiller, although if deeper sowing leads to markedly reduced plant densities, the final number of ears per plant can sometimes increase with sowing depth (da Silva, 1991). Excessive depth can result in complete emergence failure as the seed runs out of energy reserves before the coleoptile reaches the surface. Reduced emergence and subsequent grain yields with deeper drilling is reported to be particularly marked when the seed is small with lower energy reserves (Vedrov and Frolov, 1990). Such deeply sown seeds are also more susceptible to soil borne diseases and pests. Those seedlings that do emerge can be severely etiolated and often have bands of yellow running perpendicular to the leaf midrib.

There has been relatively little work on the effect of sowing depth on the quality of wheat. Campbell *et al.* (1991) found no influence of sowing at 2.5, 5.0 or 7.5 cm on grain protein concentration, while da Silva (1991) found no effect on either thousand grain weight or specific weight of sowing at incremental depths between 3.5 and 14 cm. Any results are likely to be complex, being so dependent on soil type, weather and type of wheat.

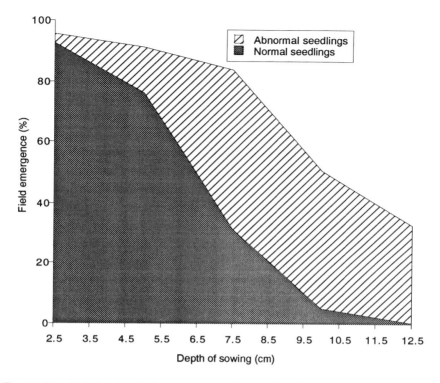

Fig. 5.4. The effect of sowing depth on emergence of winter wheat (from data presented in Petr *et al.*, 1988).

SOWING METHOD

The two main ways of sowing wheat are by broadcasting and drilling. The former is simplest, involving scattering the seed on to a seedbed which is then normally covered. In less mechanized systems of agriculture, and on difficult terrains, broadcasting by hand is still practised, e.g. Nepal. Uncomplicated mechanical devices such as seed fiddles (a shallow trough which is oscillated from side to side) can be used, but in more mechanized systems seed can be broadcast from fertilizer spreaders, or even from the air. Broadcasting is usually faster than drilling and is also less demanding on soil conditions, being more appropriate at higher soil moisture contents. Using a fertilizer spreader may also negate the need to buy a specialized seed drill. There may be benefits in scattering the crop evenly over the soil rather than concentrated in the rows produced from a drill. Interplant competition might be less, particularly in the seedling stages. The absence of the gaps between rows could also help suppress weeds. The main disadvantage of broadcasting, however, is the lack of control over seeding depth. It is almost inevitable that some seed will be poorly covered and thereby susceptible to pests, drying out and frost damage, while other seed may be too deeply covered and fail to emerge vigorously. Both of these hazards can be at least partly compensated for by increasing the seed rate by 10–15%. In between the extremes, seed at varying depths will emerge satisfactorily but at different times resulting in an uneven crop.

Drills are designed to place seed at a controlled depth in equally spaced rows. This often results in higher yields than broadcasting and is particularly beneficial when relatively deep sowing is desired due to high drought, frost or blowing risks (Martin *et al.*, 1976). Most commonly, seed from a hopper is guided into a furrow or slot created by a coulter and then covered. A combine drill is designed to apply seedbed fertilizer as well as sow the seed. This has the advantages of taking less time than applying seed and fertilizer separately, and places nutrients close to the emerging seedlings. This is particularly beneficial for relatively immobile nutrients, such as phosphate, on deficient soils. Very large fertilizer application rates should be avoided, however, due to phytotoxicity and hindered germination.

Ideal distances between rows must, in part, be influenced by seed rates. High seed rates requiring more, and therefore closer rows, to avoid excessive overcrowding within rows. Traditionally, row widths have been set at about 15–17.5 cm, but with high seed rates, high yielding winter wheat crops grown intensively have given 5% more yield when drilled in 10 cm rows (Wibberley, 1989). The narrower row drills, however, do require better standards of seedbed preparation and do not work as well when previous crop residues, stones and large clods remain on the surface. More even seed distribution can be achieved with 17.5 cm row drills by cross drilling, i.e. by drilling half of the seed in one direction and then sowing the remainder either at right angles or diagonally to the original path.

High speed drilling on poor seed beds can result in *coulter bounce* and poor distribution of seed along the drill row, and an uneven drilling depth. The latter is likely to have the greatest impact on yield, with the former being compensated for, at least in part, by greater tillering.

Conventional seed drills have changed little in principle since their inception for cereal establishment. Substantial benefits have been reported, however, from precision drills. The objective of this type of machinery is to control not only seeding depth and row width, but also the distance between seeds within a row. Gedye and Joyce (1978) report average yield increases of 12.5% of wheat established by experimental precision drills in comparison with conventional drilling on 15 cm rows. Up to 25% yield improvements were reported from some trials, where planting was on a square design rather than rows. However, the benefits have not been so apparent in commercial practice on a farm scale in the UK. The difficulties on the farm include the need for a well-prepared uniform seedbed, and sufficient time for the slower precision drilling.

Pneumatic drills with force feeding of seeds through flexible piping provide a large seed capacity and more rapid drilling potential, of particular value under variable weather conditions.

6

Crop Nutrition and Fertilizer Use

For growth to proceed wheat requires at least 17 chemicals as macro- (or major) and micro- (or trace) elements. Macroelements are required in relatively large amounts, with the crop demands expressed in terms of kg t^{-1}, while the needs of microelements are measured at g t^{-1} levels. Carbon (C), hydrogen (H) and oxygen (O) are macroelements obtained from carbon dioxide in the atmosphere and water. The crop requirements which are most met by fertilizers are nitrogen (N), phosphorus (P) and potassium (K). Other macronutrients include calcium (Ca), magnesium (Mg) and sulphur (S). Calcium is not normally considered to be a nutrient requiring application because large amounts are often applied in liming material to correct for increases in soil acidity. Similarly, specific application of magnesium fertilizers has not received much attention in the past. Increasingly though, particularly for high yielding crops and with high potassium applications, deficiency may need to be alleviated with magnesium fertilizers. Traditionally, sulphur has not been applied as a specific fertilizer to cereals as it has often been a component of chemicals supplying other nutrients, e.g. ammonium sulphate. Crop demands have also been met by sulphur deposition in rainfall after emission into the atmosphere from industrial processes. For reasons discussed later, these sources are declining in many areas and, therefore, fertilizer application is likely to be increasingly important in the future.

Micronutrients include copper (Cu), manganese (Mn), iron (Fe), boron (B), sodium (Na), chlorine (Cl), molybdenum (Mo) and zinc (Zn). The requirements of these particular nutrients are very small, and they can be toxic at higher rates. Chlorine, in particular, is hardly ever deficient in field conditions, with excesses associated with salinity causing rather more problems. Most growers rely on natural nutrient cycling from organic and rock material to satisfy demand, rather than routinely applying micronutrients as fertilizers. Deficiencies of one or more micronutrients are more likely on specific soil types under certain conditions which may warrant application. As with macronutrients, deficiencies are more likely to occur with the adoption of high yielding varieties and intensive growing systems due to the higher rates of removal from the soil over successive seasons and increased demands at peak growth periods. As a result more disappointing productivity than predicted has often occurred with long-term depletion of essential nutrients (e.g. Nambiar, 1991).

NITROGEN

Nitrogen is the most important fertilizer element determining the productivity of wheat. It is a major component of proteins, and therefore enzymes, which catalyse reactions essential for life. It is also a constituent of nucleic acids (e.g. DNA), the material carrying the genetic information from generation to generation of individuals and cells. Making more nitrogen available to deficient plants increases chlorophyll contents and leaf greenness, particularly of the older leaves; increases leaf size; delays senescence; stimulates tillering; increases height; and boosts grain site formation. Excessive availability, however, can heighten the risks of frost damage, lodging, foliar disease and water shortage, and can also delay crop maturation. It also has many influences on the various quality characteristics used to market grain. The plant itself is far more efficient at partitioning nitrogen to the grain than it is for partitioning dry matter (DM). Various studies have shown crop uptake of about 20–25 kg nitrogen per tonne of grain (at 85% DM), and around 75–80% of this nitrogen is incorporated into grain by harvest. For example, a crop yielding 5 t ha^{-1} each of grain and straw would typically contain 93 kg ha^{-1} of nitrogen in the grain but only 17 kg N ha^{-1} in the straw (Archer, 1985). This may partly reflect the evolutionary importance of seed protein to the vigour and competitive ability of the seedling (Ayers *et al.*, 1976).

Soil Availability of Nitrogen

Soils can contain large quantities of nitrogen, often far in excess of the requirements of a single crop. Usually, more than 90% of this nitrogen is in the form of organic matter (Jenkinson, 1986). Part of this organic matter will consist of decaying plant, microorganism and animal tissue, and a further contribution will be animal excreta. Nitrogen contained within this organic matter can be released slowly as ammonia by a process of mineralization. This is conducted by a range of soil microorganisms including both bacteria (e.g. *Pseudomonas, Bacillus, Clostridium, Serratia, Micrococcus* and *Streptomyces* spp.) and fungi (e.g. *Alternaria, Aspergillus, Mucor, Penicillium* and *Rhizopus* spp.) (Lewis, 1986). The nitrogen in ammonia, which can be taken up by wheat, is often oxidized by a process of nitrification by *Nitrosomonas* spp. to produce nitrite (NO_2^-) and then by *Nitrobacter* spp. to produce nitrate (NO_3^-). Both mineralization and nitrification are dependent on adequate soil moisture and only occur slowly in dry soils, reaching a peak at between 20 and 30% soil water content of a silt loam in controlled conditions (Alexander, 1977). Nitrification slows down with high water contents due to lack of oxygen, although some mineralization can continue if anaerobic mineralizing organisms are present. Mineralization is often higher under a sequence of dry : wet cycles, rather than continually moist conditions. Temperature is also very important with mineralization increasing up to temperatures of at least 40°C, and nitrification reaching a peak at between 30° and 35°C with very little occurring below 5°C and above 40°C. Nitrification is reduced in acid environments with declining rates occurring as pH falls below 6.0. In these conditions, therefore, nitrification can be increased by liming.

Release of nitrogen from organic matter will, therefore, occur at times of the year when adequate temperatures and moisture contents occur simultaneously. In the UK, nitrogen liberation reduces when soil moisture becomes too low in summer and temperatures become too low in winter, with potentially higher levels of release occurring in spring and autumn. Large amounts can also be released in summer following high rainfall, particularly following a dry spell. Total release will also depend upon the total amounts of substrate, which increases with more soil organic matter.

Availability to plants, however, will also depend on the rate of loss of nitrogen once mineralization and nitrification have occurred. Losses of released nitrogen can ensue via several processes (Powlson *et al.*, 1992), and the most important include immobilization, volatilization of ammonia, leaching and denitrification. Measured nitrogen losses at Rothamsted Experimental Station, UK, have varied between 1 and 35% (Powlson *et al.*, 1986). Immobilization occurs when ammonia released by mineralization is then utilized by soil microorganisms for their own growth. The rate at which this occurs depends greatly on the C : N ratio in the organic matter. If there is a large amount of carbon relative to nitrogen, say more than 20 : 1, all the mineralized nitrogen can be used up with fast microbial growth exploiting both available carbohydrate and nitrogen. Immobilization can increase, therefore, in soils where plant matter of low nitrogen content has been returned to the soil (Lyon *et al.*, 1929), in particular where cereal straw is incorporated.

Volatilization of ammonia occurs in alkaline conditions with insufficient H^+ ions allowing the formation of the soluble ammonium ion (NH_4^+).

Some ammonia (NH_3) can remain in solution but it sis readily lost as a gas. Alkaline soils, such as those with a high limestone or chalk content will be susceptible to this form of nitrogen loss. Ploughing can also increase nitrogen volatilization to the atmosphere (Jordan *et al.*, 1993).

Leaching of nitrogen occurs particularly as the highly mobile nitrate anion. Being negatively charged and highly soluble it is not readily held on the cationic exchange sites of clay particles and is readily carried with the mass flow of water through the soil. It can be lost from the rooting zone, therefore, on soils which drain rapidly, such as sandy soils and/or shallow soils overlying permeable rock such as limestone. Losses occur particularly when conditions for nitrification are good, but crop requirements are low, for example, in the autumn in the UK. The reasons for this are threefold. Firstly, the moisture and temperatures are adequate for mineralization and nitrification; and secondly, the rainfall exceeds evapotranspiration resulting in water drainage down through the soil profile. Thirdly, the inversion and disturbance of the soil in cultivation for autumn sown crops increases nitrification. Minimizing soil disturbance for crop establishment can, therefore, be a method of reducing losses of nitrogen when subsequent crop growth rate is low and there is a high leaching risk. Conversely, if leaching risk is low the increased frequency and aggressiveness of tillage can increase nitrate availability to the crop (Radford *et al.*, 1992). In the UK nitrate leaching can also be reduced by sowing winter wheat in September rather than October such that the plants produce more dry matter and a better developed root system to take up nitrogen during the autumn and winter period (Cosser *et al.*, 1994a). Figure 6.1. shows the amount of nitrogen captured in the shoot systems of wheat over winter to decline as the sowing date is delayed from September to December in the UK. This benefit of early sowing, however, needs to be weighed against the risks from weeds, pests and diseases which are generally increased with earlier drilling dates.

Denitrification occurs when nitrate is converted into N_2, N_2O, NO or NO_2 gas. This can arise when nitrate has been produced but then anaerobic conditions ensue. This is often the result, for example, after heavy rainfall events and is particularly serious when combined with warm temperatures (e.g. 25°C and above). High C : N ratios encourage denitrification by *Pseudomonas* spp., *Paracoccus* spp. and *Thiobacillus denitrificans*. In Canada, no-tillage systems resulted in higher levels of denitrification in wheat systems and increased the population of denitrifiers sixfold compared with conventional tillage. This was possibly due to the denser surface soil and consistently higher moisture contents of the no-tilled soil (Aulakh and Rennie, 1986).

Whatever the determinants of nitrogen availability in the soil, the amount of nitrate N remaining in the root profile at the beginning of the cropping season can have a large impact upon both grain yield and

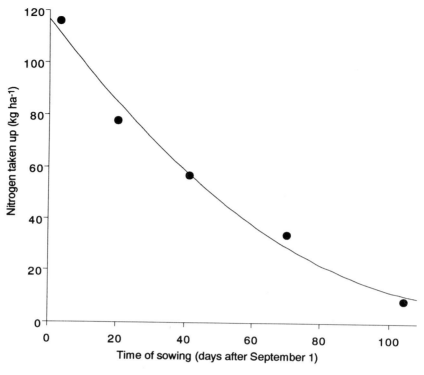

Fig. 6.1. The effect of sowing date on the amount of nitrogen accumulated in wheat shoot systems by March, before nitrogen fertilizer application (after Plant Breeding Institute, 1989).

protein concentration. In a review of 74 experiments on winter wheat after fallow in the semi-arid climate of western Nebraska, it was found that when nitrate N was less than 45 kg N ha⁻¹ in the top 180 cm, protein contents averaged about 9% DM, and more than 60 kg N ha⁻¹ needed to be applied to obtain maximum yields. Protein contents rose to above 11% DM as nitrate N increased to above 135 kg N ha⁻¹ and no nitrogen fertilizer was required to maximize yield. In less arid environments, however, the amount of nitrogen in the soil at the start of the growing season may be less important as much may be lost through leaching before large quantities are required by the crop. It has been considered by some that the amount of nitrate left in the soil in these conditions is too small and variable to warrant analysis for predictive purposes. In some systems, however, adjustment of spring nitrogen fertilizer applications for soil-available nitrogen have relied on assessing soils in February when the crop was sown in the previous October (Davies and Gooding, 1995). This avoids the complication of adjusting estimates for possible leaching during the greater part of winter based on weather conditions.

Assimilation of nitrogen

Wheat roots take up both nitrate and ammonium ions from soil solution but via different pathways. The two pathways proceed at differential rates depending on the environment. Nitrate absorption enters the root cells via permeases which require energy and is most rapid at relatively low pH conditions and high temperatures. This active uptake of nitrate allows wheat to concentrate the nitrate in the root tissue and xylem vessels compared with the levels of nitrate in soil solution. Once inside the plant, nitrate needs to be reduced to ammonium, either in the root cortex or in the leaves after transport via the xylem vessels. Reduction is catalysed by the combined action of nitrite and nitrate reductase enzymes and requires a further 347 KJ mol^{-1} of energy. Ammonium uptake occurs faster if energy is available but will also occur passively. Uptake is quicker at higher temperatures but is not as quickly reduced at low temperatures in comparison with nitrate uptake. Ammonium from direct uptake and from nitrate reduction is incorporated into organic compounds, first by incorporating the nitrogen into a glutamate molecule following the action of the glutamine and glutamate synthases. Glutamate is a precursor of other amino-acids. Ammonium becomes toxic at much lower concentrations than nitrate and, therefore, ammonium assimilation through the roots is dependent on carbohydrate transport to the roots to provide the carbon skeletons to incorporate ammonium quickly into organic compounds.

Rotation and Nitrogen Availability

The cropping rotation has a large impact upon the availability of nitrogen to wheat crops. Of most relevance are the inclusion of legume crops (*Fabaceae*). These plants form symbiotic relationships with *Rhizobium* spp. bacteria within nodules on their roots. *Rhizobium*, provided with sugars from the host, can fix nitrogen from the soil atmosphere and make it available to the legume. The nitrogen thus fixed can be made accessible to wheat plants in two main ways. The first is through degradation of the legume roots and/or shoots after the crop has been harvested or incorporated. In assessments made in each of six years, Widdowson *et al.* (1987), for example, reported average levels of 85 kg nitrate N ha^{-1} in the top 90 cm of soil in the November following a winter bean crop, but only 46 kg after winter wheat. The second is if the legume has been fed either *in situ*, or after harvest, to livestock from which farmyard manure (FYM) and slurry can be used on fields intended for wheat. It has been claimed that a large proportion of the 50% increase in average Australian wheat yields that occurred between 1960 and 1980 was due to the use of more appropriate rotations with a better exploitation of legume nitrogen.

Preceding a wheat crop with a legume can be exploited to either increase yields in extensive systems, or decrease the amount of nitrogen fertilizer required for maximum yield in more intensive systems (Narwal *et al.*, 1983). Similar effects can be achieved, however, with non-legume break crops, such as oilseed rape (*Brassica napus*) and sunflower (*Helianthus annuus*) which are less exhaustive than wheat and may, therefore, leave a greater amount of applied and/or mineralized nitrogen in the soil for use by the following crop (Vaidyanathan *et al.*, 1987; Echeverria *et al.*, 1992). Potato crops (*Solanum tuberosum* L.) have been shown to leave more nitrate N in soil than either wheat or beans (Widdowson *et al.*, 1987). Increased nitrogen availability to wheat might also be achieved if the preceding crop has (i) improved soil structure through having a vigorous and extensive rooting system; (ii) reduced the incidence of wheat root infection or stem base diseases; or (iii) has been grown with large amounts of FYM or slurry. The implications of preceding crops for recommended rates of nitrogen fertilizer to wheat in the UK are shown in Table 6.1. There is greater efficiency of nitrogen use in wheat following a break crop, probably due to more effective wheat root systems (Sylvester-Bradley *et al.*, 1987a). This means that first wheats can give higher grain and nitrogen yields at lower nitrogen rates (Table 6.2). Although soil type also has an effect, the implications of the rates shown in Table 6.1 are that a break crop such as oilseed rape or potatoes is worth about 50 kg N ha^{-1} to the wheat. It may be surprising that these appear to have a greater effect than the legume combinable break crops such as peas and beans. It must be remembered, however, that the legumes do not require nitrogen fertilizer themselves so the overall rotational requirements for nitrogen will usually be less when legume, rather than non-legume, combinable break crops are used. Non-cereal crops may also have residual effects lasting for more than one succeeding crop of wheat. This is particularly true of perennial crops. Lucerne, for example, would expect to give significant benefits for increasing nitrogen availability until at least the third successive cereal. An intensively managed permanent pasture with either a high clover content or high nitrogen fertilizer usage (250 kg N ha^{-1} year^{-1}) would expect to have residual benefits lasting at least five years after it was ended – and probably much longer (Ministry of Agriculture, Fisheries and Food, 1994).

For grain quality the interpretation of rotational effects is complex. One of the reasons for this is that it is difficult or impossible to separate the effects of rotation on increasing nutrient availability from the effects on pest, disease and weed control (McDonald, 1989). In the extensive systems of much of Australia, rotating wheat with legumes has been shown to improve yield and quality. What appears to be more important than the legume chosen, however, is the way in which the legume is utilized. Silsbury (1990) found that peas (*Pisum sativum*), vetch (*Vicia sativa*), and

Table 6.1. Recommended rates of spring nitrogen applications (kg ha^{-1}) to wheat in the UK as affected by soil type and previous crop (from Ministry of Agriculture, Fisheries and Food, 1994).

		Previous crop		
Soil type	Target yield (t ha^{-1}, 85% DM)*	• Cereal • Sugar beet • Forage crops removed • Peas and beans†	• Potatoes • Oilseed rape • Forage crops grazed • Peas and beans†	• Lucerne • Managed permanent pasture • Any crop receiving large frequent applications of FYM or slurry
Winter wheat				
Sandy soils	7	175	140	80
Shallow soils	8	225	190	130
Deep silty soils	8	180	90	0
Clays	8	190	110	0
Other mineral soils	8	210	150	70
Organic soils	8	120	60	0
Peaty soils	7	80	20	0
Spring wheat				
Sandy soils		170	130	70
Shallow soils		170	130	70
Other mineral soils		170	110	30
Organic soils		100	40	0
Peaty soils		40	0	0

* For crops with a larger or smaller yield potential the recommendation is to adjust nitrogen rates by plus or minus 20 kg ha^{-1} per t ha^{-1} yield variation.
† Recommended rates for wheat after peas and beans are intermediate between the groups headed by cereals and potatoes.

Table 6.2. Efficiency of nitrogen use in wheat, determined in 36 field experiments in the UK (Sylvester-Bradley *et al.*, 1987a).

	Previous crop		
	Cereal	Break crop	Difference
Grain yield at opt. nitrogen (t ha^{-1})	8.4	9.0	0.6
Optimum nitrogen rate (kg ha^{-1})	210	163	47
Crop N with no N fertilizer (kg ha^{-1})	71	99	28
Crop N at optimum N rate (kg ha^{-1})	192	195	3

medic (*Medicago truncatula*) produced similar grain nitrogen concentrations in following wheat, but all three gave much higher yields and nitrogen concentrations when they were ploughed in as a green manure crop after flowering, rather than after the seeds or residue had been removed. In other, more traditional, Australian rotations, Littler (1984) reported that wheat following a four year pasture of a rescue grass (*Bromus unioloides*) plus alfalfa (*Medicago sativa*) had higher yields, grain protein concentrations and baking quality compared with wheat in a continuous wheat rotation. The pasture rotation also suppressed wild oat populations. In Canada, Zentner *et al.* (1990) found that the yields and protein concentrations were higher for wheat grown in rotations which included a grass plus legume hay crop compared with the normal wheat–fallow systems. Wright (1990) reported that when evaluating legume–barley–wheat rotations, the legume increased the protein content of barley but there was no residual effect lasting to the wheat crop in year 3. In Germany, Kruez and Zabel (1989) found that a two year crop of alfalfa increased wheat grain and protein yields, thousand grain weight and baking quality.

On other occasions the effects of rotation on grain quality have been less clear cut. For example, in Australia, a crop of lupins has increased DM and nitrogen yields in the following wheat grain but effects on protein concentration have been inconsistent (Doyle *et al.*, 1988a). In Argentina and the UK a one year break crop from cereals failed to affect protein content of following wheat irrespective of whether it was a legume or not (Vaidyanathan *et al.*, 1987; Echeverria *et al.*, 1992). In both situations, however, there were increases in grain DM and nitrogen yields.

In summary it appears that rotational cropping has a much larger effect on yield than on grain quality. Increased protein concentrations are possible but are more likely after a legume-rich crop or pasture which has been growing for more than one year, and/or when the legume crop is ploughed in as a green manure.

Intercropping and undersowing

Legumes can also help increase nitrogen availability if grown in the field at the same time as the wheat. One form of intercropping that is particularly relevant to extensive systems which rely on rotating wheat with legume rich pasture is that of undersowing. Establishing the ley as an undersown crop in the wheat has a number of benefits (Thompson *et al.*, 1993c) in relation to limiting nitrogen losses. Energy and labour requirements can also be reduced when autumn cultivations are unnecessary to introduce the ley after the wheat (Bosshart *et al.*, 1986). Undersowing can be achieved in many ways. In the UK it is possible to sow a winter wheat crop in the autumn and then broadcast clover and grass seed in the spring. Going through the crop with a light finger-tined harrow can then help to cover the undersown species without causing excessive damage to the wheat. Alternatively, both wheat and undersown species can be sown in the spring. This technique is more suited to extensive systems where there are less compromises about fertilizer and herbicide use for the two crops. For example the majority, but not all, of the broad-leaved herbicides would be very detrimental to undersown clover. Additionally, large amounts of nitrogen fertilizer tend to reduce the establishment of clover and may increase the competitivity of the wheat to the detriment of the undersown species. Any nitrogen-induced lodging of wheat would be particularly troublesome for both harvest of the cereal and establishment of the ley if it lay across the clover and grass. As well as providing more efficient establishment of the following ley, undersowing can also increase the degree of competition against weeds. Thompson (1995) found that undersowing with white clover (*Trifolium repens*) reduced total weed ground cover immediately after the wheat harvest from 36 to 7%.

As well as establishing the following ley and increasing the amount of nitrogen fixed in the total rotation, undersown legumes can sometimes increase the amount of nitrogen available to the covering wheat crop (Thompson, 1995). It appears that symbiotically fixed nitrogen becomes available within the soil through the process of mineralization of sloughed off nodules. This, however, requires time and to be of noticeable benefit to the crop growing concurrently, it has been suggested that the legume would have to grow rapidly and start contributing nitrogen within three months (Charles, 1958). Excretion of nitrogen from the legume is maximized by shading of the legume after rapid growth, but limited by continuous shade (Willey, 1979). This is obviously a difficult balance to achieve within a wheat plus legume and grass mixture and may contribute to large amounts of variation in the relative effects on nitrogen availability in such systems in different regions and years.

In addition to undersowing, other forms of legume plus wheat intercropping have received attention in many parts of the world. Intercropping with a pulse can help satisfy the increasing demand for wheat at the same time as growing a crop with better protein. Wheat plus chick pea (*Cicer arietinum*) combinations have, for example, been extensively studied in India for both rainfed systems and as part of the rice : wheat rotations in irrigated rice lands (Ali, 1993).

Intercropping systems (including undersowing) require careful attention to the harvesting dates and requirements of the two crops, as well as to the relative competitiveness of each crop against the other at various growth stages. Such considerations will impact upon relative planting dates, seed rates and sowing patterns. Thompson *et al.* (1993c) and Kahurananga (1991) both found broadcast clover to reduce wheat yields, but this risk appeared to be reduced when the undersown clover was established in rows (Kahurananga, 1991).

Inorganic Sources of Nitrogen

As described in Chapter 1, the use of inorganic nitrogen fertilizers on UK wheat has increased dramatically since the mid-1970s (Fig. 1.23), with the productivity increases made possible by the adoption of potentially high yielding varieties. In addition to yield, however, nitrogen can have greatest potential impact upon the end-use suitability of grain compared with all other nutrients commonly applied in fertilizers. There are several components of fertilization that can impact upon wheat yield and quality including the nitrogen compound used, the form of material (solid or liquid), the site of utilization (e.g. soil or foliage), and application timing.

The most common forms of application are solid prills, of either ammonium nitrate or urea. Ammonium nitrate (NH_4NO_3) contains 35% N and is regularly used as a top-dressing on wheat in the UK in the spring. The nitrate ion is mobile and readily taken up by plant roots. The ammonium ion will often be partially absorbed on soil colloids so crop uptake can be delayed compared with the nitrate ion. On alkaline soils, however, there is also the possibility that the ammonium ion will be lost through volatilization. The use of ammonium nitrate has been restricted in a number of countries as it is a potential explosive. The risk of fire or explosion is reduced by prilling, i.e. coating granules, which also reduces water absorption in stored fertilizer in humid conditions. Similar beneficial effects can be obtained by mixing ammonium nitrate with calcium carbonate. These mixtures are classified as calcium ammonium nitrate (CAN) and have been marketed as Nitro-chalk. These mixtures vary in N content but typically contain about 21% N. There are, therefore, greater handling and application costs for supplying a given amount of nitrogen but, as well as reducing water adsorption and fire risk, they are less acidifying than ammonium nitrate due to the alkalinity of the calcium carbonate.

Urea CO $(NH_2)_2$ contains 46.6% N and is the most concentrated form of nitrogenous fertilizer available as a solid. Urea dissolves rapidly in soil water and is converted to ammonium carbonate. This temporarily raises the pH and, therefore, there is an increased risk of ammonia volatilization, particularly on chalk and limestone soils which are alkaline. On poorly aerated soils, however, the risk of denitrification and nitrate losses is reduced due to the absence of a nitrate ion in the fertilizer. After conversion to ammonium, the urea has a stronger acidifying effect than ammonium nitrate so, in the longer term, more lime will be required.

As well as prills, both ammonium nitrate and urea fertilizers can be dissolved in water and applied as a solution. Application to a standing crop is usually via stream jets or dribble bars to the soil, so as to avoid leaf scorch. In the UK a common solution contains a 50 : 50 mix of nitrogen from urea : ammonium nitrate to supply 37 kg N l^{-1}. Solutions have advantages in handling, compared with bags of prills, and it can be argued that more accurate application is possible. They can also sometimes be applied with herbicides and/or fungicides thus allowing savings in labour, machinery and energy costs (Palgrave, 1986; Gooding *et al.*, 1988). These benefits saw non-pressure liquid fertilizers accounting for 13% of total fertilizer usage on winter cereals in England and Wales by 1986, following their introduction in the late 1950s.

Despite the predicted interactions between the form of nitrogen and its loss on different soil types, in practice the form of nitrogen appears to be of relatively minor importance for yield and quality of wheat. Most studies have not detected significant differences in yield response between urea and ammonium nitrate (Christensen and Meintz, 1982), ammonium nitrate and urea plus ammonium nitrate solution (Hunter, 1974), or between urea and urea plus ammonium nitrate solution (Howard, 1986). Occasionally small interactions between nitrogen form and yield response have been reported (Howard, 1986) and urea plus ammonium nitrate solutions have been less effective when they have caused leaf scorch (Penny and Freeman, 1974). What is much more important, however, is the amount of nitrogen applied, irrespective of product used, and timing.

Yield, specific weight and thousand grain weight

Because of the risks of nitrogen loss from commonly used fertilizers, it is important that the timing of application coincides with the ability of the wheat plant to utilize the largest amounts efficiently. Nitrogen demand by the crop is greatest during rapid periods of growth, commonly from the start of stem elongation through to flag leaf emergence, then it declines gradually as the crop matures. In the UK it was once common practice to apply nitrogen to the seedbed and/or young seedlings of wheat. The efficiency of uptake by the plant, however, can be low due in part to the scanty root system and, in the case of autumn sown wheat, the slow growth rates possible due to low temperatures during winter. Nitrogen could, therefore, be lost through leaching and volatilization (Jain *et al.*, 1971). In the UK the proportion of nitrogen given to winter wheat in the autumn steadily declined during the 1980s (Table 6.3). Dressings in spring, before the reproductive stage has started, can be taken up by the crop, but are at risk of stimulating excessive tillering. Relatively small amounts, of up to about 40 kg N ha^{-1} can be applied, however, to stimulate

Table 6.3. Timings and proportions of total nitrogen application to winter wheat in the UK (after Davies, 1990).

Year	% of total nitrogen			
	August to December	January to February	March to May	June to July
1983/84	8.0	11.3	79.9	0.8
1988/89	1.9	7.7	88.5	1.9

tillering of poorly established and/or backward crops. Assuming adequate moisture, the greatest and most reliable utilization of nitrogen and yield responses derive from applications at the start of the reproductive stage, i.e. approximately from the pseudo-stem erect stage to the second node detectable stage. In crops of spring wheat, or short season wheat in hot environments, the timing of nitrogen application for most efficient utilization is often less critical as crop development is much more rapid and long periods of slow growth are less likely. In such situations, however, the efficacy may be improved by applying nitrogen at irrigation timings (Singh, 1988). Timing may also be less important when leaching risk is diminished. Sander *et al.* (1987), for instance, cited work indicating that autumn applications of nitrogen were generally as effective at improving yield as spring applications to winter wheat grown in central USA.

Even when applied at recommended timings the response of wheat to nitrogen can be highly variable. For example, Fig. 6.2 illustrates some significant ($P < 0.05$) effects of nitrogen applied as ammonium nitrate prills at the first node detectable growth stage on Avalon winter wheat in two contrasting seasons in the UK (Smith *et al.*, 1990). Linear and quadratic responses have been fitted where appropriate. The replicated field experiment was at Coates Manor Farm, Cirencester, on shallow free draining stony soil of the Sherborne series. The 1989 season was much drier during grain filling than in 1988, with rainfall in July only amounting to 19 mm in 1989, compared with 112 mm in 1988. Figure 6.2 demonstrates, therefore, the importance of moisture for optimizing yield responses to nitrogen fertilizer. In 1988, with adequate moisture, the yield at first increased rapidly with incremental increases in nitrogen application. The rate of increase then declined as yield approached a maximum. In other experiments further additions of nitrogen have caused yield reductions, particularly if nitrogen applications have increased lodging and/or disease severity. These types of yield response can be described by quadratic (Johnson *et al.*, 1973; Holbrook *et al.*, 1983; Howard, 1986) or linear plus exponential curves (Sylvester-Bradley *et al.*, 1984). In 1988 in the Avalon

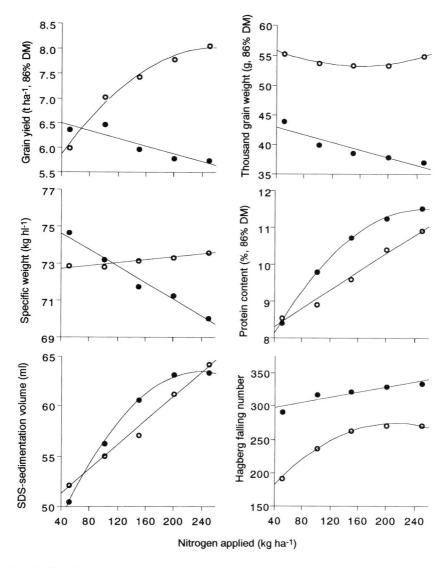

Fig. 6.2. The effect of nitrogen applications at the first node detectable growth stage on the grain yield and bread-making quality of Avalon winter wheat in 1988 (○) and 1989 (●) (after Smith *et al.*, 1990).

study at Coates Manor Farm, the yield improvements were closely correlated with grain numbers per ear, confirming the role of nitrogen in stimulating grain site formation, when applied at this growth stage. In contrast, there was no improvement in thousand grain weight. In other experiments, thousand grain weight has been reduced by nitrogen fertilizer applications, even when yields have been increased (Bayles *et al.*, 1978; Howard, 1986; Gallagher *et al.*, 1987). Thousand grain weight can sometimes be improved by later applications, particularly when flag leaf senescence is delayed by nitrogen applications after flag leaf emergence (Gooding and Davies, 1992). This is presumably because these later applications are unlikely to have a dramatic effect on grain numbers and, therefore, the nitrogen does not increase the intergrain competition for assimilate in the same way as earlier applications. Delaying nitrogen application beyond the second node detectable growth stage, however, often reduces the yield response (Finney *et al.*, 1957).

The slight, but statistically significant, increase in specific weight in 1988 in Avalon was possibly due to the increased protein content and, therefore, density of the grain (Pushman and Bingham, 1975) as it cannot be assigned to better filling of the grain. Bayles *et al.* (1978) also found specific weight generally to increase with nitrogen application rate even though large amounts caused reductions in thousand grain weights and an increase in grain shrivelling as determined visually.

Results from 1989 in Fig. 6.2 demonstrate the risks of applying nitrogen to a drought prone soil in a season with considerable moisture deficits during grain filling. In this season the nitrogen application stimulated a greater amount of tillering, and therefore foliage. This may have increased the total amount of transpiration, thus exacerbating the moisture deficits and leading to inadequate grain filling and reduced yields (Smith *et al.*, 1990). Excessive nitrogen availability early in the season can, therefore, lead to reduced thousand grain weights. The associated grain shrivelling could also contribute to reduced packing efficiency and specific weights as observed in Fig. 6.2. This type of yield response may be common in certain areas such as the High Plains of North America (Sander *et al.*, 1987) but is relatively rare in the UK, as rainfall is usually more evenly distributed throughout the year. For example, Sylvester-Bradley *et al.* (1984) report yield reductions with low levels of nitrogen fertilizer application in only 16 out of 121 widely distributed UK field experiments. They also reported that an application of 175 kg ha^{-1} of nitrogen would have been economic on about half of the crops tested. Recommended rates, however, may be made more precise by consideration of soil type, previous cropping, variety, lodging risk and stem base disease, i.e. greater applications appear to be justified on sandy soils, after cereals, when there is a low lodging risk, and where it is desired to alleviate some of the symptoms of take-all (Ministry of Agriculture, Fisheries and Food, 1994). Also of importance is the availability of other essential elements. Increased growth arising from nitrogen applications can lead to deficiencies in other nutrients. For example, copper requirement increases when nitrogen fertilizer is applied and complete crop failure has occurred following large nitrogen dressings to copper deficient soil (Fleming and Delany, 1961).

In summary, nitrogen fertilizer applied at rates and timings to optimize yield response does not necessarily give comparable improvements in thousand grain weight and specific weight (Kosmolak and Crowle, 1980; Gallagher *et al.*, 1987). Reductions in all three measurements can be accentuated in drought conditions when competition for water limits grain growth. Others have also frequently found the nitrogen effect on specific weight to be small and inconsistent (McClean, 1987; Withers, 1987; Kettlewell, 1989). A further complication is that varieties can interact differently with nitrogen with respect to grain shrivelling, thousand grain weight and specific weight even when grown under identical conditions (Fig. 6.3)

Grain nitrogen concentration

The most dramatic influence of nitrogen fertilizer on grain quality is through effects on grain protein concentration. Figures 6.2 and 6.4 demonstrate that, in UK conditions, increasing nitrogen inputs can often upgrade grain from having inadequate to sufficient protein content for inclusion in the Chorleywood bread-making process. The shape of the response around the levels commonly used to obtain optimum yields has been found to be approximately linear in a wide range of growing environments (Johnson *et al.*, 1973; Benzian and Lane, 1986). The rate of response differs with site and season but, in the UK, the increase in protein for nitrogen applied at the main time of top-dressing around early stem extension is on average 1% (at 86% DM) for every 80 kg ha^{-1} N applied, up to a total of about 200 kg N ha^{-1} on crops yielding about 8 t ha^{-1} at 85% DM. Figure 6.2 shows that the same increase at Coates Manor Farm required 80 and 53 kg N ha^{-1} in 1988 and 1989, respectively, indicating that responses can be steeper for lower yielding crops when there is less dilution of the accumulated nitrogen by carbohydrate. Confirming this trend, Penny *et al.*, (1978) obtained 1% increases in protein concentration with only 46 kg N ha^{-1} but here the average yield was only 5 t ha^{-1}. In countries where yields are much lower, protein increases can be even more striking. The results of Kosmolak and Crowle (1980), for instance, suggest that a 1% increase in protein concentration could be achieved with an application of only 26 kg N ha^{-1} to a crop yielding about 2.5 t ha^{-1}. Much of this variation is due to water availability. Barber and Jessop (1987), for example, showed that the yield and protein concentration responses to increased nitrogen application rates were, respectively, increased and decreased as the number of water applications increased. Whether soil moisture content is adequate or restrictive for maximum yield, the grain protein content response continues to rise with nitrogen application rates after yields have stopped increasing (Fig. 6.2; Sander *et al.*, 1987).

Despite these variations, the response of protein to nitrogen fertilizer is more stable and predictable over sites and seasons than yield effects.

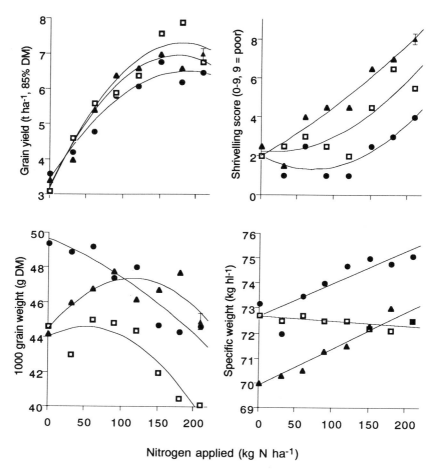

Fig. 6.3. The effects of nitrogen fertilizer applied in the spring on the grain yield and grain characteristics of winter wheat varieties Capelle-Desprez (●), Maris Fundin (□), and Maris Huntsman (▲). Bars represent standard error of difference for comparing all points (plotted from data presented in Bayles *et al.*, 1978).

Similarly, the protein concentration response does not interact greatly with variety (Fig. 6.4), despite significant nitrogen × variety interactions for grain yield.

Although the response of protein to nitrogen fertilizer is approximately linear around application rates commonly used to maximize yield, in high yielding intensive systems, non-linear relationships can be observed at both very low and very high levels of nitrogen availability. At low levels of availability, small increases in nitrogen fertilizer rate can

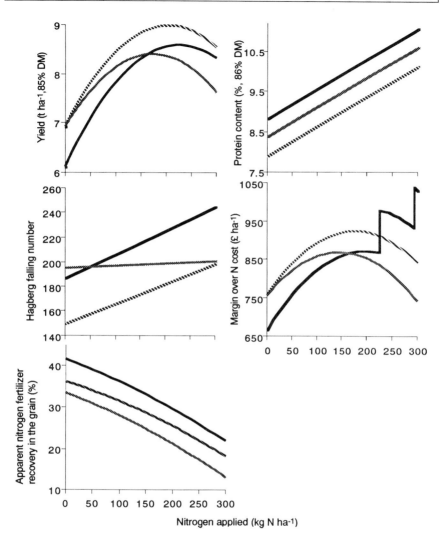

Fig. 6.4. The effect of nitrogen applications at the first node detectable growth stage on the grain yield, bread-making quality and nitrogen recovery of Avalon (solid line), Brimstone (hatched line) and Mission (shaded line) winter wheat in 1985 on a sandy loam (after Gooding, 1988).

cause grain protein to be depleted as the increase in yield can outstrip nitrogen accumulation when severe nitrogen deficiency is relieved (Benzian and Lane, 1986). No consistent positive effects of nitrogen fertilizer on protein content were, therefore, reported from work conducted in the UK before 1940, because these studies usually only investigated applications up to 30 kg N ha^{-1} (Sander *et al.*, 1987). As nitrogen is applied at rates higher than that necessary to achieve maximum yield, the protein response does not increase indefinitely but, as is shown for 1989 in Fig. 6.2, tends to flatten off. The modelling of protein responses to large ranges of nitrogen fertilizer application, therefore, requires the use of non-linear functions, as described by Murray and Nunn (1987).

As well as amount, the effect of nitrogen on protein content depends on timing. In contrast to yield, the largest increases have been obtained by applications at, and after, ear emergence (Gooding and Davies, 1992). A contributory factor to this trend is that yield increases are less likely at these later timings and, therefore, there is less dilution of extra grain N by increased amounts of carbohydrate. The precise timing for optimum effect, however, has varied with different systems. There is some evidence that when water availability is low at, and after anthesis, or the crop senesces early for other reasons, the greatest protein responses are from applications at anthesis (Finney *et al.*, 1957). When nutrient uptake occurs later into the season due to extended leaf and root life with sufficient soil moisture, however, applications as late as the milky-ripe stage can be particularly beneficial (Dampney, 1987; Rule 1987; Astbury and Kettlewell, 1990; Dampney and Salmon, 1990). At timings as late as this, however, root activity would still be expected to be impaired, and greater nitrogen uptake may be achieved through application to the foliage (Strong, 1982; Curic, 1988).

The most common fertilizer used for foliar application is urea dissolved in water to produce products containing up to 0.2 kg N l^{-1}. These may be diluted further before application to the crop. Urea is favoured for this type of application, having a low salt index compared with ammonium nitrate and it is, therefore, less likely to desiccate leaf cells through osmosis (Gray, 1977). Even at relatively low rates (e.g. 15 kg N ha^{-1} in 220 l ha^{-1}), however, flag leaves may still show tipping and scorch symptoms (Gooding, 1988), particularly if leaf tissue remains wet with urea solution for an extended period. In addition to desiccation of leaf cells, damage may occur through aqueous ammonia and urea toxicity, biuret contamination, and the disruption of carbohydrate metabolism (Gooding and Davies, 1992). Damage can be reduced by applying the urea in more dilute sprays and using additives which reduce the rate of accumulation into the leaf. Conversely, combining urea with certain wetters has increased damage. Spraying late-season crop protection chemicals in the same mixture with urea would be a major advantage (Gooding *et al.*, 1988), but this can often increase the risk of crop damage (Gooding and Davies, 1992). Typical amounts of nitrogen supplied as late season foliar urea, specifically for increasing grain protein concentration, vary between 30 and 50 kg N ha^{-1}.

Protein quality

As nitrogen fertilizer increases grain protein concentration the composition of the protein changes. Usually this does not have deleterious effects on baking quality because it is the storage proteins implicated in dough formation, i.e. the glutenins and gliadins, which increase more than the cytoplasmic (water-soluble) proteins. This shift is also apparent in effects of nitrogen fertilizer on protein quality for ruminants, where increased applications reduce the immediately rumen soluble N but increase the soluble but degradable N (Givens, 1994). Unfortunately, improvements in nutritional value are not as great because the storage proteins are poor in essential amino-acids. Hence increased nitrogen applications often depress the percentage content of lysine, threonine and valine and increase the content of glutamic acid, proline and phenylalanine (Larson and Nielson, 1966; Abrol *et al.*, 1971b). Some studies, however, suggest that a depreciation in lysine content is not inevitable (Sander *et al.*, 1987).

Returning to baking quality, Fig. 6.2 shows that nitrogen influences on crude protein content can be associated with similar effects on sodium dodecyl sulphate (SDS) sedimentation volume. This is in agreement with most previous tests showing that the amount of protein conferring good baking potential increases in line with protein concentration and nitrogen application rate (Petrakova *et al.*, 1974; Rodriguez-Bores and Bushuk, 1975). Sometimes, especially at high rates of nitrogen application, the SDS test is not improved despite increases in crude protein content (Dampney and Salmon, 1990) suggesting that protein quality is reduced. This is supported by others showing that the ratio of gliadins : glutenins is increased as grain N increases (Triboi *et al.*, 1990) and it is particularly the high molecular weight (HMW) glutenins that impart resistance to a dough and are found in the SDS sediment (Fullington *et al.*, 1987). In Canadian experiments where both foliar urea and soil ammonium nitrate applications gave large increases in protein concentration (e.g. from 12.1% to $> 16\%$), dough strength and the quality of loaf crumb texture was reduced (Tipples *et al.*, 1977). Other disappointing quality responses following large increases in protein concentration deriving from considerable applications of nitrogen have been attributed to incomplete synthesis of gluten (Finney *et al.*, 1957). Gooding *et al.* (1987b) found the improvements in loaf quality and dough strength following late-season urea sprays were less reliable when the urea was applied with a fungicide.

Increased nitrogen applications often increase grain sulphur contents (McGrath, 1985), but not always at the same rate as nitrogen increases. Increased nitrogen applications have, therefore, sometimes been associated with a higher ratio of nitrogen : sulphur in the grain, reduced proportions of the sulphur containing amino-acids cysteine and methionine, increased proportions of sulphur deficient ω-gliadins, and decreased HMW-glutenin

(Timms *et al.*, 1981; Cressey *et al.*, 1987; Gooding *et al.*, 1991). These reductions in protein quality may, therefore, have been alleviated with better late-season sulphur nutrition.

There has been speculation as to how late nitrogen can be applied and still be incorporated into grain proteins of use for baking. Salmon *et al.* (1990) found that foliar urea applications even as late as the milky-ripe growth stage were better at improving both protein content and bread-making performance than ammonium nitrate prills applied much earlier, when the second node was detectable. However, this result contrasted with earlier studies (Hook *et al.*, 1989) and even Salmon *et al.* (1990) found the late-season urea to increase the proportion of ω-gliadins present relative to earlier applications. This effect was partly corrected by simultaneous application of micronized elemental sulphur, again indicating that nitrogen effects on protein quality are dependent on the adequate availability of sulphur. Applications after the end of grain filling have not improved SDS-sedimentation volumes, but have not increased grain protein contents either (Sylvester-Bradley *et al.*, 1987b). There has been some concern that these very late applications may result in urea remaining on the surface of the grain, thus contributing to assessments of crude protein but being of no value to baking. This does not, however, appear to occur even for applications one hour before harvest, presumably because the grain is shielded sufficiently by the lemma and palea (Sylvester-Bradley *et al.*, 1987b).

The effect of nitrogen fertilizer applications on protein quality can interact with genotype. For example, the response of SDS-sedimentation volume to foliar urea applications can vary among varieties despite responses of protein concentration being similar (Grama *et al.*, 1987; Gooding *et al.*, 1991). Kosmolak and Crowle (1980) also reported contrasting effects of soil applied ammonium nitrate on the dough strength of five Canadian hard red spring wheat varieties despite similar responses in protein concentration. These interactions, however, were not consistent over seasons (Grama *et al.*, 1987; Gooding *et al.*, 1991).

Mineral content

It might be expected that the mineral content of wheat grain would tend to decline with increased nitrogen fertilizer applications as they become diluted with the increased DM accumulation. In an analysis of nine field experiments, however, McGrath (1985) found the grain concentrations of iron, zinc and copper to increase significantly with nitrogen application rate in six, five and five cases, respectively. In contrast, significant reductions were only recorded once for both copper and zinc, and not at all for iron. The author suggests that the increased nutrient concentrations and nutrient offtakes were due to increased root biomass developing in response to the nitrogen fertilization. Manganese, however, was not increased by nitrogen fertilizer, and was significantly reduced at one site, which was particularly poor in soil manganese availability, where plants developed manganese deficiency symptoms.

Flour characteristics

Effects of nitrogen fertilizer on flour extraction rates are usually small and inconsistent (Tipples *et al.*, 1977; Bayles *et al.*, 1978; Kosmolak and Crowle, 1980; Salmon *et al.*, 1990), presumably reflecting the variability of specific weight, thousand grain weight and grain shrivelling responses described above. Both Canadian and UK experiments have shown, however, that flour tends to get darker as nitrogen application rate increases (Fuller and Stewart, 1968; Tipples *et al.*, 1977; Salmon *et al.*, 1990). This is often associated with effects of different nitrogen rates on protein contents (Fig. 6.5), or in the case of the UK experiments, with protein effects

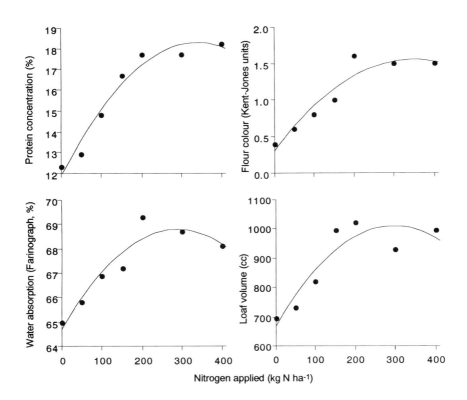

Fig. 6.5. The effect of increasing nitrogen application rate on the grain protein content, flour characteristics and loaf volume of wheat (from data presented by Tipples *et al.*, 1977).

induced by different rates, forms and timings of nitrogen application (Salmon *et al.*, 1990). These reports are, therefore, consistent with the view that protein contributes directly to the inherent greyness of the flour, although increased bran contamination from shrivelled grain cannot always be ruled out. In contrast to the effects on greyness, increased nitrogen applications have been associated with reduced amounts of yellow pigment in the flour (Tipples *et al.*, 1977).

Water absorption is also often increased with extra nitrogen fertilizer, again possibly as a result of the extra protein (Fig. 6.5; Bayles *et al.*, 1978; Kosmolak and Crowle, 1980).

Hagberg falling number

Increasing applications of nitrogen to wheat have decreased Hagberg falling numbers in contrasting field environments (Brun, 1982; Tabl and Kiss, 1983; Oskarsen, 1989). This can occur when nitrogen increases lodging, and therefore grain moisture content (Brun, 1982; Oskarsen, 1989). This would be expected to increase the risk of pre-harvest sprouting and hence reduce Hagberg falling numbers. Conversely, since nitrogen application can also delay maturity, it has been suggested that this influence could help maintain high falling number (Anon., 1985). Pushman and Bingham (1976) found that increased nitrogen application decreased α-amylase activity, lending support to this suggestion, but contrasting with the results of Brun (1982), Tabl and Kiss (1983) and Oskarsen (1989). Latterly, there have been many findings, similar to those shown in Figs 6.2 and 6.4, that nitrogen fertilizer can sometimes increase Hagberg falling number, commensurate with reduced α-amylase activity (Gooding *et al.*, 1986b; McDonald and Vaidyanathan, 1987; Astbury and Kettlewell, 1990). Doubt has, however, been cast on the original suggestion that this was due to delayed maturity. Firstly, late-season urea sprays, which are unlikely to have a dramatic effect on crop maturity, have increased Hagberg falling number as effectively as applications of nitrogen during stem extension (Rule, 1987). Secondly, the effect of nitrogen on Hagberg falling number interacts strongly with variety in ways not apparently related to maturity effects (Fig. 6.4).

Compared with the variety × nitrogen interactions that occur on SDS, the interactions on Hagberg falling number are relatively consistent over site and season. For example, Gooding *et al.* (1986b), Rule (1987) and McDonald and Vaidyanathan (1987) all found various nitrogen treatments to increase the Hagberg falling number of cv. Avalon, but not of Mission. The reason for this interaction has not been determined although its explanation would surely help explain the positive effects of nitrogen.

Financial implications

The financial implications of nitrogen fertilizer application to bread-making quality wheats are complicated by seasonal variations and variety interactions. Figure 6.4 shows an analysis of the response of three comparable bread-making wheats available to UK farmers in 1985, on free draining sandy loam in Shropshire. There was a significant variety × nitrogen interaction with regards to yield response but all three were close to their highest yields at 200 kg N ha^{-1}. The nitrogen effect on protein concentration did not interact with variety although cv. Avalon was nearer the bread-making thresholds of 10.5 and 11% protein than the others. Avalon also reached the Hagberg falling number threshold of 220 at 176 kg N ha^{-1}. Mission had comparable Hagberg falling numbers to Avalon at low nitrogen application rates, but failed to reach threshold levels because it failed to respond to nitrogen. Hagberg falling number of Brimstone did respond to nitrogen but, because this was a very low scoring variety to start with, it never reached threshold levels required for bread-making. Avalon, therefore, was the only variety that could receive a premium payment for bread-making. Margins over cost of nitrogen application were calculated on the basis of feed wheat fetching 108.9 £ t^{-1}, intermediate bread wheat 121.6 £ t^{-1} (10.5% protein, 220 Hagberg falling number), bread wheat 134.3 £ t^{-1} (11% protein, 220 Hagberg falling number), and the cost of nitrogen being 0.3 £ kg^{-1}. The return on Avalon, therefore, showed a marked increase at about 225 kg N ha^{-1} as its protein was increased above 10.5%, and again at 295 kg N ha^{-1} when 11% was reached. This type of analysis demonstrates the difficulties in being able to predict the economic optimum for nitrogen on bread-making wheat varieties when one or two characters are close to threshold levels. In the analysis presented here, the nitrogen required for economic optimum for Avalon is nearly twice that required for Mission despite similar requirements for optimum yield. Comparisons of other varieties have shown similar variability with step functions for describing economic return when just yield and protein concentration have been considered (Holbrook *et al.*, 1983).

In an analysis of 58 experiments on bread-making varieties of winter wheat over nine years in the UK, Sylvester-Bradley and George (1987) calculated that on average 25 kg N ha^{-1} extra over and above what would be optimum on the basis of feed prices, could be justified with premiums for bread making as low as 6 £ t^{-1}. This was true whether or not the premiums were paid in one step as protein exceeded 11%, or as a continuous sliding scale from 10 to 12%. With premiums of over 24 £ t^{-1} an extra 75 kg N ha^{-1} could be justified. These average results must, however, hide a great deal of site and variety variation as exhibited in Fig. 6.4. In practical terms, improved advice to farmers will require better prediction systems to determine how close the wheat will be to threshold levels without additional application, and the likely responses when influenced by variety, site and season.

Better advice is also required to reduce potential impact upon the environment, particularly with regards nitrogen losses. Figure 6.4, for example, shows the apparent recovery of fertilizer nitrogen in the grain. This is calculated by subtracting the nitrogen in the grain at nil application from the nitrogen recovered in the treated plots and expressing the remainder as a percentage of how much was applied. Encouraging farmers to increase nitrogen application rates above that required for optimum yield to help achieve quality thresholds, therefore, exacerbates problems associated with poor nitrogen recovery. This is mostly a problem in areas where the climate confers high yield potential. Such areas will also be associated with low grain protein, unless large quantities of nitrogen fertilizer are used. Latterly, however, even in the UK the usefulness for the end-user of applying nitrogen above the rates necessary for optimum economic grain yield in order to achieve higher quality have been cast into doubt. Although protein contents will be increased, improvements in protein quality are much more variable and benefits for the baker of this environmentally risky strategy may be minimal. Sylvester-Bradley (1990) and Webb and Sylvester-Bradley (1995) found no consistent benefits for loaf volume or SDS-sedimentation volume when applying nitrogen above the rates necessary to achieve optimum economic grain yields in a large number of experiments. This inconsistency may be improved in the future by advances in sulphur nutrition. At present, though, the encouragement of high nitrogen inputs to combine high yields with high quality increases the risk of nitrate leaching, N volatilization and support energy requirements, without necessarily giving a consistent benefit to the end-user.

Organic Sources

In many areas of the world, particularly when livestock and arable systems run side by side, the utilization of organic wastes from animals is important for maintaining soil fertility for cropping. Where intensification has occurred, however, the ready availability of affordable inorganic nitrogen fertilizers has led to many farmers neglecting the nutrient value of organic manures. Instead such products are looked upon more as a disposal problem rather than as a valuable resource and are sometimes applied to soil at inappropriate times for the environment and efficient crop nutrition. During the 1970s and 1980s, for example, slurry became the most important farm waste product in areas of intensive livestock farming in north west Europe, which led to serious problems of nitrate leaching to ground water and to ammonia and odour emissions (Lorenz and Steffens, 1992). On the other hand, failure of farmers to account for the nutrient value of organic wastes applied to cropping areas can lead to overdosing, particularly with nitrogen, increasing risks of lodging and disease. There is, therefore, a clear need to understand how slurry and farmyard manure (FYM) can be used to maximize crop potential, reduce reliance on synthetic fertilizers, and reduce harmful effects on the environment.

Nitrogen from organic sources, such as FYM and slurry, can be used as a sole nitrogen source or as a substitute for part of the inorganic fertilizer. These benefits, however, are not always easy to optimize. This is partly because of nitrogen content and its subsequent release being difficult to predict. Also because organic wastes contain much more than just nitrogen, and may affect crops in ways not just related to nitrogen response. Other notable constituents include both phosphorus and potassium (Table 6.4). Application of FYM also increases soil organic matter, improving both soil structure and biological activity.

FYM consists of dung, urine and the bedding, usually straw, from stock. Its value depends on the type of animal (Table 6.5), the nitrogen content of the food fed, the straw content, and the manner of storage, transport and application method in the field. Much nitrogen is lost through volatilization of ammonia, particularly when composted. This can be reduced, however, by minimizing the amount of times that FYM is moved and also by ploughing in immediately after application. Manure piles exposed to rain will also lose considerable amounts of nitrogen by leaching and volatilization.

Table 6.4. Nutrient content and availability to following crop of phosphate and potash in livestock wastes as estimated from UK experiments (source: Ministry of Agriculture, Fisheries and Food, 1994).

	Dry matter (%)	P_2O_5		K_2O	
		Total	Available	Total	Available
Fresh manures (kg t⁻¹)					
Cattle	25	3.5	2.1	8.0	4.8
Pig	25	7.0	4.2	5.0	3.0
Poultry (layer)	30	13.0	7.8	9.0	6.8
Broiler litter	60	25.0	15.0	18.0	14.0
Slurries (kg m⁻³)					
Dairy	6	1.2	0.6	3.5	3.2
Beef	6	1.2	0.6	2.7	2.4
Pig	6	3.0	1.5	3.0	2.7

Table 6.5. Typical values for nitrogen content and availability to the following crop in livestock wastes in the UK assuming 750 mm annual rainfall (calculated from Ministry of Agriculture, Fisheries and Food, 1994).

			Availability when applied in:				
			Autumn (Aug–Sep)		Winter (Nov–Jan)		Spring (Feb–April)
	Dry matter (%)[a]	Total nitrogen content	Sandy/ shallow soils	Other[b] mineral soils	Sandy/ shallow soils	Other[b] mineral soils	All soils
Surface application							
Fresh manures (kg t^{-1})[c]							
Cattle	25	6.0	0.3	0.3	0.6	0.9	1.2
Pig	25	7.0	0.4	0.4	0.7	1.1	1.4
Poultry (layer)	30	15.0	1.5	2.3	2.3	3.8	5.3
Broiler litter	60	29.0	2.9	4.4	4.4	7.3	10.2
Slurries (kg m^{-3})							
Dairy	6	3.0	0.2	0.3	0.3	0.6	0.9
Beef	6	2.3	0.1	0.2	0.2	0.5	0.7
Pig	6	5.0	0.3	0.5	0.5	1.3	1.8
Rapid incorporation/injection							
Fresh manures (kg t^{-1})[c]							
Cattle	25	6.0	0.3	0.6	0.6	0.9	1.5
Pig	25	7.0	0.4	0.7	0.7	1.1	1.8
Poultry (layer)	30	15.0	1.5	2.3	2.3	4.5	7.5
Broiler litter	60	29.0	2.9	4.4	4.4	7.3	11.6
Slurries (kg m^{-3})							
Dairy	6	3.0	0.2	0.3	0.5	0.9	1.5
Beef	6	2.3	0.1	0.2	0.3	0.7	1.2
Pig	6	5.0	0.3	0.5	0.8	1.8	3.0

[a] DM contents above and below those stated should invoke a similar adjustment in the nitrogen content and availability.
[b] Deep silty and clay soils will have greater availability as less nitrate will be leached.
[c] When FYM has been stored in the open or outside values may be reduced by half.

With the introduction of intensive livestock systems, the rising cost of straw and mechanization of cleaning yards, many farmers now deal with manure in a liquid state (slurry). For environmental reasons this usually has to be stored. Its nutrient value will depend strongly on its dilution with rainwater or water from cleaning hoses. As with FYM, losses will be affected by method of application and the degree to which the product is incorporated, or injected into the soil, or sprayed onto the crop. Table 6.5 shows the amounts of nitrogen that might be expected to be made available to the following crop from manures and slurries from relatively intensive livestock systems found in the UK. This table emphasizes the additive effects of source, DM content, storage, soil type, timing and method of application on subsequent nitrogen availability. Greatest amounts being recovered in the crop when storage periods are low; DM contents are high; soil types and application timings are less conducive to leaching; and when slurries and manures are not left on the soil surface for any significant period of time.

Despite the complicated nature of nutrient release, many field experiments have shown that application of organic wastes can have analogous effects to inorganic nitrogen fertilizers. FYM manure applications have, for example, increased crude protein and gluten concentrations, and the degree of vitreousness (Lui *et al.*, 1987; Lebidinskaya *et al.*, 1988; Shatilov and Sharov, 1992; Bagrintseva and Khodzhaeva, 1992). Additionally, applications have sometimes increased thousand grain weight and specific weight (Bagrintseva and Khodzhaeva, 1992). Applications of pig slurry at sowing have also increased protein concentrations (Salomonsson *et al.*, 1994). However, others have not found FYM manures to increase grain protein concentrations, although yield improvements occur more frequently (Ionescu *et al.*, 1988; Zhou and Li, 1991). The somewhat small effects of FYM and slurry on quality characters may derive from most applications to arable land being made before sowing in the autumn and winter to allow incorporation by ploughing. Increased nitrogen availability this early may result in higher risk of nitrogen loss through volatilization, leaching and denitrification. As discussed previously, the earlier availability of nitrogen is more likely to favour yield improvements than grain protein concentrations (Smith and Chambers, 1993). Hayward *et al.*, (1993a,b) have shown that both cattle and pig slurry applied to growing crops in the spring can be more efficient at increasing both yield and protein content than applications in the autumn or winter. Figure 6.6 shows the results from one experiment where two applications of cattle slurry (29 m^3 ha^{-1}; 4.2% DM; 2.8 kg total N m^{-3}; 1.6 kg ammonium N m^{-3}) were made to winter wheat receiving different levels of inorganic fertilizer at the beginning of March (four tillers) and April (pseudo-stem erect). As can be seen, the slurry gave responses in yield, lodging, specific weight and protein concentration comparable to 50–75 kg N ha^{-1} applied as ammonium nitrate, which equates roughly with the ammonium N applied in the slurry.

Another source of organic nitrogen is human sewage sludge. Its use would appear to offer a degree of sustainability with respect to nitrogen nutrition of crops as nitrogen being exported off farms for human

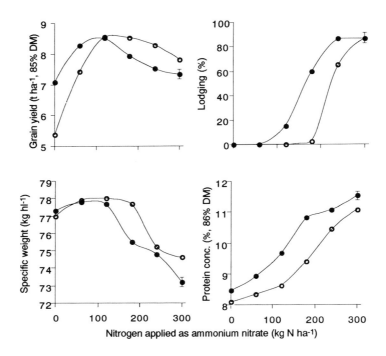

Fig. 6.6. The effect of applying cattle slurry and ammonium nitrate to winter wheat in the spring on grain yield, lodging, specific weight and grain protein concentration (●,○ = with and without slurry, respectively. Error bars represent SED for comparing slurry treatments at same level of ammonium nitrate (from Hayward *et al.*, 1993b).

consumption would be returned. The use of sewage sludge has increased yields of wheat in a large number of experiments due, in particular, to increased uptake of nitrogen, phosphorus, calcium and magnesium. In dry conditions sewage sludge may also increase yields through increasing the water holding capacity of the soil (Al-Mustafa *et al.*, 1995). Some concern, however, remains as to the concentrations of toxic heavy metals present in such products, due to industrial discharges. These contaminants, which may build up in soil, limit its long-term use, both in relation to crop growth and the utilization of crop products. For example, heavy applications of sewage sludge have occasionally depressed crop yields (Mitchell *et al.*, 1978) and produced cereal grains containing levels of cadmium in excess of recommended tolerances (Chaney and Giordano, 1977). Nonetheless, in a long-term study, Lerch *et al.* (1990) found that an application of 7.5 t ha^{-1} of sewage sludge per year increased financial returns relative to the normal practice of applying 60 kg N ha^{-1} as inorganic fertilizer due to the higher levels of protein content achieved with

Table 6.6. Effect of sewage sludge application (t ha^{-1}) on grain yields (t ha^{-1}) and element contents (%) of wheat averaged over three seasons and two irrigation treatments (calculated from data presented by Al-Mustafa *et al.*, 1995).

Rate of sludge application[a]	Yield	Protein	P	K	Fe	Zn	Mn	Cu
0	6.19	12.89	0.35	0.33	0.016	0.0047	0.0054	0.0013
20	7.38	13.24	0.38	0.34	0.014	0.0046	0.0052	0.0014
60	7.16	13.35	0.37	0.35	0.015	0.0050	0.0050	0.0014

[a] Average composition (%) of sludge was 36.2 organic matter, 1.32 N, 1.12 P, 0.46 K, 0.35 Fe, 0.054 Zn, 0.0198 Mn, and 0.0148 Cu.

the sludge application. In this work contamination with cadmium, nickel and lead remained at acceptably low levels. Others have also found sewage sludge to increase yields and grain protein contents simultaneously (Kasatikov and Runik, 1989) and without the concentration of metals contained within the sludge increasing in the grain (Table 6.6). The particular potential hazard of cadmium contamination may be alleviated by liming as its availability is greatest on acid soils and declines markedly as pH rises above 6.5.

SULPHUR

Sulphur is an important component of plant proteins and some oils. In protein, sulphur containing amino-acids, such as cysteine and methionine, are responsible for forming sulphydryl (S–H) and disulphide (S–S) bonds among proteins and hence contribute to their three-dimensional configurations and subunit associations. A 5 t crop of grain at 11% protein (DM) will remove approximately 6 kg of sulphur in the grain and 4.5 kg in straw. Leaf symptoms indicating poor availability are very similar to those of nitrogen, while deficiency in grain is deemed to have occurred if the N : S ratio is 17 : 1 or above. Wrigley *et al.* (1984) argue that deficiency only occurs if the grain has a higher N : S ratio than 17 : 1 <u>and</u> a grain sulphur concentration of less than 0.12%. In many parts of the world, sulphur deficiency in wheat is becoming of increasing importance. For example, a survey of wheat grain in the UK revealed that no samples were deficient in 1982 but up to 10% were by 1992 (McGrath *et al.*, 1993). Average grain sulphur concentration declined from 0.172 to 0.143% over the same period, a decline which has been even more marked in Germany (Haneklaus *et al.*, 1992). This is because of four main factors. Firstly, the use of sulphur containing fertilizers has been steadily declining. Ammonium sulphate was once used as a common nitrogen source but has now been largely replaced by ammonium nitrate and urea. Secondly, with greater concerns for the environment, sulphur emissions from power stations and industry are being curtailed in many areas, such as northern Europe. Thirdly, with the increasing wheat yields, achieved through increased nitrogen application, there is a greater demand for sulphur by the crop. Fourthly, intensive arable production systems have tended to deplete the organic matter content of soils and, therefore, reduced the amount of sulphur that would otherwise be slowly released from this material.

Sulphur deficient plants will have reduced grain yields and thousand grain weights (Byers et al., 1987) and application of sulphur containing fertilizers has alleviated these problems in many cropping areas (Haneklaus et al., 1992). The most effective fertilizers appear to be those containing SO_4 (such as ammonium- or magnesium sulphate) applied to the soil. Foliar applications are thought to be less effective due to the relative immobility of SO_4 within the plant. Application of sulphur containing fertilizers can sometimes increase grain protein concentration (Byers et al., 1987; Dampney and Salmon, 1990), particularly, it appears, at low nitrogen fertilizer applications (Camblin and Gall, 1987).

The greatest effect of sulphur deficiency on wheat is on protein quality, which can be adversely affected in the absence of obvious effects on yield and leaf chlorosis. This is because the bonds formed from the sulphydryl groups of cysteine are essential for the formation of gluten and, therefore, the viscoelasticity of the dough (Wall, 1971). Inadequate sulphur concentrations in grain can be associated with reduced dough extensibility, leading to reduced loaf volumes with poor crumb structure (Moss et al., 1981, 1983). With high nitrogen fertilization, inadequate sulphur leads to increased levels of ω-gliadins because these have a comparatively low sulphur content with no cysteine and little or no methionine and, therefore, minimal disulphide bonding capacity (Byers et al., 1987; Haneklaus and Schnug, 1992). Application of sulphur can, therefore, reduce the proportion of ω-gliadins present (Salmon et al., 1990) and negate some of the negative effects of nitrogen fertilizer on protein quality discussed previously. From an analysis of German data, Haneklaus et al., (1992), recommended a standard rate of 50 kg S ha^{-1} to be applied as SO_4 to the soil in early spring to satisfy the needs for quality wheat. This would be expected to increase grain sulphur concentrations by 0.1 mg g^{-1} and increase loaf volumes by 50 ml per 100 g. More precise recommendations, however, rely on access to sulphur deposition maps to determine the crop requirement that could be satisfied from the atmosphere. It has, for example, long been known that sulphur deficiency is much more likely to occur in rural areas compared with areas close to, or down-wind, of urban industrial areas.

POTASSIUM

Potassium plays critical roles in metabolism and, in particular, water regulation through controlling the salt and osmotic balance of cells. Deficient plants show areas of bleaching or dieback, particularly at the tips of the older leaves. Increasing availability to deficient plants can improve tolerance to frost and drought, and can increase resistance to a number of diseases including powdery mildew. In contrast to nitrogen, increasing availability helps maintain straw strength and thereby reduce the risk of lodging. Also as distinct from nitrogen, less potassium is translocated to the grain compared with what is retained by the straw. A 5 t grain ha^{-1} crop would, for example, be expected to have about 25 kg K ha^{-1} in the grain and 33 kg K ha^{-1} in straw.

Potassium is absorbed by the plant as K$^+$ ions which, being positively charged, are readily held on the surface of clay and humus in the soil, or more tightly between clay plates within soil aggregates. The implications of this are, firstly, that the ion is much less likely to be leached through the soil compared with the negatively charged nitrate ion. As a result the timing of application is less significant, particularly on soils rich in clay. Secondly, clayey soils tend to have a reserve of potassium which may help offset fertilizer applications. Deficiency is more likely to occur on sandy soils.

Table 6.7 emphasizes the impact of both straw utilization and soil availability on potassium requirements of high yielding crops. This is determined by standard laboratory techniques (Ministry of Agriculture, Fisheries and Food, 1986) and converted to a soil index. Indices of 0 and 1 indicate deficiencies. In these cases it is wise to apply more potassium than the crop requires so that, over time, deficiencies can be avoided. In

Table 6.7. Recommended application rates of potash (kg K$_2$O ha^{-1}) to wheat crops in the UK (Ministry of Agriculture, Fisheries and Food, 1994) in relation to soil availability as determined by standard methods (Ministry of Agriculture, Fisheries and Food, 1986).

	Target yield (t ha^{-1}, 85% DM)	Available potassium (mg l^{-1}): Potassium index:	0–60 0	61–120 1	121–240 2	241–400 3
Straw of previous crop ploughed in/incorporated						
Winter wheat	8		95	70	45	0
Spring wheat	6		85	60	35	0
Straw of previous crop removed						
Winter wheat	8		140	115	90	0
Spring wheat	6		120	95	70	0

these situations it is also advisable to apply the fertilizer to the seedbed and, if possible in a combine drill with the seed, so as to enhance its availability to the young plants. A soil index of 2 indicates that soil release will be adequate for the crop so the amount recommended for application roughly equates to crop removal such that levels are not depleted to deficient situations. At indices above 2 potassium levels are in excess and, can be allowed to drop over time so no additional fertilizer need be applied.

For traditional reasons, the potassium content of fertilizers is usually expressed as % K_2O (83% K) or potash although it is regularly applied as potassium chloride (KCl; 60% K_2O), potassium sulphate (K_2SO_4; 48–50% K_2O), or in organic forms (see Tables 6.4 and 6.6). Potassium sulphate is more expensive per unit of potassium than the chloride salt and is, therefore, less frequently used. It does, though, also supply sulphur and hence gives additional benefits for yield and quality as described above. The best form of potassium to use has been found to depend on rotation (Bakhsh *et al.*, 1986) and possibly, therefore, relative sulphur availability.

The effect of potassium fertilizers on the quality of wheat has not received much attention compared with nitrogen and sulphur. In glasshouse pot experiments, large increases in grain protein content have been attributed to increased potassium nutrition (Saad *et al.*, 1986; Singh *et al.*, 1992) and a similar effect has been achieved with potassium chloride seed treatments (Sen and Misra, 1987). In spite of this, large-scale field experiments have not detected additional potassium effects on protein concentration, despite improvements in yield and, sometimes, thousand grain weight (Hagras, 1985; Jackson *et al.*, 1991).

Increasing availability of potassium above levels required for optimum yield can result in *luxury* uptake, i.e. plants absorb more than is necessary to maximize production. This may help the plant to guard against future disease and drought stresses but it can also have a deleterious effect on magnesium uptake.

PHOSPHORUS

Phosphorus plays a crucial role in energy storage and transfer within cells. In particular, inorganic phosphate is essential for photosynthesis (Herold and Walker, 1979). High energy phosphate to phosphate bonds are created in the formation of adenosine triphosphate. The energy can be released when the bonds are broken to form adenosine diphosphate. In contrast to nitrogen, wheat plants that are deficient in phosphorus keep their green colouring, mature late and have increased susceptibility to a number of diseases (Wiese, 1987). In extreme cases, leaves have a dull, bluish-green colour with a purple or bronze hue, and die back from the tips. Increasing phosphorus availability to deficient plants can aid establishment, speed up the growth of seedlings, increase root development, and encourage tillering. At harvest, most phosphorus is partitioned to the grain, and 5 t grain typically contains about 19 kg P with 5 t straw containing only 3 kg P.

Phosphorus can be applied in many different forms but once in the soil it is comparatively immobile, only becoming available to the plant slowly as $H_2PO_4^-$ in weak soil solution. A low supply is more likely in acidic (pH < 5.5), alkaline/calcareous (pH 7.5–8.2) and/or low organic matter soils. It may also be in short supply in wet or compacted conditions which limit root growth, and where applications to previous crops have been inadequate (Archer, 1985; Wibberley, 1989). Due to its poor solubility, phosphorus is not leached readily and, therefore, timing of application is less critical. It is, however, difficult to correct deficiencies quickly. The strategy should, therefore, be one of building up or maintaining reserves over several years. Recommended rates for high yielding crops in the UK are shown in Table 6.8. As with potassium, requirements depend on straw utilization and availability in the soil. Soil indices of 0 and 1 indicate deficiency when application to the seedbed is likely to be most effective.

In fertilizers, the nutrient content is usually expressed in terms of P_2O_5, which contains 43% P. Availability, however, varies greatly amongst different fertilizer products. The most suitable forms for wheat production, particularly when in deficient situations, are water soluble. These include *superphosphate* which is produced by treating ground rock phosphate with sulphuric acid and contains 18–21% water-soluble P_2O_5. Alternatively, *triple superphosphate* contains about 47% water-soluble P_2O_5 and is produced by treating the rock phosphate with phosphoric acid. Availability in deficient soils can be improved by applying the fertilizer with the seed

Table 6.8. Recommended application rates of phosphate (kg P_2O_5 ha^{-1}) to wheat crops in the UK (Ministry of Agriculture, Fisheries and Food, 1994) in relation to soil availability as determined by standard methods (Ministry of Agriculture, Fisheries and Food, 1986).

Target yield (t ha^{-1}, 85% DM)	Available phosphorus (mg l^{-1}): Phosphorus index:	0–9 0	10–15 1	16–25 2	26–45 3
Straw of previous crop ploughed in/incorporated					
Winter wheat 8		110	85	60	60
Spring wheat 6		95	70	45	45
Straw of previous crop removed					
Winter wheat 8		120	95	70	70
Spring wheat 6		100	75	50	50

in a combine drill, thus ensuring closer placement of the phosphate to the seedlings.

As well as maintaining yield, increasing phosphate availability has improved thousand grain weight and specific weight (Hagras, 1985). In India, grain protein contents have also been improved in some field experiments (Sen and Misra, 1987; Singh _et al._, 1991). In other experiments, however, this was not the case (Hagras, 1985). Dubetz (1977) found that phosphate application to irrigated wheat in Canada gave significant increases in the total yield of grain protein, but these did not keep pace with yield improvements so protein concentration declined slightly. Phosphate applications to a deficient soil would be expected to improve root development, facilitating greater nitrogen uptake and higher protein yields.

MAGNESIUM

Magnesium is a constituent of chlorophyll and is also associated with phosphorus metabolism. Poor availability results in yellow or chlorotic patterns on the leaf from low chlorophyll. Magnesium deficiency is most likely to occur on light sandy or chalk soil, or where there is poor soil structure. Poor uptake can also be exacerbated by large applications of potassium and/or calcium. Crop removal in 5 t of grain and straw equates to about 6.5 and 4.0 kg, respectively. Magnesium can be applied as magnesium sulphate, but it is also contained within FYM and keiserite (16% Mg). It is also possible to apply magnesium containing limestone (e.g. dolomite) for pH adjustment.

CALCIUM AND LIMING

Many intensive wheat production systems tend to reduce pH, due partly to the use of acidifying fertilizers such as ammonium nitrate and urea. This needs to be rectified by the application of lime (calcium carbonate). However, lime can be lost through leaching of calcium carbonate which may amount to between 250 and 1750 kg ha^{-1} year^{-1} in the UK (Wibberley, 1989). In addition, removal of 5 t of grain and straw would also expect to extract 10 and 30 kg of calcium carbonate equivalents, respectively. For the vast majority of wheat varieties, the ideal pH for vegetative and reproductive growth is between 6.2 and 6.8. Maintaining pH within this range ensures that many plant nutrients are available while many toxic elements are in insoluble forms and therefore not adversely affecting the plant. Wibberley (1989) quotes a critical level of 5.3 below which performance is significantly impaired while alkaline soils, above pH 7.5, can also give unsatisfactory development. The amount of lime necessary to alter pH depends on soil type. Approximately 1.6, 2.5 and 4.0 t ha^{-1} of lime (or calcium carbonate equivalents) are needed to to give a 0.5 pH rise on sandy, medium loam, and heavy or organic soils, respectively.

TRACE ELEMENTS

Several trace elements influence the activity of enzymes. This effect is often dependent on the micronutrient forming stable complexes with naturally occurring ligands. For example, micronutrient–enzyme associations include zinc in carbonic anhydrases and dehydrogenases; iron in catalases, peroxidases and some cytochromes; manganese in pyruvate carboxylase; copper in cytochrome oxidase and ascorbic acid oxidase; and molybdenum in nitrate reductase and nitrogenase (Katyal and Friesen, 1988).

Poor availability of manganese results in grey–white spots on both young and old leaves. When spotting is severe the leaves often kink or droop from the base of the blade. Manganese deficiency is most likely to occur in cool dry weather on light sandy or chalk soil where there is a pH greater than 6.0. Relieving deficiency can increase resistance to both take-all and powdery mildew infection (Graham, 1991).

Copper deficiency is most likely to occur on high organic matter soils and also when phosphorus is in excess. Copper is required for calcium translocation from old to new tissues. Reduced calcium movement results in pale green leaves which are darkened, dry and twisted at the tips. Roots are often stunted and profusely branched. When the ears emerge they often appear as white heads and develop only shrivelled grain. Wheat varieties differ in their sensitivity to copper deficiency and durum wheats appear less tolerant than bread wheats (Katyal and Friesen, 1988).

Iron deficiencies are most likely on calcareous soils and cause either overall or interveinal yellowing, starting with the younger leaves. Availability can be complicated by iron forming complexes with other elements in the soil such as copper, manganese, aluminium and phosphorus.

Zinc deficiencies are most likely in calcareous soils and when phosphorus is in excess. Poor availability results in stunted plants with reduced tillering. Leaves become yellow, although the central area may remain green. Foliar applications can correct deficiencies in the crop and may increase resistance to foot rot (*F. graminearum*) (Graham, 1991).

Boron deficiency is widespread in wheat producing areas of China, Brazil, Thailand, Nepal and Bangladesh, as evidenced by sterility (Rerkasem *et al.*, 1991). Grain set may fail in any floret from the base of the ear upwards. The palea and lemma appear normal but the anthers are absent or poorly developed, with few and/or deformed pollen grain.

Molybdenum deficiency can occur on acid soils (pH < 6.0). The amounts required to prevent deficiency are only of the order of 70–200 g ha^{-1} and may be achieved simply by sowing seeds enriched with molybdenum.

In the scientific literature, there are many instances of trace element applications having both yield and quality effects on wheat. For example, Hemantaranjan and Garg (1988) report yield and thousand grain weight improvements following applications of iron plus zinc. Comparable effects have also been reported for applications of boron (Mitra and Jana, 1991).

Molybdenum and zinc applications have increased yield and/or grain protein concentration in a number of experiments (Zeleny, 1986; Zhmurko, 1992). Application of molybdenum, in particular, has been shown to improve the ability of nitrogen containing fertilizers to increase protein concentration on a number of occasions, commensurate with its role in influencing key enzymes of nitrogen metabolism (Grifanov and Davydov, 1972; Butorina *et al.*, 1991).

Nearly all of these instances have occurred when soil pH has not been ideal, being either very high or very low for wheat, and/or confirmed low soil availability. Although showing very desirable effects, therefore, the significance to farmers in the majority of wheat growing systems is questionable. The potential adverse effects of deficiency, however, emphasize the requirements for an appropriate soil pH and periodic soil and plant analyses. Commonly used sources of micronutrients are shown in Table 6.9. Application can be made to the soil or foliage (Table 6.10). Soil applications often need to be greater than foliar sprays due to less efficient uptake by the target crop. However, soil applications can have residual beneficial effects for following crops. It is unwise, however, to repeat large soil applications in successive years as this can lead to surpluses, and possible toxicity problems (Katyal and Friesen, 1988).

Table 6.9. Common sources of micronutrients (from Katyal and Friesen, 1988).

Element	Source	Element concentration (%)
Zinc	Zinc sulphate ($ZnSO_4,7H_2O$)	23
	Zinc sulphate ($ZnSO_4,H_2O$)	36
	Zinc oxide (ZnO)	60–80
	Zinc chelate (Na_2–ZnEDTA)	14
	Zinc frits	4 (variable)
Iron	Ferrous sulphate ($FeSO_4,7H_2O$)	20
	Ferric sulphate ($Fe_2(SO_4)_3$)	20
	Fe–EDDHA	6
Manganese	Manganese sulphate ($MnSO_4,3H_2O$)	26–28
	Manganese oxide (MnO_2)	63
	Manganese chelate (Mn–EDTA)	12
Copper	Copper sulphate ($CuSO_4,5H_2O$)	25
	Copper sulphate ($CuSO_4,H_2O$)	35
	Cuperous oxide (Cu_2O)	89
	Copper chelate (Na–CuEDTA)	9
	Copper chelate (Na_2–CuEDTA)	13
Boron	Borax ($Na_2B_4O_7,10H_2O$)	11
	Sodium tetraborate ($Na_2B_4O_7,5H_2O$)	14
	Sodium tetraborate ($Na_2B_4O_7$)	20
	Solubor($Na_2B_4O_7,5H_2O+Na_2B_{10}O_{16},10H_2O$)	20
Molybdenum	Sodium molybdate ($NaMoO_4,2H_2O$)	39
	Ammonium molybdate(($NH_4)_6Mo_7O_{24},4H_2O$)	54
	Molybdenum trioxide (MoO_3)	66

Table 6.10. Methods and rates of micronutrient application for wheat (from Katyal and Friesen, 1988).

Element	Source	Suggested application rates (kg element ha^{-1})		
		Broadcast	Band	Foliar
Zinc	Zinc sulphate	5–20	3–5	0.015–0.250
Copper	Copper sulphate	4–15	1–4.5	0.1–0.5
Manganese	Manganese sulphate	20–130	6–11	0.5–2
Iron	Ferrous sulphate	–	–	5–10
Boron	Sodium tetraborate	0.6–1.2	–	–
Molybdenum	Sodium molybdate	0.07–0.2	–	0.1–0.15

7

Biology and Control of Diseases, Weeds and Pests: Effects on Grain Yield and Quality

Throughout the life cycle of a wheat crop, it is susceptible to a wide range of stresses imposed by biological agents comprising weeds, pests and diseases. These undoubtedly have a major impact on wheat productivity but the difficulties of estimating crop losses are well known. As a result estimates vary widely, particularly if a global perspective is attempted. One of the earliest studies of global losses in wheat was by Cramer (1967) who concluded that 5.0% loss was from insects, 9.1% from diseases and 9.8% from weeds – giving a total of 23.9% worldwide. Russell (1981) considered these figures to be an underestimate as a result of neglecting postharvest losses and damage caused by nematodes, molluscs and vertebrates. It has been suggested that crop losses have increased substantially since Cramer's estimates in the mid-1960s. Dr E.C. Oerke, of the University of Hanover, considers such factors as greater crop susceptibility, higher fertilizer use, changes in cropping patterns, and the build up of populations of pathogens, pests and diseases to be some of the reasons (British Agrochemicals Association, 1993). Oerke estimates that loss in total world wheat production between 1988 and 1990 was probably one and a half times greater than the estimates of Cramer (Fig. 7.1). The most important losses are attributed to disease (13.3%) and weed (13.1%) problems (British Agrochemicals Association, 1993). Even Oerke's figures, however, underestimate total impact as losses are only kept to this level by the use of control practices which themselves can incur a significant cost. In the absence of crop protection measures, total losses would be expected to rise to over 50%.

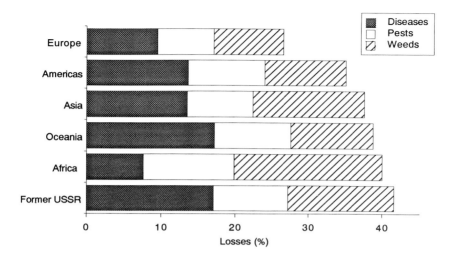

Fig. 7.1. Estimated losses in wheat production in different regions due to weeds, pests and diseases 1988–1990 (from British Agrochemicals Association, 1993).

Control has traditionally relied on cultural practices, for example, by choosing rotations and cultivations which interfere with the life cycle of the offending organism, by exploiting varietal resistance, or removing alternative hosts. Latterly, however, the use of agrochemicals has become widespread, permitting more intensive, and often more reliable production. For example, reduced losses in European production (Fig. 7.1) can be attributed to the particularly intensive and developed use of chemicals for crop protection in this area. Overreliance on agrochemicals can, though, lead to several problems, not least the build up of resistance to the agrochemical in populations of the target organism, potential pollution of the environment, and a reduction in non-target organisms such as natural predators of crop pests, or wild flowers in hedgerows. A balance is, therefore, required whereby the benefits of agrochemicals can be exploited, but non-chemical methods of control are also utilized. Modern approaches commonly exploit agrochemicals in the framework of an integrated crop protection strategy which relies equally, if not more heavily, on genetic and cultural components.

This chapter reviews the main biological agents causing yield and quality reduction of wheat, highlighting the principal control methods in each case, and examining the implication of control on grain characteristics. The major diseases which cause problems in all or most of the major wheat growing areas are covered. In the case of weeds, examples are drawn from important plant families, e.g. the *Poaceae*, while the major taxonomic orders of pests causing problems on wheat worldwide are introduced, followed by a number of examples of specific species. Differential treatment arises, however, as it has been attempted to be comprehensive with regards to the effects on quality. More minor weeds, pests and diseases are included, therefore, if their effects on quality have been well documented. In addition, the implications of using plant growth regulators will be assessed in this section as they are often grouped with fungicides, herbicides and pesticides as agrochemicals by farmers.

INFECTIOUS DISEASES

Infectious diseases of wheat can be caused by fungi, viruses, bacteria, mycoplasmas and nematodes. By far the most important group of diseases in wheat production systems are those caused by fungi.

Fungal Diseases of Wheat

In a compendium of wheat diseases, Wiese (1987) describes over 40 fungal diseases of wheat attacking the seedlings, roots, stem bases, leaves and stems, and/or the ear structures. The reader is referred to this account for more detail concerning this large range of pathogens and symptoms. Amongst this group, however, Oerke *et al.* (1994) identified eight particular genera and specific species of pathogen which can be considered to have particular importance. These comprised *Tilletia* spp. (bunts), *Ustilago* spp. (smuts), *Puccinia* spp. (rusts), *Septoria* spp. (leaf blotch, leaf spot and glume blotch), *Erysiphe graminis* f.sp. *tritici* (powdery mildew), *Pseudocercosporella herpotrichoides* (eyespot), *Gaeumannomyces graminis* var. *tritici* (take-all) and *Cochliobolus sativus* (seedling blight, root rot and leaf spot). It is these species together with *Fusarium* spp. (foot rot, ear scab), *Rhizoctonia solani* (root rot), *R. cerealis* (sharp eyespot), and *Pyrenophora triticirepentis* (tan spot) that will be dealt with here.

Fungi mostly exist as branched chains of cells forming filaments known as hyphae which are commonly 0.5–100 μm wide. The hyphae of most wheat pathogens are partitioned by cross-walls (septa) and make up the vegetative mass of the fungus (the mycelium). Reproduction is chiefly by means of spore production. This can be asexual, for example, with spores such as conidia being borne on specialized hyphae, the conidiophores. The conidia may be simply formed in exposed pustules as in the case of powdery mildew (*Erysiphe graminis* f.sp. *tritici*), or produced in flask-like structures known as pycnidia as in *Septoria* spp. (Parry, 1990). The production of masses of asexual spores allows the fungi to multiply and disperse rapidly and it is this form that is usually encountered by plant pathologists. Most pathogenic fungi of wheat, however, are also known to reproduce sexually. Ascospores, for example, are produced from within small sacs, known as asci (singular, ascus), in the sexual phase of take-all (*Gaeumannomyces graminis*) and powdery mildew. Other fungi, including the rusts (*Puccinia* spp.), bunts (*Tilletia* spp.) and smuts (*Ustilago* spp.) can have much more complicated life cycles, sometimes involving up to five different spore types. These fungi all produce a basidium, a club-shaped structure on which sexually produced basidiospores or hyphae derive. Epidemics of the rust fungi are, however, mostly associated with asexual reproduction and the dispersal of uredospores. They also commonly produce teliospores, a spore in which sexual fusion occurs. It is this type of spore which is mostly responsible for the dispersion of the smuts and bunts.

The principal diseases of wheat can be broadly divided into those attacking mainly the leaves and stem (e.g. powdery mildew, the rusts, tan spot and *Septoria*), those damaging mainly the roots and stem base (take-all, eyespot, *Rhizoctonia* spp. and *Fusarium* spp.), and those causing injury to the grain and ear diseases (the bunts, loose smut, glume blotch and *Fusarium* spp.). The foliar diseases have attracted most attention with regards to wheat yield and quality and will, therefore, be dealt with first.

Foliar diseases of wheat caused by fungi

Powdery mildew

Powdery mildew is one of the most common diseases of wheat, occurring throughout the major wheat growing areas of the world, including humid and semi-arid areas (Wiese, 1987). Oerke *et al.* (1994) identified powdery mildew as a particular problem of western and eastern Europe. Its economic importance has increased with the intensification of wheat production as exemplified by the use of high yielding varieties, increased seed rates and greater amounts of nitrogen fertilizer (Engels, 1995). Symptoms can be found on all aerial parts of the plant but they are most commonly observed on the upper surface of the lower leaves (Wiese, 1987; Parry, 1990). The disease is first seen as yellow flecks which are replaced by a white fluffy pustule. These then become powdery as masses of asexual conidia are released. The mycelium is entirely located above the leaf surface, except for the growth of specialized intracellular hyphae, known as *haustoria*, into the epidermal cells. The haustoria are the route by which nutrients can flow from the plant to the fungus (Mount and Ellingboe, 1969). In order for the haustoria to function, the invaded host cell must remain alive. *Erysiphe graminis* is, therefore, classed as a *biotrophic* pathogen, i.e. it can only live and reproduce by feeding on living plant material. The difficulty of growing such fungi on media in the laboratory has also led to the biotrophs being referred to as obligate parasites.

Conidia blown by the wind onto healthy plants can germinate at between 5 and 30°C, but 15°C with a relative humidity of 95% is optimal (Parry, 1990). Free water on the leaf surface inhibits conidial germination but in drier conditions the next generation of conidia can start to be produced in about seven days. Powdery mildew infection pressures can, therefore, increase rapidly with alternating wet and dry conditions, and breezes to disperse the spores. Wheat is most susceptible during periods of rapid growth. For UK grown winter wheat this commonly occurs in April and May, with spring wheat being especially vulnerable about a month later (Gair *et al.*, 1987).

As summer temperatures increase above 25°C the growth and development of the pustules is retarded. The increase in temperature, and/or moisture stress is accompanied by the production of black cleistothecia embedded in the mycelial mat as the pustule loses its white fluffy appearance and becomes brown. The cleistothecia are the sexual stage in the life cycle of *E. graminis* and allow recombination of genetic material and the development of new races (Gair *et al.*, 1987). The cleistothecia are also more resistant to desiccation than conidia and can thereby be a means of pathogen survival as the green tissue of the crop senesces. They can also persist on the crop debris after harvest. Each cleistothecium usually contains between 15 and 20 asci, which themselves contain eight ascospores (Wiese, 1987). These are released as the cleistothecia imbibe water after rain or heavy dews. In the UK ascospores can infect volunteer plants after harvest. Conidia from these, together with any surviving conidia produced on late tillers, can then infect autumn sown crops. In North America, the cleistothecia may survive the winter to permit ascospore infection of crops in the spring, but this is not thought to occur in the UK (Jones and Clifford, 1983).

Powdery mildew can cause significant losses in yield with field epidemics reducing yields by up to 25% in the USA and 40% in the UK (Gair *et al.*, 1987; Parry, 1990). Surveys in England between 1970 and 1980 suggest that annual yield losses range between 1.5 and 4.4%, with an annual national average loss of 3%. These percentage yield losses (y) have been related to the percentage area of leaf three (third leaf below the ear) covered at ear emergence (x) in equation 7.1 (Large and Doling, 1962; King, 1973):

$$y = 2.0 \sqrt{x} \qquad [7.1]$$

Yield reductions may result in part from reductions in the amount of light energy intercepted and therefore the amount of dry matter (DM) produced. Specific mechanisms include premature leaf senescence, reduced size of emerging leaves, and with heavy infections, tiller or even whole plant death. The efficiency with which the plant uses light energy to produce yield can also be reduced by infection. For example, respiration can be increased (Allen and Goddard, 1938) representing a loss of DM. Obligate pathogens, such as powdery mildew, are thought to block the non-cyclic electron transport chain and adversely affect photophosphorylation. Rates of photosynthesis are reduced as a result (Buchanan *et al.*, 1981). Various adverse reactions have been reported including altered source–sink relationships in infected plants such that export of sugars from the leaves is reduced (Farrar, 1995), possibly implicating effects on growth regulator metabolism.

The rusts

Rust diseases of wheat can cause significant damage throughout the world (Oerke *et al.*, 1994). The principal rust diseases which attack wheat are yellow rust (also known as west stripe or stripe rust) (*Puccinia striiformis* f.sp. *tritici*), brown rust or leaf rust (*P. recondita*), and black stem rust (*P. graminis*). These are all biotrophic pathogens but, unlike powdery mildew, the mycelium ramifies within the leaf tissue and pustules (uredia) rupture through the epidermis to release pigmented uredospores. Like powdery mildew the rusts increase transpiration and respiration, and reduce photosynthesis and the export of assimilates from the leaves. They can reduce plant vigour, seed filling, and root growth. The poor root activity renders infected plants more drought susceptible. Premature desiccation of rusted leaves in hot dry weather can cause serious yield losses.

Yellow rust is usually first seen as discrete patches within fields with yellow/orange uredia, about 0.5–1 mm in diameter occurring on leaves. In the UK, epidemics tend to start in May and continue into early summer. On mature tissue the uredia are often aligned, causing characteristic yellow stripes between the leaf veins. On younger tissue, pustules may appear to be more randomly scattered. The optimum conditions for uredospore production, dispersal by wind, penetration of new tissue and the production of new uredia are temperatures between 10 and 15°C and a relative humidity of 100%. In such environments the whole cycle may be repeated in as little as seven days (Parry, 1990). Development is hindered by temperatures above 20°C. Yellow rust is, therefore, the most important rust disease in the cooler areas of wheat production such as north-west Europe and north-western USA (Jones and Clifford, 1983). Within these areas epidemics are most common in coastal areas typified by cool summer temperatures and regular mists, or as in North America, at high elevations between mountain ranges. Yellow rust is also found in South America, Kenya, China, the Mediterranean, the Middle East, India and Oceania.

At the end of the season the yellow uredia are often replaced by black stripes of telia producing teliospores, mostly seen on the leaf sheaths. The teliospores can give rise to basidiospores but these are not known to infect wheat or any other host, and are currently thought to be insignificant in yellow rust epidemiology. Yellow rust can survive over the autumn and winter period as mycelium and/or uredospores on volunteers, early sown autumn crops, and on late tillers in stubble. This is assisted by the mycelium within leaves being able to survive temperatures as low as −5°C. Uredospores arriving from nearby fields can infect healthy crops, but dispersal over hundreds of miles is also possible in air currents (Gregory, 1973). Because *P. striiformis* cannot withstand high temperatures, during the summer in California it is found only in cooler coastal areas in the Sierra Nevada above 1830 m. In autumn, however, wind-blown uredospores infect early sown wheat and volunteers over a wider range, and from these sources it later attacks wheat crops more generally in spring (Tollenaar and Houston, 1967).

Crop losses in susceptible varieties can be very high, as much as 50% in South America and Oceania. The widespread use of resistant varieties has, however, contributed to low overall annual yield losses. In England and Wales these are not thought to exceed 1% regularly. Yield losses have been related to infection using equation 7.2 (King, 1976):

$$y = 0.4 \times x \hspace{4cm} [7.2]$$

where y is the percentage yield loss and x is the percentage area of the flag leaf infected when the caryopsis is milky-ripe. More complex, but possibly more robust, models have recently been derived relating yield loss from yellow rust infection to an estimate of disease effects on the accumulated radiation intercepted by green tissue after flag leaf emergence (Bryson *et al.*, 1995).

Brown rust (also known as leaf, dwarf or orange rust) is first seen as uredia rupturing the epidermis as orange/brown pustules, 1×1–2 mm in diameter (Parry, 1990). The disease occurs virtually everywhere the crop is grown, but is particularly evident when the wheat matures late, as is the case in many predominantly spring wheat growing areas (Wiese, 1987). Due to its widespread occurrence, it may well be the single most important disease of wheat on a worldwide basis (Jones and Clifford, 1983). The uredia can occur on all aerial parts of the plant but are most common on leaves. Unlike yellow rust the pustules are randomly scattered over the surface rather than in stripes. The temperatures for rapid development are warmer than for yellow rust (15–22°C with 100% relative humidity), so epidemics occur later in the season – mid- to late summer in the UK. Breezes to disperse the spores, followed by heavy dews to promote infection favour the rapid build up. As the crop senesces, darkly coloured telia are evident, releasing teliospores. In certain countries further spore stages are then known to infect other species in the genera of *Thalictrum* (meadow rue), *Anchusa*, *Anemonella*, *Clematis* and *Isopyron*. These stages, however, are not thought to be of major significance to the epidemiology of the disease, as new infections seem to occur from mycelium and uredospores that have overwintered as is supposed with yellow rust. Nonetheless the sexual stages that progress on the alternative hosts may assist the pathogen to develop new races.

Yield reductions by brown rust on susceptible varieties can be as high as 50% (Gair *et al.*, 1987) but, in the UK, infection is often too late to cause significant damage. In warmer areas with adequate humidity such as central and southern Europe, the former USSR, Australia, and south, central and mid-western states of the USA, annual yield losses of between 5 and 10% have been common, and may be much higher in certain years. Lower levels may be achieved following the widespread adoption of resistant varieties. Yield reductions have been related to yield loss in equations very similar in coefficients to yellow rust (see equation 7.2), although Burleigh *et al.* (1972) give a more complex analysis relating yield loss to brown rust at three separate growth stages.

Black stem rust produces pustules which are initially brown but turn much darker as telia develop. Unlike yellow and brown rust, the pustules are quite large (2×5 mm) and, although all aerial parts can be affected, they occur most frequently on the stems. The surrounding frayed epidermis is also more conspicuous (Wheeler, 1969). Also, unlike the other rusts, the spore stages on the alternate host can be very important to its epidemiology. The fungus overwinters either as telia or uredia. The spores germinate in the spring after repeated freezing and thawing. Basidiospores are produced from the teliospores which in turn infect an alternate host, which is most commonly barberry (*Berberis* spp.), but also can be *Mahonia* spp. After further spore stages on the barberry, aeciospores are released in late spring and blown on to wheat from where uredospores are produced which are capable of being blown over large distances to cause secondary infection and widespread dispersal. The infrequent infections that do occur in the UK are thought to derive from uredospores blown northwards from the western Mediterranean (Jones and Clifford, 1983). Disease development is favoured by warmer temperatures than yellow and brown rust (i.e. optimal at about 26°C with 100% relative humidity), being significantly hampered below 15°C and above 40°C. The disease is, therefore, only regularly serious in the warmer areas of wheat production such as parts of North America, India, Australia, North Africa and the Mediterranean. In these areas, however, black stem rust is often the most serious disease.

Annual yield losses in vulnerable areas may regularly exceed 10% on susceptible varieties. As well as yield losses through mechanisms mentioned above, severe infection can cause water loss and prevent assimilate transport up the stem resulting in white heads. The stem can also be weakened, increasing the risk of lodging (Jones and Clifford,

1983). Equations relating yield loss in spring wheat to disease severity include 7.3 (Romig and Calpouzos, 1970):

$$y = -25.53 + 27.17 \log x \qquad [7.3]$$

where y is the percentage yield loss and x is the percentage disease at the milky-ripe growth stage.

Diseases caused by Septoria *spp.*

Septoria tritici (leaf blotch, sexual stage *Mycosphaerella graminicola*) and *S. nodorum* (leaf and glume blotch, sexual stage *Phaeosphaeria* (syn. *Leptosphaeria*) *nodorum*) are prevalent in all major wheat growing regions of the world, of which probably North America and western Europe are most significantly affected (Oerke *et al.*, 1994). In the UK and western Europe, there have been many seasons when the septorias have been the most important foliar pathogens in terms of both disease severity and yield loss (Royle *et al.*, 1995). This has been particularly true for *S. tritici* since the early 1980s.

Septoria nodorum is distinct from powdery mildew and the rusts as the pathogen kills the host tissue with the aid of toxins and then utilizes the dead host tissues as a substrate on which to feed and sporulate. It is, therefore, known as a necrotroph, and due to its ability to grow easily on media, or on dead and decaying tissue, it can be classed as a facultative pathogen. *Septoria tritici* is not as specialized at biotrophy compared with powdery mildew and the rusts, but neither is it as efficient at necrotrophy as *S. nodorum,* and it possesses an extended biotrophic phase following infection (Royle *et al.*, 1995). This distinction in infection strategy contributes to the different symptoms associated with the two pathogens. Both initially appear as chlorotic flecks on the lower leaves which develop into lesions. The centres are initially water-soaked, then become yellow and finally red–brown, often surrounded by a chlorotic ring. Asexual conidia are then released from flask-like structures embedded within the lesion. The conidia are splash dispersed vertically up the infected plant and horizontally from plant to plant. In the case of *S. nodorum* the lesions are lens-shaped and spread into each other, often interacting with physiological leaf senescence. The pinky-brown translucent pycnidia are not necessarily produced until long after lesion development, increasing the chance of confusion with other processes causing leaf death. The cycle of conidial dispersal, infection and conidia production can be completed in 10–14 days but this requires temperatures of 20–27°C with 100% relative humidity (Polley and Clarkson, 1978).

Septoria tritici lesions are parallel sided and usually remain discrete for longer. They contain more recognizable black pycnidia (Royle *et al.*, 1995). The conidial cycle of development is optimal at about 15–20°C. In the UK *S. tritici* is, therefore, often evident earlier in the season than *S. nodorum,* but requires longer periods of high humidity to initiate infection (Parry, 1990).

During the UK summer the *Septoria* spp. are capable of entering the sexual stage, producing ascospores from pseudothecia containing asci (Cunfer, 1994). Ascospores, together with conidia, can be produced by surviving mycelium on stubble and other crop debris. The ascospores in particular are prevalent in air during late summer and autumn, and can travel large distances causing primary infection of emerging crops. Infection of the ear can lead to grain infection by *Septoria nodorum* which can also, in turn, generate infection of subsequent autumn sown crops.

Yield loss through *Septoria* infection is thought to account for nearly 2% of the world's wheat (Wiese, 1987). Losses can reach at least 20% on susceptible varieties when weather is conducive. Even in countries where the routine use of fungicides is widespread, favourable weather patterns have given substantial losses in national yields. For example, 8% of the English wheat crop is thought to have been lost to *Septoria* in the wet summer of 1985. Damage is primarily through the destruction of photosynthetic tissue and yield loss has been equated to severity in equations 7.4 to 7.7.

$$y = -2.6943 + 0.6366\ x_e \qquad \text{(Eyal, 1972)} \qquad [7.4]$$
$$y = 5.412\ x_{m1}^{0.6} \qquad \text{(King } et\ al.,\ 1983) \qquad [7.5]$$
$$y = 1.011\ x_{m1} \qquad \text{(King } et\ al.,\ 1983) \qquad [7.6]$$
$$y = 0.551\ x_{m2} \qquad \text{(King } et\ al.,\ 1983) \qquad [7.7]$$

where y is the percentage yield loss, x_e the percentage of disease on the foliage at ear emergence, and x_{m1} and x_{m2} the percentage area of disease at the milky-ripe growth stage on the flag leaf and leaf below it, respectively. In addition, *Septoria* infection often interacts with other diseases. *Septoria nodorum* and brown rust can be much more damaging when they occur together than the additive effects of the two diseases (Gair *et al.*, 1987).

Tan spot

Tan spot (syn. yellow leaf spot, *Pyrenophora tritici-repentis*, asexual stage *Drechslera tritici-repentis*) occurs throughout the world but, together with brown rust and black stem rust is often considered more important in hot and warm areas (coolest month mean temperature > 17.5 and 12.5°C, respectively) compared with the *Septoria* spp., powdery mildew and yellow rust (Dubin and van Ginkel, 1991). In such areas, chemical control trials have indicated that yield losses can exceed 20% (Dubin and van Ginkel, 1991). The fungus overwinters on stubble and other crop debris from which ascospores are released and dispersed by wind. Infection on the leaves results in brown flecks which develop into oval lesions with yellow borders. Lesions coalesce and develop darker, central areas from where conidia are released to cause secondary infections. Infection, lesion development and conidia production are all significantly increased by prolonged wet periods resulting from rains and/or dews (Wiese, 1987).

Control of foliar fungal diseases

Non-chemical methods

The principal means of non-chemical disease control are similar for all the major foliar pathogens of wheat. They include the removal of stubble, crop debris and volunteer cereals which assist the pathogens' survival from one year's crop to the next. Sowing autumn crops later also extends the time that the fungus needs to survive in the absence of the host and reduces the time during which the pathogen can multiply in the crop before winter. Conversely, in some climates late sowing may make crops more vulnerable to brown and black stem rust in spring by extending the season into warmer conditions. This effect can be exacerbated when some crops have been sown early to allow build up of inoculum. Rapid transfer of pathogens to spring crops can be reduced by isolating them as much as possible from infected autumn crops. Avoiding excessive seed rates, double drilling and excessive nitrogen fertilizer application rate can also reduce the build up of most of the diseases.

The removal of alternative hosts may reduce foci for new epidemics. This strategy has been notably exploited in the case of black stem rust, where national programmes of barberry removal have attempted to hinder disease progress in a number of countries. For septoria using seed free of disease can reduce inoculum and seedling death (Wiese, 1987).

All of these techniques, however, will only give partial reductions in disease severity. The most important means of non-chemical control is through the breeding and use of resistant varieties. Mechanisms of resistance can include the host's cuticle being a mechanical barrier, or the host cell quickly dying to prevent haustorial development (hypersensitivity). Some varieties slow rates of conidia production. Unfortunately, however, new races can develop which overcome resistance based on simple, particularly single gene, changes in the host. This is particularly true for powdery mildew and the rusts. It is hoped that reduced rates of mildew and rust development caused by the additive effects of several genes in some varieties will remain durable (Wiese, 1987). Differences in variety susceptibility may also be exploited by diversifying the varieties grown in different fields within a particular area. To be successful it is important that the genetic basis of resistance is known in the varieties chosen so that the overall vulnerability of crops to the spread of one race of the disease is reduced. This information forms the basis of variety diversification schemes for yellow rust in the UK (Anon., 1995b). Different varieties may also be grown together as a blend on the same field. In some experiments, suitable variety combinations have been shown to reduce the spread of certain foliar pathogens (Jeger *et al.*, 1981). Attention needs to be paid, however, to the relative maturation rates and market suitabilities of the different genotypes involved.

Control of foliar diseases with fungicides

Despite important advances made in resistance breeding, control of foliar pathogens by non-chemical means is rarely total. Resistance to the *Septoria* spp. is, for instance, only partial. With the rusts and powdery mildew, varietal disease resistance is still regularly overcome by new races of the pathogens. Combining durable resistance with high yielding, and high end-use value in the same genotype still remains a long-term objective. For example, in 1995 in the UK a new race of yellow rust resulted in potentially severe infestations of the disease in two of the most widely grown and highest yielding wheat varieties. The effective use of fungicides limited potentially damaging reductions in productivity, and also ensured the future use of these otherwise very useful genotypes (Orson, 1995).

Examples of fungicides available for cereal foliar diseases are shown in Table 7.1. They can be broadly divided into those which are surface protectant and those which are systemic. Protectant fungicides include most of the traditional products, such as elemental sulphur, which protect the tissue on which they land from fungal infection. As a generalization, their main disadvantages are that they do not move within the plant and, therefore, do not have curative activity and do not protect new emerging leaves or other tissue not easily targeted. Most modern fungicides have the ability to move within the plant, i.e. they are systemic. They can, therefore, have curative activity and may safeguard newly emerging tissue. Against this, however, is the fact that most of them interfere with a single biochemical reaction within the fungus. They are thus also known as site-specific fungicides. This means that only relatively small changes in the target pathogen need to occur and be selected before the population contains significant numbers of resistant individuals, which may lead to failure of the fungicide to control the disease in the field. Within the systemic group there is a large degree of variation in mobility. Eradicant and internal protectant activity with iprodione, in particular, shows very little systemic activity.

Modern fungicide control of powdery mildew and the rusts relies heavily on translocated chemicals which inhibit ergosterol synthesis in the fungus, i.e. the triazoles, morpholines and piperidines (Engels, 1995). The importance of triazoles and morpholines is emphasized in surveys of fungicide use in England and Wales (Fig. 7.2). The heavy reliance on compounds with related modes of action is highly undesirable due to the repeated selection pressures on the pathogen to become resistant. Resistance to triazoles has developed over a number of years in *E. graminis* f.sp.

Table 7.1. Fungicides available for wheat disease control in Europe (compiled from Parry, 1990; Leprince, 1995).

Chemical group	Probable mode of action	Foliar pathogen(s) particularly sensitive	Examples
Protectant fungicides			
Sulphur	Many potential targets including protein metabolism, respiration and heavy metal chelation	Powdery mildew	Elemental sulphur
Dithiocarbamates	Inhibition of thiol groups in dehydrogenase enzyme	*Septoria* spp., powdery mildew, rusts	Mancozeb, maneb
Phthalonitriles	Inhibition of thiol groups in dehydrogenase enzymes	*Septoria* spp.	Chlorothalonil
Systemic fungicides			
Benzimidazoles	Disruption of tubulin formation	Eyespot	Benomyl, carbendazim
Imidazoles	Inhibition of ergosterol biosynthesis	Eyespot, powdery mildew, *Septoria* spp.	Prochloraz
Triazoles	Inhibition of ergosterol biosynthesis at the production of 4,4-dimethylergosta-trienol	Powdery mildew, *Septoria* spp., rusts, eyespot	Triadimefon, propiconazole, triadimenol, flutriafol, flusilazole, hexaconazole, tebuconazole, cyproconazole, difenoconazole, diniconazole, fenbuconazole, epoxiconazole, bromuconazole, metconazole, fluquinconazole
Morpholines and piperidines	Inhibition of ergosterol biosynthesis at the production of 4,4-dimethyl-dienol and ergosta-trienol	Powdery mildew, rusts	Fenpropimorph, tridemorph, fenpropidin
Dicarboximides	Unclear	*Septoria nodorum*	Iprodione

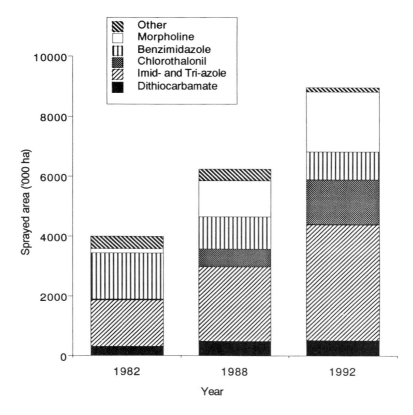

Fig. 7.2. Sprayed area in England and Wales of the main groups of fungicides (after Orson, 1995).

tritici and these products may sometimes be combined with a morpholine to increase efficacy and reduce further selection pressure on the pathogen population (Heaney *et al.*, 1988). There has been much slower build up of resistance to morpholine fungicides, possibly because they appear to inhibit two sites of the ergosterol synthetic pathway. Nonetheless, a small decrease in the sensitivity of powdery mildew populations to morpholines has been recorded in several western European countries (De Waard *et al.*, 1992), although field performance of these fungicides does not appear to have been reduced.

In several countries, rates of fungicide application have been reduced with benefits for profitability and risks of environmental damage. The degree of control is not necessarily reduced, particularly if the fungicide is applied at optimum timing. It has been argued, however, that reduced dose rates may encourage the build up of resistance. Examples of this occurring in *E. graminis* have not been well documented but splitting the application rate into two separate timings can increase the development of reduced sensitivity (Engels, 1995).

The threat of resistance in pathogen populations increases the need to develop new fungicides with different modes of action. Engels (1995) lists four new compounds to be marketed in the near future from the chemical groups of anilinopyrimidines, benzamides and strobilurins. Future chemical control may also incorporate systemic acquired resistance (SAR), i.e. a long-lasting whole plant disease resistance following an initial challenge by microorganisms or certain chemicals such as salicylic acid or 2,6-dichloroisonicotinic acid. Results on barley have shown, for example, that resistance to *E. graminis* f.sp. *hordei* can be induced by the prior application of bacterial metabolites (Engels, 1995).

Fungicides to control *Septoria* include the surface protectant, non-translocated chemicals such as the dithiocarbamates (e.g. maneb, mancozeb and zineb) and chlorothalonil. Translocated fungicides include the benzimidazoles (e.g. carbendazim and thiabendazole). Important levels of resistance to the benzimidazoles are, however, present in *S. tritici* populations in the UK (Gair *et al.*, 1987; Clark, 1995). More recent programmes have, therefore, included triazoles.

The effect of fungicides for foliar disease control on grain quality

The fungicide applications receiving most attention for their effects on quality have been those sprayed at flag leaf and ear emergence. Fungicide treatments applied at flag leaf emergence impact most on grain yield in the UK (Cook and King, 1984). The main aim of these applications is to extend the life of the flag leaf, from which at least 45% of the grain carbohydrate is derived (Lupton, 1972). A way of quantifying effects on flag leaf survival is to calculate green leaf area duration (GLAD) by computing the area under the percentage green area curve plotted against time from flag leaf emergence until complete senescence (Fig. 7.3). This is a useful approach as diseases rarely appear in isolation and assessments of the severity of individual disease problems can underestimate the full impact of combined infections and hence the benefit of fungicide application.

To facilitate a review of the effects of foliar disease control on grain quality results are presented from a series of experiments on wheat grown on a free draining sandy loam at Harper Adams Agricultural College, Shropshire, UK, between 1983 and 1987. These field trials assessed the impact of propiconazole, or propiconazole plus tridemorph, fungicide applications at flag leaf and ear emergence on GLAD, grain yield, thousand grain weight, specific weight, nitrogen yield and protein content, sodium dodecyl sulphate (SDS) sedimentation volume, and Hagberg falling number. It was attempted to supply enough nitrogen fertilizer to achieve maximum yields, and this was confirmed in most experiments by neighbouring

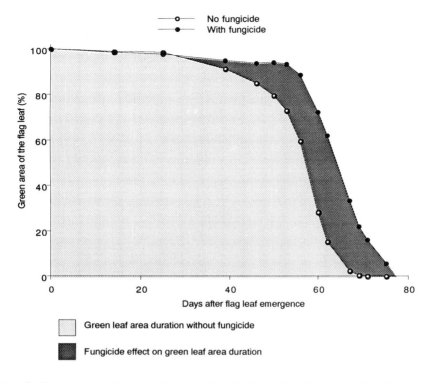

Fig. 7.3. The calculation of green leaf area duration (GLAD) and the effect of fungicide (after Gooding, 1988).

nitrogen response trials (Gooding, 1988; Lawson, 1989). Contrasting diseases and infection pressures resulted from contrasting weather patterns over the five seasons, and by testing treatments on varieties with differing disease susceptibilities.

Figures 7.4 to 7.10 show the relationships between the effect of the fungicide programmes on flag leaf GLAD and the fungicide effects on grain yield and quality. These results reveal the clear benefits of extending the life of the flag leaf for improving grain yield (Fig. 7.4) which is primarily a result of increased grain weight (Fig. 7.5). On average, the improved grain filling also appears to benefit specific weight (Fig. 7.6). The relationship of specific weight with GLAD is not, however, as close as with grain weight and is not necessarily related to improved packing efficiency of plumper grain. Kettlewell (1989), for example, found that significant increases in specific weight following fungicide application were related to improvements in individual grain density, rather than packing efficiency. The coefficients presented here suggest that extending the flag

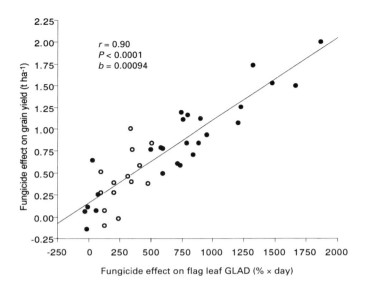

Fig. 7.4. The relationship between the effect of fungicide applied at flag leaf and ear emergence on the green leaf area duration (GLAD) of the flag leaf and grain yield when either powdery mildew (○) or *Septoria tritici* (●) has been the principal disease controlled.

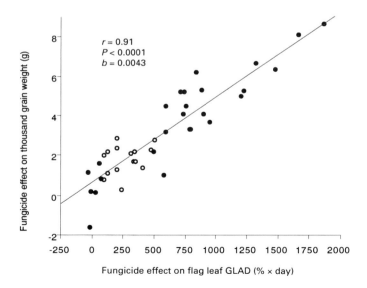

Fig. 7.5. The relationship between the effect of fungicide applied at flag leaf and ear emergence on the green leaf area duration (GLAD) of the flag leaf and thousand grain weight when either powdery mildew (○) or *Septoria tritici* (●) has been the principal disease controlled.

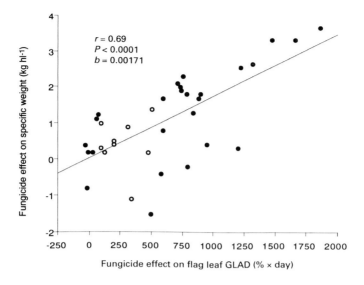

Fig. 7.6. The relationship between the effect of fungicide applied at flag leaf and ear emergence on the green leaf area duration (GLAD) of the flag leaf and grain specific weight when either powdery mildew (○) or *Septoria tritici* (●) has been the principal disease controlled.

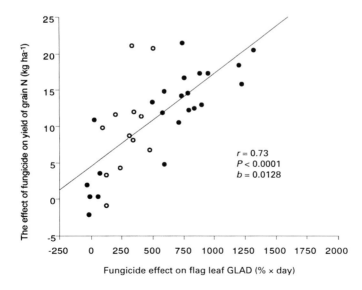

Fig. 7.7. The relationship between the effect of fungicide applied at flag leaf and ear emergence on the green leaf area duration (GLAD) of the flag leaf and yield of nitrogen in the grain when either powdery mildew (○) or *Septoria tritici* (●) has been the principal disease controlled.

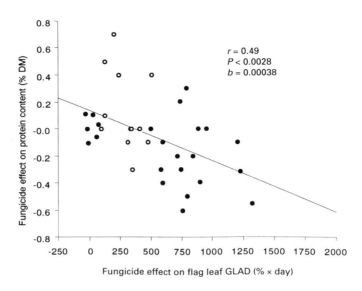

Fig. 7.8. The relationship between the effect of fungicide applied at flag leaf and ear emergence on the green leaf area duration (GLAD) of the flag leaf and grain protein concentration when either powdery mildew (○) or *Septoria tritici* (●) has been the principal disease controlled.

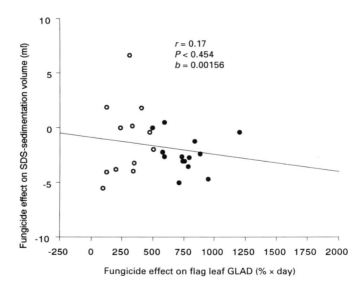

Fig. 7.9. The relationship between the effect of fungicide applied at flag leaf and ear emergence on the green leaf area duration (GLAD) of the flag leaf and SDS-sedimentation volume when either powdery mildew (○) or *Septoria tritici* (●) has been the principal disease controlled.

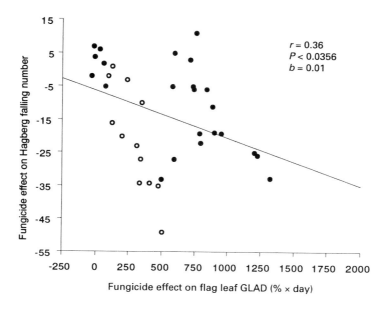

Fig. 7.10. The relationship between the effect of fungicide applied at flag leaf and ear emergence on the green leaf area duration (GLAD) of the flag leaf and Hagberg falling number when either powdery mildew (○) or *Septoria tritici* (●) has been the principal disease controlled.

leaf life by one week through controlling foliar diseases will result in about 0.65 t ha^{-1} extra grain yield, 3 g higher thousand grain weight and about 1 kg hl^{-1} extra specific weight. Average values in the absence of fungicide treatment were 6.73 t ha^{-1}, 39.1 g and 73.3 kg hl^{-1}, respectively. Similar improvements in mean grain weight and specific weight by flag leaf and ear sprays have been reported (McClean, 1987; Cook, 1987b; Proven and Dobson, 1987; Gooding *et al.*, 1994) with a wide range of fungicides including several triazoles, morpholines, benzimidazoles and surface protectants.

As well as improving grain yield, late-season disease control can benefit the total yield of nitrogen in the grain. In these experiments delaying flag leaf senescence by one week was associated with about 9 kg ha^{-1} extra nitrogen being harvested (Fig. 7.7) on top of the average yield of 135 kg N ha^{-1} harvested from untreated plots. Fungicides can increase nitrogen yield in many ways. The leaves are a major site of nitrate reductase activity and they also play a major role in aiding translocation from, and activity by, the roots. Controlling yellow rust, for example, has increased protein synthesis in flag leaves (Abdel *et al.*, 1987) and flag leaf removal can reduce the amount of nitrogen in the grain by a greater amount than the nitrogen actually in the leaves detached (Wardlaw *et al.* 1965). Specific diseases may also have more direct influences. Severe infection with powdery mildew can increase the loss of nitrogen from leaves as gaseous ammonia (Sadler and Scott, 1974); reduce the remobilization of nitrogen from senescing leaves (Finney, 1979); and greatly reduce the uptake of nitrogen by the roots (D. Walters, Scottish Agricultural Colleges, 1995, personal communication).

Despite the increased nitrogen uptake by the grain following disease control, this rarely keeps pace with grain weight improvements such that controlling severe disease on the flag leaf can result in small reductions in protein concentration (Fig. 7.8). Lisoval *et al.* (1991) also found that propiconazole applications leading to large yield increases (+17%) gave reductions in both protein and gluten concentrations. This is consistent with other reports showing that severe epidemics of *Septoria* spp. (Shipton *et al.*, 1971; Karjalainen and Salovaara, 1988) and yellow rust can lead to increased grain protein concentration (O'Brien *et al.*, 1990). This is not, however, inevitable and simultaneous increases in grain yield and protein concentration following fungicide application have occasionally been reported, particularly when powdery mildew and brown rust have been controlled (Jordan, 1992; Gooding *et al.*, 1994). The relationship between late-season disease control and grain protein concentration is, therefore, complex and must depend on the relative availability of nitrogen for uptake and/or remobilization compared with photosynthetic activity and carbohydrate partitioning, as well as the biology and severity of the pathogen concerned.

Measures of protein quality such as the SDS-sedimentation volume do not appear to be adversely affected by extending the life of the flag leaf (Fig. 7.9). Indeed, there have been several reports of disease control benefiting overall protein and/or loaf quality. Gooding *et al.* (1994) report improvements in loaf quality score following control of *Septoria tritici* and powdery mildew with propiconazole plus tridemorph in successive seasons despite reductions in protein concentration. SDS-sedimentation volume has also been improved in the absence of effects on protein concentration when powdery mildew has been controlled with fenpropimorph (Gooding *et al.*, 1993b). Peltonen and Karjalainen (1992) also reported improvements in protein quality following control of the same diseases in Finland with propiconazole. Fullington and Nityagopal (1986) found that severe brown rust infection resulted in a lower proportion of HMW-glutenin subunits being stored in the grain, while yellow rust infection has been associated with reduced dough resistance and a deterioration in mixing tolerance (O'Brien *et al.*, 1990). It appears, therefore, that even when protein concentration has been reduced by fungicide use, this can sometimes be more than compensated for by an improvement in protein quality. The impact upon the farmer depends on what characteristics of the grain are measured at the point of sale. Specifications relying on protein content and variety alone for assessing protein quality would appear to be inadequate and to the disadvantage of the producer.

As well as sometimes reducing grain protein concentration, controlling severe disease can give a reduction in Hagberg falling number (Fig. 7.10). This has been shown to be due to increased α-amylase activity in the grain (Salmon and Cook, 1987) and has occurred in response to controlling *Septoria tritici,* powdery mildew, brown rust and yellow rust with a wide range of fungicides (Gooding *et al.,* 1986b, 1994; Salmon and Cook, 1987; Cook and Hims, 1990). The basis of the effect is not yet fully understood but α-amylase in grain is known to increase when the drying of the grain during maturation is delayed (Olered and Jonsson, 1970; Gale *et al.,* 1983), and the degree to which fungicide delays senescence does appear to relate to delayed grain drying and reduced Hagberg falling numbers (Gooding *et al.,* 1987b). Additionally, increased ear weight resulting from disease control can sometimes lead to increased lodging and, therefore, contribute to pre-harvest sprouting and the production of α-amylase. Also, there appears to be a relationship between the nitrogen content of grain and Hagberg falling number, which has yet to be explained, but reductions in protein content by fungicides might contribute to effects on Hagberg falling number. Whatever the mechanism, the effect appears to be small. The results presented here suggest that delaying senescence of the flag leaf for a week is only associated with a reduction in Hagberg falling number of about seven seconds. This, together with other surveys (Clark, 1993), demonstrates that there is also a large degree of scatter around the best fit line and in individual experiments disease control has sometimes actually led to significant improvements in Hagberg falling number (Gooding, 1988).

The financial implications of these varied impacts upon quality are shown in Fig. 7.11. The value of each wheat sample was calculated by multiplying the yield by average spot prices for December in the appropriate year (Anon., 1984–1988). Bread-making quality and livestock feed was worth (£ t^{-1} at 85% DM) 139.9 and 120.2 in 1983, 119.9 and 107.3 in 1984, 134.3 and 108.9 in 1985, 121.7 and 109.2 in 1986, and 149.0 and 111.8 in 1987. The value for bread-making milling wheat was applied if the grain satisfied requirements stated in those years as being typical, i.e. a protein concentration, Hagberg falling number and specific weight of at least 11% (at 86% DM), 220 and 72 kg hl^{-1}, respectively. When protein concentration was below 11%, but above 10.5%, the mid-price between bread-making quality and feed wheat was used. Margins were then calculated by subtracting the value per hectare achieved in the absence of late-season fungicide, as well as the cost of the fungicide product itself. To evaluate the implications of the fungicide treatments on quality, margins were also calculated on the basis of feed prices alone, with no consideration of

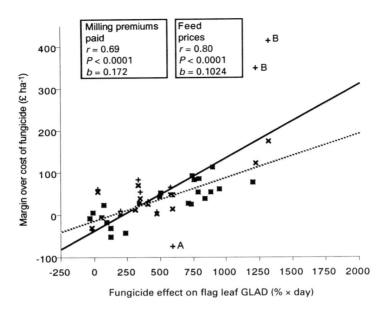

Fig. 7.11. The relationship between the effect of fungicide applied at flag leaf and ear emergence on the green leaf area duration (GLAD) of the flag leaf and the financial margin over the cost of the fungicide when premiums are paid for grain satisfying threshold requirements for bread-making (+, solid line) or for when these are not available (×, hatched line). Letters denote when fungicide altered the grading of grain either by pushing grain below (A) or above (B) thresholds compared with no application.

quality. This analysis demonstrates a number of points. Firstly, fungicide rarely changed the grading of wheat (only three out of 35 comparisons). In one case, wheat was downgraded as fungicide caused a reduction in protein concentration below the 11% threshold (A in Fig. 7.11), and in two cases, the fungicide upgraded the wheat because specific weight was increased above the 72 kg hl^{-1} threshold (B in Fig. 7.11). Secondly, because of this upgrading, and also because fungicide increased financial gain by protecting the yield of higher value wheats, on average, fungicide use showed a higher financial response when quality was taken into account. Clark (1993), therefore, rightly argues that increased yield and specific weights following late-season fungicide use will generally have a greater impact on gross margins than effects on Hagberg falling number. It is also apparent that fungicide use is potentially more valuable on wheat varieties with potential for bread-making than varieties suitable for livestock feed alone. Thirdly, the fungicide only needed to control disease which would lead to about 2–3 days reduction in the life of the flag leaf in order to justify the cost of the product, irrespective of whether quality was taken into account.

In addition to fungicide effects on the grain characters commonly assessed at the point of sale detailed above, late-season foliar disease control has been shown to have effects on other attributes of relevance to the end user. Flour extraction rates often increase when foliar disease is controlled (Salmon and Cook, 1987). However, work in Shropshire, UK, shows that the effects are not necessarily large or consistent over seasons (Table 7.2) and are sometimes only achieved with a simultaneous increase in grade colour figure. Fungicide application often decreases the amount of small particles in a milled sample, signifying an increase in hardness (Table 7.2; Myram, 1984), but this is not always associated with an increase in water absorption (Table 7.2). Effects on loaf volume have also been inconsistent, possibly reflecting contrasting effects on Hagberg falling number, protein content and protein quality (Table 7.2).

Table 7.2. The effect of propiconazole plus tridemorph fungicide applied at flag leaf emergence and again at ear emergence on flour characteristics of winter wheat in three years (after Gooding, 1988).

Principal pathogen controlled:	1985 S. tritici	1986 Powdery mildew	1987 S. tritici	Mean
Flour extraction rate (%)				
No fungicide	73.0	75.5	71.1	73.2
With fungicide	74.0*	75.2	71.5*	73.6
Texture (% flour passing through 75 μm)				
No fungicide	43.2	45.8	41.7	43.6
With fungicide	42.3	44.6*	39.2*	42.0*
Grade colour figure				
No fungicide	2.42	2.27	4.5	3.06
With fungicide	2.36	2.33	5.2*	3.30
Water absorption (%)				
No fungicide	15.4	15.2	15.5	15.4
With fungicide	15.4	15.0*	15.7	15.4
Loaf volume (cm³)				
No fungicide	1555	1510	1649	1571
With fungicide	1520*	1542*	1589*	1550

* Fungicide had a significant (*P* < 0.05) effect.

Root and stem base diseases of wheat caused by fungi

Eyespot

Eyespot (*Pseudocercosporella herpotrichoides*) occurs on the stem bases of many cereal and grass species, but the main hosts are winter wheat and winter barley. It is particularly severe in Western Europe on heavy wet clay soils with frequent cereal cropping, high nitrogen fertilizer applications, high seed rates, and early autumn sowing. Initial symptoms consist of a brown smudge on the leaf sheaths at the stem base. As the plant and lesion mature, an eye-shaped lesion develops with a diffuse margin and a black mass of hyphae at its centre (Parry, 1990). Grey mycelium can often be observed in the internal cavity of straw when cut open (Jones and Clifford, 1983). The pathogen overwinters on infected stubble, early-sown autumn crops, volunteers and grass weeds. Survival on stubble can be as long as three years. Conidia are produced during late autumn and winter. These are splash-dispersed by rain over short distances from stubble to healthy stem bases. Conidia production appears to be optimal at about 10°C with new infection occurring between 6 and 15°C (Wiese, 1987). Infection is virtually halted at temperatures above 16°C and the pathogen is least active or dormant during summer. As well as producing asexual spores, a sexual stage has been reported in Australia, New Zealand, Europe and South Africa. This stage has been named *Tapesia yallundae* (Wallwork and Spooner, 1988) which is evident as apothecia producing ascospores on stubble. In the UK, viable ascospores capable of causing infection appear to be released between December and May, with a peak between January and March when mean daily temperatures range between 3 and 8°C (Lucas, 1995). Unlike the conidia in water droplets, the ascospores are genuinely airborne and, therefore, capable of wider dispersal. The sexual recombination necessary in ascospore production is also likely to increase variability within populations which may increase risks of fungicide resistance.

Early infection can kill young seedlings and tillers. Later, severe infections of eyespot penetrate the stem and cause lodging. The movement of water and nutrients through the stem and leaf sheaths is also disrupted, contributing to early senescence and the production of white ears containing shrivelled and reduced numbers of grain. Yield losses on individual tillers can approach 36% but average losses in the UK probably range between 5 and 10% (Parry, 1990). Clarkson (1981) related yield reductions to eyespot severity as shown in equation 7.8:

$$y = 0.1x_m + 0.36x_s \qquad [7.8]$$

where y is the percentage yield loss, and x_m and x_s are the percentage tillers with moderate and severe symptoms, respectively, at the milky-ripe growth stage.

Eyespot infection is less in rotations which have at least a two year break from wheat or barley. Avoiding excessive nitrogen applications, seed rates and early drilling is also important, as is effective disposal of stubble. The most effective method of non-chemical control when a two year break from cereals is uneconomic is through the growing of resistant varieties. Eyespot frequently produces lesions on these varieties but the rate of penetration through the stem tissue is reduced (Gair *et al.*, 1987). This form of resistance in many modern UK varieties was derived from cv. Capelle Desprez and has remained durable. The resistance, however, is far from total and the disease may still cause significant losses if conditions are particularly favourable for infection. Other forms of resistance are now entering breeding programmes, for example from the goat grass *Aegilops ventricosa*. These additional genes have heightened resistance in certain varieties used in the UK, e.g. cv. Rendezvous. As a generalization, short, stiff strawed varieties should also be relatively tolerant of the disease as they are less likely to lodge when the stem is partially weakened.

Chemical control of eyespot in wheat has primarily relied on fungicides applied at the start of stem extension, i.e. when the first node is detectable. Delaying application after this makes it very difficult to target the stem base through a rapidly rising canopy. Control was initially achieved with benzimidazole fungicides. Because they were relatively cheap, and often gave yield responses when eyespot was not damaging, they were used extensively such that by 1982 about half of all wheat in the UK received a benzimidazole fungicide. Surveys in 1983 and 1984, however, revealed that resistant isolates were present in about half of the crops so that benzimidazole could no longer be safely recommended (Gair *et al.*, 1987). When benzimidazole resistance became a problem, fungicidal control of eyespot was switched to prochloraz. Prochloraz was, however, not as effective as the benzimidazoles against benzimidazole sensitive strains so mixtures of benzimidazole plus prochloraz were also formulated. Prochloraz has remained the most effective fungicide for controlling benzimidazole resistant eyespot. There is recent evidence, however, of reduced sensitivity to the fungicide, sometimes associated with reduced field performance (Lucas, 1995). Triazole fungicides, such as flusilazole and propiconazole, also show activity against eyespot but these are not as effective as prochloraz (Lucas, 1995).

Eyespot control and grain quality

Premature ripening and lodging resulting from severe eyespot would be expected to reduce Hagberg falling number but this does not appear to have been reported. Yield increases following eyespot control are often associated with increased thousand grain weight, but only rarely with improved specific weight (Jones, 1994). Others have also found that increases in specific weight are less likely with fungicides applied at the first node detectable growth stage than at flag leaf and ear emergence (Proven and Dobson, 1987). Similarly the earlier applications have not had a significant effect on grain protein concentration (Clare *et al.*, 1990).

Sharp eyespot

Sharp eyespot (*Rhizoctonia cerealis*) occurs in most temperate wheat growing regions of the world, particularly on coarse textured, neutral to acid soils. Overwintering occurs as mycelium in host debris or as sclerotia in the soil. Infection of healthy tissue is encouraged by cool conditions near the base of the plant. Similarities exist with the symptoms of eyespot but lesions at the base of stems are usually surrounded by a thinner (sharper) brown margin, are less regularly oval shaped, and tend to occur higher up the stem (up to 30 cm above soil level) (Wiese, 1987). Unlike the black pupil developing at the centre of true eyespot lesions, dark brown mycelium developing in sharp eyespot lesions can be easily scraped away. Damage can be similar to that caused by eyespot, including the death of seedlings, development of whiteheads and lodging, but is rarely as severe. Control of sharp eyespot is difficult given that varietal differences in susceptibility are not well documented. In addition, fungicides applied primarily for eyespot control, such as prochloraz, have sometimes increased the severity of sharp eyespot (Hoare and Jordan, 1984). Conversely, benadonil has reduced sharp eyespot and increased eyespot in experiments where both pathogens were present (Hoare and Jordan, 1984).

Take-all

Take-all (*Gaeumannomyces graminis* var. *tritici*) attacks the roots and stem base of cereals and a number of grasses (e.g. *Bromus* and *Agropyron*). It is most severe in temperate climates, on neutral or alkaline soils and in successive wheat crops. Infection often initiates from mycelium that survives on the roots and stubble debris. Hyphae reaching roots of a healthy plant appear as black runners on the root surface. Infection pegs are then produced which lead to the colonization of the internal structures of the root, including the endodermis and vascular tissue. The roots, and sometimes the stem base, become blackened. Water and nutrient uptake is, therefore, restricted and patches of stunted, poorly competitive plants develop, which can allow weeds to proliferate. Ears are bleached (whiteheads) and contain shrivelled and/or few grain. The senescent ears then regularly become invaded with sooty moulds (e.g. *Cladosporium* spp.), turning them black.

Annual grain yield losses in England and Wales due to take-all are estimated at 1–4% for second and subsequent wheat crops (Yarham, 1995). Take-all severity has also been inversely related to thousand grain weight, specific weight, yield of nitrogen (but not protein concentration) and sometimes Hagberg falling number (Bateman *et al.*, 1990). Severely infected crops have been found to yield darker flour, possibly due to contamination by sooty moulds after colonization of prematurely ripe grain (Jones, 1987b). There is no recommended method of controlling take-all with fungicides but a degree of suppression has been achieved with triadimenol, and triadimenol plus fuberidazole seed treatments. This suppression has been related to improvements in both thousand grain weight and specific weight (Jones, 1987b). Other fungicidal methods attempted include drenching the soil with nuarimol or applying fungicide pellets at sowing (Parry, 1990), but these techniques are still experimental. In Australia, flutriafol applied in the seed furrow, coated onto granules of super-phosphate, has shown promising results (Brennan, 1989).

The most effective control practices are non-chemical. The disease can build up rapidly as two to four successive wheat crops are grown. In the UK, peak take-all losses usually occur in about the third crop, although they may occur as early as the second wheat in some rotations (Yarham, 1995). A one year break from a host crop will, however, generally reduce disease severity to minimal levels as long as alternative hosts such as volunteers and particular grasses are well controlled. Alternatively, after four or five successive cereal crops take-all levels can start to diminish (take-all decline). Continuous wheat production can, therefore, result in tolerable take-all levels. Take-all decline may be the result of increasing numbers of microorganisms in the soil that are antagonistic to *G. graminis*. Attempts have been made to achieve biological control by inoculating wheat seed with potential antagonistic bacteria. *Pseudomonas fluorescens* can, for example, reduce take-all infection (Leggett and Sivasithamparam, 1986). The disease can be partially suppressed by ensuring rapid stubble decomposition with thorough tillage. High pH encourages take-all so high levels of lime application are ill-advised. Some of the effects of poor root growth from infection may be compensated in part by application of additional nutrients, some of which may stimulate new root growth.

Foot rot and associated leaf and ear diseases caused by Fusarium *spp.*
Many *Fusarium* spp. are relatively unspecialized pathogens which can cause foot or crown rot, seedling blights, leaf spots and ear blights (scab) (see Chapters 2 and 4) of a range of small grained cereals and grasses. The fungi can be either soil or seed borne. Foot rot associated with *Fusarium* spp. is expressed as brown discoloration occurring on the roots, and up to three nodes and internodes above the soil. Severely affected plants mature early, produce whiteheads containing shrivelled grain, and may have a bronzed appearance (Wiese, 1987). In wet environments *Fusarium* spp. infection may also lead to light brown lesions on the leaves. It has been estimated that foot rot reduces annual wheat production from most of North America by 3–4% (Wiese, 1987). In severe cases losses of up to 50% have been recorded, particularly when drought stress has exacerbated

symptoms. In most regions, losses are thought to be much less. Most serious losses derive from seed borne inoculum, which is likely to kill seedlings. Many fungicides applied to the seeds can limit this type of damage. Control of foot rot symptoms from soil borne inoculum is difficult to achieve although systemic fungicides applied at the start of stem extension to control eyespot can have a useful suppressive effect. Fungicides targeted to control the ear scab symptoms have also been successful. Mielke and Meyer (1990) found that a number of fungicides, but particularly guazatine and tebuconazole, improved flour yield, dough characteristics and loaf volumes associated with the control of *F. culmorum*. Gareis and Ceynowa (1994) also found that tebuconazole plus triadimenol controlled *Fusarium* ear blight caused by *F. culmorum*. In this latter study, however, the authors also detected an increase in the concentration of nivalenol mycotoxin following the use of the fungicide. The authors concluded that tebuconazole plus triadimenol application had stimulated mycotoxin production by *F. culmorum*. Cultivars can vary greatly in their tolerance of or resistance to certain *Fusarium* spp. Cultivars with resistance to ear scab caused by *F. graminearum*, for example, are available in a number of wheat growing areas (Luzzardi, 1985; Liu, 1985; Wiese, 1987). Crop rotation can help reduce the number of alternative hosts but significant levels of viable inoculum can survive for more than one year in the absence of a susceptible crop.

Foot rot, leaf and ear diseases caused by Cochliobolus sativus
Cochliobolus sativus (conidial stage, *Bipolaris sorokiniana*, syn. *Helminthosporium sativum*) occurs in the warmer cereal growing regions. The combination of foliar, ear and blackpoint symptoms associated with its conidial stage result in its being considered one of the most economically damaging diseases in very hot (coolest month mean temperature (CMMT) > 22.5°C), hot (CMMT > 17.5°C) and warm environments (CMMT > 12.5°C) (Dubin and van Ginkel, 1991). Foot rot symptoms are often difficult to diagnose but a dark brown/black discoloration of the subcrown internode (i.e. the stem between the seed and the soil surface) which may extend a short way above ground level is most symptomatic. Infected plants are stunted and yellow. Leaf infections by *C. sativus* (spot blotch), encouraged by warm (24–28°C) damp weather, give rise to elongated, dark brown lesions that rarely exceed 1 cm in diameter, most commonly found after ear emergence on the lower leaves. The fungus spreads up the plant by dispersal of conidia. When stem nodes are attacked, brownish black spots are produced which may coalesce and become velvety with the production of large numbers of conidia (Jones and Clifford, 1983). Control strategies are similar to those for *Fusarium* spp. Leaf symptoms can be suppressed with a number of fungicides (Mehta and Igarashi, 1985; Raemaekers, 1985; Lapis, 1985) and breeding can result in resistant varieties (Gilchrist, 1985;

Wiese, 1987). Seed borne inoculum can be reduced by treatment of seed with fungicides, and crop rotation can reduce soil borne inoculum (Gair *et al.*, 1987).

Root rot caused by Rhizoctonia solani

Root rot can occur in both temperate and warmer areas, extending through parts of Europe, Australia, northwestern USA, as well as semi-tropical regions such as in Brazil (Dubin, 1985) and Southeast Asia (Saari, 1985). The pathogen is widespread in soils, existing on crop residues as mycelium and sclerotia (Wiese, 1987). The root is attacked in the top 10–15 cm of soil where roots can be severed. If only a small proportion of roots are affected the above ground plants can appear normal. Lightly affected plants may outgrow the disease by producing more roots. Other-wise, patches of bare ground or stunted plants are evident. Stunted plants may also exhibit purple leaves and lodge easily and/or produce white-heads. Varieties of wheat can differ in their tolerance and/or suscepti-bility to *Rhizoctonia solani* but there are no truly resistant varieties. One reason for this is that *R. solani* consists of several groups which vary in host range (Dubin, 1985). Despite the wide host range of *R. solani*, crop rotation can sometimes help suppress the disease. Conditions conducive to good root growth will also help plants tolerate the disease. The disease appears particularly severe in systems relying on minimal or zero cultiva-tions, possibly because more rigorous tillage helps to disturb mycelium and/or the decay of crop residue.

Control of grain and ear diseases

Many grain and ear diseases contribute directly to wheat quality as con-taminants, or through the production of mycotoxins. Their prevalence, importance and life cycles have, therefore, been described previously, together with the effects of climate and variety on their severity (Chapters 2–4). This section concentrates on the impact of fungicide treatments.

Blackpoint

Blackpoint has been difficult to control with fungicides. Conner and Kuzyk (1988) were unable to obtain reliable control of blackpoint on soft white spring wheat from the foliar application of propiconazole, triadi-mefon, fenpropimorph, oxycarboxin, chlorothalonil or mancozeb, nor from seed treatment with triadimenol. In other studies, prochloraz ap-plied at flag leaf and ear emergence increased blackpoint severity, while fenpropimorph applied at the same timings increased and decreased severity depending on previous nitrogen fertilizer application rate (Gooding *et al.*, 1993c). In recent studies a comprehensive fungicide programme with applications at the start of stem extension, flag leaf emergence and ear emergence has dramatically increased blackpoint severity, particularly on varieties which benefited most from the fungicide programme with respect to foliar disease control, thousand grain weight and yield (Ellis *et al.*, 1996). This variety × fungicide interaction was also associated with grain dimensions lending support to the idea that severity might increase when larger, wider grain lead to a more open floret allowing greater access to the germ end of the grain. Other explanations are, however, possible. Fungicide programmes which control severe foliar disease, leading to large increases in thousand grain weight, can also delay crop maturity and slow the rate of grain drying (Gooding *et al.*, 1987a, 1994). This may change and/or extend the period over which grain is susceptible to infection. The growth of subepidermal fungi in the grain, often dominated by potential causal organisms of blackpoint, such as *Alternaria* spp. (Hyde and Galleymore, 1951), is highly dependent on grain moisture content. Variety and fungicide effects on grain moisture and drying rate may, therefore, greatly alter the quantity of mycelium present. Fungicide applications after anthesis may give better blackpoint control but this has yet to be confirmed. Such applications, however, would be very unlikely to give a yield response and, therefore, the farmer would need to know that there was a high risk of blackpoint, and that blackpoint severity would significantly reduce the marketability of his grain, before such applications could be recommended.

Ergot

Ergot control has been reviewed by Yarham (1993a). As with blackpoint, the incidence of ergot (*Claviceps purpurea*) has been increased with fungicide applied at flag leaf and ear emergence to control foliar pathogens, possibly because it reduced competition from other microorganisms or because it had a growth regulatory effect on the crop. Benzimidazole fungicides applied at anthesis have reduced severity and this has been exploited particularly on highly susceptible durum wheats. Nonetheless, the fungicide can only reduce severity on those florets which are open, allowing access to the stigmatic surfaces and ovaries, and only a certain proportion will be at a suitable stage of development at any one time. Seed treatments, particularly those containing triadimenol or bitertanol, which also cover ergot sclerotia, caused a reduction in subsequent sclerotial germination and infection of blackgrass. However, treated sclerotia were better preserved and more germinated in following years, although ascocarp production was still reduced.

Tilletia *spp.*

Attacks of common bunt can originate from both seed and soil borne inoculum. On the seed, the fungi is borne externally so it has often been sufficient to apply a protectant, non-systemic, fungicide such as organomercury. These compounds are relatively cheap and their wide-spread use has, in the past, led to common bunt being reduced to minor importance (Wainwright, 1995). Such compounds, however, are now restricted in many regions due to environmental considerations. When fungicide use has been curtailed, bunt often recurs. Greater survival of soil inoculum, possibly due to bans on straw burning, may also be favouring the disease in some areas (Yarham, 1995). More modern, systemic fungicides applied to the seed, such as triadimenol plus fuberidazole, may therefore be necessary to protect the seed from both seed and soil borne inoculum (Wainwright, 1995; Yarham, 1995).

Most inoculum for dwarf bunt originates from soil borne spores. Fungicide applied to the soil surface can therefore be effective. Some protection against karnal bunt can be afforded by fungicide sprays at heading.

Loose smut

Seed treatments for loose smut control need to be systemic to reach the mycelium. Active ingredients have included carboxin, triadimenol and fuberidazole, although there is some evidence of resistance to carboxin in *Ustilago nuda* (Parry, 1990).

Diseases Caused by Viruses

Viruses are relatively simple structures consisting of nucleic acid (RNA or DNA) surrounded by a protein coat (i.e. a capsid). They are obligate parasites, capable of development only within host cells. They cause damage by modifying the host genome, replicating within cells and altering metabolism. Wiese (1987) lists 32 virus or virus-like diseases that occur naturally on wheat. Most are spread by insects but some depend on other vectors such as mites, soil borne fungi or nematodes. Many of them cause only minor economic damage, or occur in limited areas of the world.

Barley Yellow Dwarf Virus (BYDV)

BYDV is the most widely distributed virus disease of the *Poaceae* attacking over 100 species of cereals and grasses. It is the most economically important virus disease of wheat. There are at least 20 aphid vectors of the disease, including the bird cherry aphid (*Rhopalosiphum padi*) and the grain aphid (*Sitobion avenae*) in the UK, and a number of strains of the virus which differ in severity. The most damaging attacks occur in early autumn sown wheat, particularly after a grass ley when transmission from surviving grass to young wheat plants can occur. Early autumn infection can cause seedling death, or severe stunting and yellowing. Early aphid feeding can increase tillering but this effect is less consistent when the virus is also transmitted (Thirakhupt and Araya, 1992), as BYDV affected plants sometimes exhibit reduced tillering. Late autumn infection may only produce slight stunting with obvious symptoms not being exhibited until flag leaf emergence when the upper leaf becomes pale yellow, often with a pink/purple colouring extending from the tip of the leaf. Discoloured flag leaves on otherwise normal plants are indicative of post-seedling infection.

Ears from infected plants contain shrivelled and/or few grain. Wiese (1987) quotes common yield losses of between 5 and 25%, while maximum yield loss can be 50% or more (Gair *et al.*, 1987). Average yield losses in the UK may be less than 1% but this is probably much higher in parts of North America. Surveys in Victoria, Australia, suggest overall yield losses of about 2% (Sward and Lister, 1987). Quality assessments appear to have been limited to thousand grain weights. El-Yamani and Hill (1991), for example, found that inoculating bread wheat could reduce mean grain weight by 24%.

In the UK, non-chemical control relies on delayed drilling in the autumn, or spring cropping, to avoid the most damaging autumn infections. In other countries, however, spring cropping can be very damaging if affected by spring flights of aphids. Varieties of both bread and durum wheat differ in tolerance to the different virus strains (El-Yamani and Hill, 1991), or may differ in suitability for the vectors. When following grass it is important to ensure a good kill of the ley before sowing the wheat. Chemical control of the vector is most economic when it controls early autumn infections. The trend towards drilling early in autumn has, therefore, increased reliance on autumn aphicide applications. Monitoring of aphid populations in the atmosphere in the UK, and identification of risk factors, has provided a better basis for insecticide recommendation. Advocacy of later drilling in integrated cropping systems will also reduce spraying requirements (Jordan and Hutcheon, 1995).

Diseases of Wheat Caused by Bacteria

Bacteria are unicellular organisms. Unlike the viruses, they can multiply and persist without exploiting the replicative mechanisms of host cells. They are, however, relatively simple in structure compared with the fungi. They are classed as prokaryotic, i.e. the nucleus is not separated from the cytoplasm with a nuclear membrane and they lack the membrane bound organelles and microtubule systems found in the fungi (classed as eukaryotic). All plant pathogenic bacteria are rod shaped, 0.5–3.5 μm × 0.3–1.0 μm. They often lack specialized mechanisms for invading host tissue and enter through stomata, wounds, or other gaps in the epidermis.

Compared with the fungal diseases and BYDV, bacterial diseases of wheat are less important. Wiese (1987) describes eight diseases but the economic importance of most is either very limited or unknown. Three are, however, worthy of mention here because of their effects on both yield and quality.

Black chaff

Black chaff or bacterial stripe is caused by *Xanthomonas campestris* pv. *translucens*. Severe epidemics can occur in warm wet conditions, typified in subtropical regions, or when sprinkler irrigation is used. The disease is particularly evident at ear emergence when dark brown, water soaked lesions appear as streaks or blotches on the glumes, awns and leaves. In wet weather, viscous exudates appear on the diseased tissue. Yield reductions are particularly severe if preanthesis infection causes sterility. Infected grain may become shrunken and have reduced germination capacity. Control methods rely on using clean seed and resistant or tolerant varieties. Frequent sprinkler irrigation should also be avoided (Wiese, 1987).

Basal glume rot

Basal glume rot is caused by *Pseudomonas syringae* pv. *atrofaciens* and occurs in most wheat producing areas of the world, particularly when moist conditions occur during ear emergence and grain maturation. Symptoms expressed are usually dark streaks on the lower halves of the glumes which may extend to the rachis and grains. Yields are reduced through a reduction in mean grain weight. Severe infection will also result in blackpointed grain. Control methods are limited to using clean seed, although subsequent infection may still occur from windborne inoculum in moist weather during summer (Wiese, 1987).

Bacterial leaf blight

Bacterial leaf blight has only been recognized as a wheat disease since 1972, but has occurred frequently in the north central USA on bread and durum wheats. Symptoms are particularly evident on the flag leaves. Lesions initially appear as small, water soaked spots which turn from grey-green to tan-white. These may then coalesce to form streaks and blotches such that in severe cases 50% of the foliage is destroyed. The pathogen is dispersed in rain droplets and possibly on seed. Disease development is favoured by high relative humidities and spring temperatures of 15–25°C. Control appears to reside mainly with the use of resistant or tolerant varieties (Wiese, 1987).

WEED CONTROL AND HERBICIDE APPLICATION

Weeds reduce the yield and quality of wheat in all areas of the world. The effects of weeds, and weed control measures, are often more difficult to delineate compared with the effects of fertilizers and fungicides. Weeds are known to reduce yields through competition, particularly in the early stages of crop development (Wilson *et al.*, 1985), but weed competitivity will depend on the limiting factor(s) to growth, i.e. whether competition is for light, nutrients and/or moisture. Certain weeds may also reduce wheat yields through increased lodging while some can host wheat diseases and pests. They may also interfere with crop operations, i.e. large amounts of green weeds during grain maturation and at harvest can delay harvesting and reduce harvesting efficiency. Alternatively, water weeds can interfere with irrigation and drainage systems, and weed growth on fallow land reduces the amount of moisture conserved within the soil. The viability of weed control not only depends on the effects of weeds on a single crop, but also the effect of letting weeds disperse seed and therefore proliferate in future crops.

Grass Weeds

There are large numbers of different weed species that commonly exist within wheat fields. Weeds present will depend on previous cropping, time of sowing, soil environment and climatic conditions. As a generalization, however, the most problematic are those from the same family as wheat, i.e. the *Poaceae*. Being closely related, the grass weeds are often well adapted to the same ecological niches and farming systems commonly used for wheat production. Their requirements in terms of nutrients and canopy structure are similar to wheat and they tend, therefore, to be very competitive. Many also act as alternative hosts for wheat pests and diseases and can, therefore, aid their survival and be a source of epidemics. In addition, they tend to be more expensive and/or difficult to control with herbicides compared with other weeds. This is partly because herbicides need to be highly selective if they are to control the weeds while the wheat is growing, and also because there is greater financial benefits in controlling them, thereby justifying a greater market price. The estimated relative costs of herbicides to control many weeds common to wheat in the UK are shown in Fig. 7.12. This reveals that grass weed control has been commonly three to four times more expensive than control of dicotyledonous, broadleaf weeds. Common examples of grass weeds occurring in many regions of the world are couch grass (*Agropyron repens*), wild oats (*Avena* spp.), blackgrass (*Alopecurus myosuroides*), bromes (*Bromus* spp.), ryegrasses (*Lolium* spp.), meadow-grasses (*Poa* spp.), and loose

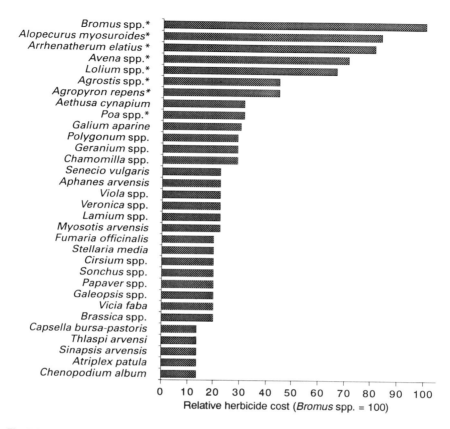

Fig. 7.12. The relative estimated costs of herbicides for controlling weeds commonly found in UK cereal fields (after Keen, 1991); *, a grass weed.

silky-bent (*Apera spica-venti*). Relative prevalence of grass weeds in the UK is shown in Fig. 7.13 but economic importance depends not only on incidence, but also on ease of control and impact upon yield and quality.

Couch grass

Couch grass (*Agropyron repens* syn. *Elymus repens*) is known by many other names including twitch, quick, quickens, wrack, scutch and quack grass. It is distributed throughout the UK, most of Europe, and temperate areas of Asia and America. It is one of the few perennial species that is well adapted to annual cropping. This is in part due to the fact that reproduction can be via underground runners (rhizomes), as well as by seed. The weed is highly competitive against wheat and dense tillering of

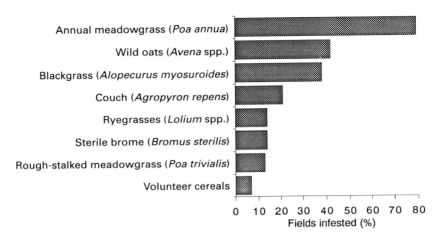

Fig. 7.13. The incidence of major grass weeds in UK winter cereals in 1989 before herbicide had been applied.

the weed can increase lodging, and hamper harvesting of the crop. It is also a host for the take-all fungus and thereby may carry over the disease from one cereal crop to another, reducing the beneficial effects of crop rotation. The rhizomes might also hinder some cultivations. Rhizomes are produced in the UK between May and November. Axial buds at nodes along the rhizome can potentially allow generation of a new tiller from every 2 cm length. Many of these buds normally remain dormant but breaking the rhizome stimulates them to develop. The success of these tillers is favoured by the rhizome acting as a storage organ, supplying assimilate to the emerging tiller in its early growth. Rhizome development can result in a mat of underground stems, usually in the top 15 cm of soil, particularly when the wheat is a poorly competitive crop.

Traditional control of couch has depended on the stimulation of tiller production from the rhizomes followed by their destruction. For example, non-chemical techniques relied on stubble cultivations to break the rhizomes followed by further cultivations to kill the subsequent leaf growth after it had reached about 5 cm. This needed to be repeated at least four times before the assimilates and nodes of the rhizome had been exhausted. It was a technique, therefore, that was expensive in terms of labour, fuel and time. It also carried a high degree of risk if the required number of operations could not be completed due to unfavourable weather or other factors, as many tillers would survive and proliferate. The procedure was more suited in the UK to following winter barley than wheat due to the greater window of opportunity in the autumn. Alternatively, a degree of suppression could sometimes be achieved by using certain tines to collect rhizomes from close to the soil surface so that they could be gathered and burnt. Burial of the rhizomes below 15 cm can also check the plant, so deep ploughing can sometimes be beneficial while minimum cultivation and direct drilling favours the weed. Seed shed should also be avoided and, as hedgerows can be a major source of seeds, careful hedgerow and headland management is required (see under sterile brome).

Until very recently, there has been little opportunity for controlling couch selectively in wheat while the crop was actively growing. A few contrasting herbicide strategies can, however, be employed. An early approach involved the use of paraquat, a non-selective, non-translocated (contact), foliar herbicide which inhibits photosynthesis. Initial cultivations were used to stimulate tiller growth from the rhizomes, followed by repeated spraying each time regrowth reached the two leaf stage, until growth stopped. As with stubble cultivations, this repeated spraying technique was time consuming but necessary as, although the paraquat gives relatively rapid destruction of above ground green tissue, it is a contact material which is not mobilized to the rhizomes. Couch control with herbicides was, therefore, revolutionized with the introduction of glyphosate. This was also a non-selective, foliar herbicide but could be translocated and kill the rhizomes directly. On stubble, applications could be made to plants 10–15 cm high followed by ploughing or other cultivations as soon as the leaves turned red/yellow. A disadvantage to the use of glyphosate is that, like most translocated herbicides, it requires the plant to be actively growing to work effectively. In cold late autumn and winter conditions it could take 50 days before the leaves turned colour. An alternative approach has been the application of glyphosate to wheat before harvest, but after grain moisture content has dropped below 30%. Applications this late, but before one week before harvest, do not adversely affect grain yield and conditions are such that couch, as well as many other weed species, can be controlled effectively. Recent reports have suggested that compounds may shortly be available that could control couch selectively in an actively growing wheat crop (Parrish *et al.*, 1995).

Wild oats (Avena *spp.*)

Two species of wild oat are common in the UK. Spring (or common) wild oat (*A. fatua*) has been long established and is also found throughout most of Europe, North Africa and Central Asia, and has been introduced to North America. In the UK it germinates mainly between March and May, with a smaller 'flush' between September and November. In contrast, winter wild oat (*A. ludoviciana*) has only been reported in the UK since 1917; it germinates mainly in October and November with small amounts in February and March; is found mainly in central, southern England, particularly on heavy soils in the South Midlands; and has been particularly favoured by the increased area of winter cereals. It is probably a native of Central Asia but is also found throughout south and central Europe and has been introduced to many other areas.

Wild oats are annual plants which can be distributed widely in crop seed (see Chapter 2). In the UK they spread rapidly during the 1960s and 1970s such that 85% of the area infested in 1972 had become so since 1957. They are very competitive with the crop and yield improvements from control have been as high as 70%, although 25–30% are more normal through controlling severe infestation. Damage coefficients (crop loss (kg ha^{-1}) by one plant m^2) for *A. fatua* in winter and spring cereals have been calculated as 4 and 8 respectively (Hurle, 1993). The relative competitive ability of wild oats against wheat, however, is influenced by nitrogen availability. Wright and Wilson (1992), for example, found that *A. fatua* responded more than wheat to treatment with nitrogen fertilizer in terms of vegetative growth and nitrogen uptake. Yield losses of wheat caused by *A. fatua* were consistently greater at high nitrogen fertilizer application rates. Increasing nitrogen availability also increased seed production of *A. fatua*. These findings emphasize the increased reliance on herbicides that can accrue from high usage of nitrogen containing fertilizers.

Because of their effect on quality of seed crops, complete eradication of wild oats is often the aim. This is made difficult because the seeds can remain dormant, but viable, in the soil for many years. Each surviving plant can return 10–500 seeds to the soil which leads to a persistent seed bank. It has been calculated that five successive years of reducing seed return by at least 95% is required before infestations reach economically roguable limits in the UK (one plant in every 20–30 m^2), depending on the cost of labour and value of the crop.

In the past, non-chemical control has relied on roguing and the sowing of certified seed. Using chaff collectors on combine harvesters to minimize weed seed spread can also be useful (Morrison and Bourgeois, 1995). Burying the seed with ploughing can increase dormancy and long-term survival whereas leaving the seed on the surface can reduce the total numbers of viable seed. If stubble burning is permitted, this can also reduce seed viability and seed dormancy.

Herbicides to control wild oats in wheat need to be highly selective given the close biology of weed and crop. The use of certain broad-spectrum annual grass weed killers which only stunt the wild oat plants can be counterproductive as the oat inflorescence may remain below or close to the height of the wheat canopy, making hand roguing more difficult and less effective. Older chemicals for the more specific control of blackgrass include triallate. This is a soil acting, residual herbicide that can be sprayed before drilling and incorporated into the top 1.5–2.5 cm of soil. Alternatively, granules may be applied pre-emergence or during the early seedling stages. More modern active ingredients include post-emergence herbicides, many of them working in the same way, by inhibiting acetyl coenzyme A carboxylase (ACCase). Diclofop-methyl, fenoxaprop-P-ethyl and clodinafop-propargyl are classed as aryloxy-phenoxypropionates while tralkoxydim is a cyclohexanedione. The first of these, diclofop-methyl, was introduced in the mid-1970s but all the others have been introduced relatively recently in the 1990s, and further introductions are expected. Where ACCase inhibitors are used repeatedly in the same field, the chances of the wild oat becoming resistant is greatly increased. Morrison and Bourgeois (1995), for example, report hundreds of resistant populations in western Canada, particularly in Manitoba. ACCase inhibitor use had increased fourfold during the 1980s such that by 1981 50% of sprayed fields were treated with these products, greatly increasing the risk of resistance. Other herbicides that have been pre-dominantly targeted at wild oats include difenzoquat and flamprop-isopropyl. Unfortunately, wild oats resistant to triallate are also often resistant to difenzoquat.

Blackgrass

Blackgrass (or black twitch) (*Alopecurus myosuroides*) is an annual grass weed which germinates mainly in the autumn. It is found throughout the UK, but most abundantly in Southeast England. It is also widespread in Europe and temperate Asia, and has been introduced to many other temperate areas including North America. It causes most yield loss in early sown winter cereals (Moss, 1985). Unlike wild oats it is relatively small seeded and shows little viability over extended periods of time. Its spread in the UK has, therefore, been favoured by an increasing winter wheat area, earlier sowing of winter wheat, the adoption of continuous cereal cropping, and the use of reduced cultivation techniques, particularly on heavy land. It can tiller profusely, reducing yield through competition with a damage coefficient estimated to be 3 and 1.5 in winter and spring cereals, respectively (Hurle, 1993). The economic impact is greater as a result of blackgrass being a host and source of ergots.

Non-chemical control includes delayed sowing of winter wheat (Moss, 1985), using competitive varieties (Cosser *et al.*, 1995), spring cropping and deep ploughing. A two year fallow also almost completely eliminates the species. However, many herbicides, mostly soil acting residuals, have been used to reduce populations. Early examples included terbutryne and methabenzthiazuron, but greatest reliance has been placed on the phenyl ureas chlortoluron and isoproturon (IPU), the latter being the most widely used herbicide in the UK on wheat during the early to mid-1990s (Fig. 7.14). Some ten years after their introduction in the early 1970s, however, *A. myosuroides* biotypes resistant to chlortoluron were

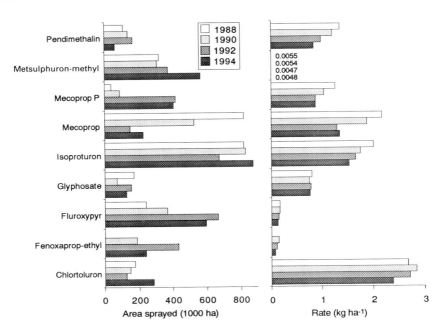

Fig. 7.14. The area sprayed and application rate of the eight most widely used herbicides
($>$ 100,000 ha in 1992) on wheat in the UK between 1988 and 1994 (after Cooke and Burn, 1995).

identified. By 1990 phenyl urea resistance was present on 46 farms in
19 UK counties, which increased to 71 farms in 22 counties by 1992. Clark
et al. (1994) classified blackgrass populations on a 1* (least resistant) to 5*
(most resistant) scale and found that in random surveys 7% of samples
were at least partially resistant (2* or more). The same authors also report
that resistance to the much newer fenoxaprop-ethyl was found on 90
farms. In the face of increasing blackgrass herbicide resistance, Mills and
Ryan (1995) stress the importance of non-chemical methods of blackgrass
control such as deeper cultivations and later sowing date, but also
advocate a system incorporating several herbicides differing greatly in
their mode of action, i.e. trifluralin (inhibits microtubule polymerization),
triallate (inhibits fatty acid synthase), and clodinafop-propargyl (ACCase
inhibitor). Sulphonylureas (inhibit acetolactate synthase) active against
blackgrass can also contribute to the diversity of modes of action available
to reduce selection pressures against any one group of herbicides.

Brome species

There are several brome species of economic importance as weeds in
wheat. Sterile brome (*Bromus sterilis*) has been particularly problematic

in lowland, southern England on well-drained soils. It is also found throughout most of Europe and southwest Asia, and has been introduced to North America. Downy brome (*B. tectorum*), also known as drooping brome or downy chess, is a native of the Mediterranean region. It has, however, been introduced into America, Australia and New Zealand. In the USA, for example, it has become a serious problem on minimum tillage grain fields in the north west. The great brome (*B. diandrus*) is also of Mediterranean origin and is a serious weed of wheat in Spain. It has also been introduced into North and South America, Australia and New Zealand.

Sterile brome in the UK increased dramatically during the 1970s and 1980s. It shares similarities with blackgrass in having small seeds, low levels of innate dormancy, and an autumn germination period (Firbank *et al.*, 1985). It is, therefore, particularly associated with winter cereal and reduced cultivation techniques (Froud-Williams, 1983). Conversely, ploughing to 20 cm in successive years can give virtually complete control of the weed (Bowerman *et al.*, 1993). Heavy infestations can reduce winter wheat yields by more than 50% depending on crop competitivity (Cousens *et al.*, 1985). In comparison with blackgrass, however, sterile brome has been much more difficult to control with herbicides (Henly *et al.*, 1985). Expensive application rates of metoxuron (Bulmer *et al.*, 1985) or multiple combinations of other herbicides have been necessary. This has increased interest in methods which reduce reliance on herbicides. Sterile brome is a common component of hedgerows (Froud-Williams *et al.*, 1980). The frequent intentional or accidental drift of sprays of broad-leaved weed killers into hedgerows, however, is thought to have reduced natural competition against sterile brome, allowing it to proliferate. With much greater numbers of seeds being produced in the hedgerow, the chances of ingress into the headland and rest of the field increases. To prevent this requires special attention to the headland areas. For example, even if reduced cultivations are desired for the majority of the field, ploughing the headlands could be desirable. If possible, compaction in the headland with subsoiling should be considered to help maximize crop competitivity. More radical measures can include sowing the headland to a spring crop, so as to allow the use of a total herbicide to kill emerging weeds in the autumn. Similarly a sterile strip of about 1 m width could be established between the hedgerow and crop and maintained with repeated cultivations or with a soil acting, residual, non-selective herbicide. In reality, very few hedgerow species are capable of surviving in an arable field. Ideally, therefore, the use of broad-spectrum herbicides close to the field boundary should be avoided. In such areas selective herbicides, targeted at the specific problem weeds, should be used so that competition of the natural flora in the hedgerow against potential crop weeds can be safe-guarded (Marshall, 1985).

In the future, reliance on ploughing and other non-chemical methods are likely to remain important in sterile brome control, but some of the new sulphonylurea herbicides may also allow more opportunities for herbicide usage (Parrish *et al.*, 1995).

Meadow-grasses (Poa *spp.*)

Annual (*P. annua*) and rough-stalked meadow-grass (*P. trivialis*) (also known as annual and rough bluegrass, respectively) are very common grass weeds, and are widespread in temperate wheat production. They appear most prevalent in higher rainfall areas, associated with non-ploughing, heavy soils, alternate grass ley arable rotations, poorly competitive crops, and winter cropping (Harvey, 1985). Many herbicides targeted for annual grass weed control including chlortoluron, isoproturon, pendimethalin and trifluralin will control meadow-grasses, but doubt has been expressed concerning the economic justification. Harvey (1985), for example, found no yield increases attributable to control of *Poa* spp. despite populations exceeding 3000 panicles m^{-2}. In other UK experiments, where *P. annua* was sown, a population of 100 plants m^{-2} equated to an average yield reduction of 0.035 t ha^{-1} (Woolley and Sherrott, 1993), i.e. a damage coefficient of only 0.35. Both of these reports suggest, therefore, that the competitive impact of the meadow-grasses against wheat is relatively low. Rough-stalked meadow-grass, however, is a host for *Septoria tritici*, so control could have benefits for disease suppression. In addition control of *Poa* spp. may reduce harvesting problems, particularly in wet areas (Sherrott, 1988).

Ryegrasses (Lolium *spp.*)

Ryegrasses are common weeds of wheat in many countries. Italian ryegrass (*L. multiflorum*) is a native of central and southern Europe, northwest Africa and southwest Asia, but has been introduced into many temperate countries as a forage grass. In comparison with the meadow-grasses they are much more competitive, with populations of Italian ryegrass (*L. multiflorum*) of only 40 m^{-2} reducing yield by as much as 30% in certain environments. Other annual ryegrasses including darnel (*L. temulentum*), *L. persicum* and, in particular, *L. rigidum* can be serious weeds of wheat in warmer growing areas such as Saudi Arabia (Alshallash and Drennan, 1993). In addition to competition effects, ryegrasses can be a host for barley yellow dwarf virus.

Loose silky-bent (Apera spica-venti)

Loose silky-bent (or wind grass) is widely distributed in England, particularly on light soils in southern and eastern regions. It also occurs throughout Europe and northern Asia, and has been introduced to North America. It has increased in prominence in several European countries including Germany (Hurle, 1993) and Denmark (Melander, 1993). The plant germinates almost exclusively in autumn so, like blackgrass and sterile brome, its spread has accompanied increased autumn cropping. Even in competition with wheat, each plant is capable of producing 1900–2000 small seeds which are shed before harvest, facilitating rapid spread under successive cereal crops, particularly on sandy and sandy loam soils. The damage caused by the plant varies widely between reports, with damage coefficients ranging between 1 to 8 in winter cereals (Hurle, 1993).

Broad Leaved Weed Control

As well as the grass weeds, there are some annual, dicotyledonous broad leaved weeds that can be very competitive and/or difficult to control in wheat. Cleavers (or bedstraw, *Galium aparine*) and chickweed (*Stellaria media*) are probably the most indicative, occurring throughout Europe and are common in the UK.

Cleavers is common in hedgerows but has grown in importance as a weed of arable crops. Froud-Williams (1985) noted that cleavers collected from arable fields differed substantially from those collected from hedgerows. In particular, differences in seed biology rendered hedgerow populations ill-suited as arable weeds. Germination in the field is concentrated in the autumn, particularly at temperatures between 5 and 15°C, and is stimulated by nitrate. Establishment is favoured by shallow tine cultivations as opposed to ploughing, while straw burning can significantly reduce the number of viable seeds.

Cleavers has, therefore, been of particular, but not exclusive, concern in winter cereals established with shallow cultivations. In UK crops, it has been the most prevalent broad leaved weed at harvest in winter wheat, occurring in 12% of crops surveyed (Froud-Williams, 1985). Its competition against wheat and seed production increases with nitrogen fertilizer application rates (Wright and Wilson, 1992). Yield losses in winter wheat of 30, 47 and 52% have been related to cleaver densities of 25, 100 and 520 plants m^{-2} (Peters, 1984), while others have found yields to be more than halved by as few as 26 plants m^{-2} (Wilson, 1984). In addition to yield losses, uncontrolled cleaver populations reduce harvesting efficiency because they can both increase lodging and reduce the proportion of the above ground biomass which is grain (Froud-Williams, 1985). A large proportion of cleaver seeds produced (30–70%) can also be collected by the combine harvester, reducing seed purity and increasing the chances of contaminated seed being sown (Froud-Williams, 1985).

Chickweed is a highly prevalent weed in UK cereal production, favoured by high nitrogen fertilizer application rates (Grundy *et al.*, 1992). In each of three surveys conducted between 1967 and 1980, incidence of chickweed was higher than any other broad leaved weed in both winter and spring sown cereals. It is able to germinate at any time of the year, sets seed rapidly, and can survive cultivation (Makepeace, 1982). It can be very competitive but perhaps less so than cleavers on a per plant basis (Froud-Williams, 1985).

Selective control of broad leaved weeds in winter wheat with herbicides has been possible since the 1950s. The first chemicals to be used widely, and often routinely, were those which mimicked the action of auxin, i.e. the 'hormonals', causing growth imbalances within the weeds. This group is typified by 4-chloro-*o*-tolyloxyacetic acid (MCPA) and 2,4-dichlorophenoxyacetic acid (2,4-D). These are inexpensive, but unfortunately, many of the present-day problem broad leaf species, including cleavers and chickweed are not adequately controlled by them. Cleavers and chickweed can be suppressed by mecoprop (CMPP), a later addition to the list of hormonal herbicides. However, mecoprop is only effective against seedlings of these species. Additionally, because the hormonals are translocated herbicides they need to be applied while the weed is actively growing and are less effective in cold temperatures during winter. They can also cause crop damage if applied after the first node detectable growth stage (Tottman *et al.*, 1988). Foliar herbicides with contact action for the control of many broad leaf weeds including chickweed, but not cleavers, is possible with the nitrile herbicides ioxynil and bromoxynil, but again only until the start of stem extension. Selective control of mature chickweed, until flag leaf emergence of the crop, is now possible with metsulphuron-methyl, a sulphonylurea herbicide with both contact and residual activity. Control of mature cleavers, until flag leaf emergence of the crop, is possible with fluroxypyr, a herbicide with both contact and translocated activity. Chickweed can also be controlled by many of the herbicides designed primarily for grass weed control, i.e. the soil acting residual herbicides chlortoluron, isoproturon and terbutryne can be effective, often at lower rates than those recommended for grass weeds. Similarly metoxuron will suppress cleavers as will some of the newly discovered sulphonylureas. Soil acting, residual activity against cleavers, chickweed and annual meadow-grass is also possible with diflufenican (Orson, 1988; Cahill, 1988).

Effects of Weeds and Weed Control on Wheat Quality

Compared with fertilizer and fungicide effects, relatively little is known about weed and herbicide influences on the quality of wheat. Allowing certain weeds to grow and compete with the crop sometimes reduces thousand grain weight. Several experiments have, for example, found cleavers to reduce yield through reductions in grain weight (Rola and Rola, 1987; Hofmann and Pallutt, 1989; Wright and Wilson, 1992). Experiments with grass weeds, however, often find yield reductions are more closely associated with reductions in ear numbers and/or grain numbers per ear (Moss, 1987; Wright and Wilson, 1992). Relative impacts upon different yield components can be largely explained by the timing and type of competition. Competition early in the season, for example by blackgrass and wild oats, is more likely to reduce yields through reducing tiller development and increasing tiller mortality such that ear numbers decline. Conversely, weeds that grow above the crop and shade it during grain growth are prone to cause reductions in thousand grain weight. Such effects in the UK are typified by cleavers which grow rapidly from May onwards, exploiting available solar radiation above the wheat canopy (Wilson *et al.*, 1985).

Grain protein concentration

Competition for nitrogen and moisture would be expected to reduce total crop recovery of nitrogen in the grain. Influences on protein concentration would depend, however, on relative effects of weeds on DM and nitrogen accumulation and partitioning, and may well differ depending on weed species. The complexities are demonstrated in work from Australia (Taylor and Lill, 1986) which compared weeded and non-weeded plots in 167 crops which contained natural mixed infestations of short annual grasses, wild oats and annual legume weeds. They found that weeding increased wheat DM production, nitrogen concentration in the whole crop, nitrogen and phosphorus uptake, and the number of fertile tillers. By harvest, however, although grain DM and nitrogen yield was increased by weeding, thousand grain weight and grain protein concentration were unaffected. In contrast, Cosser *et al.* (1996) found wheat grain dry matter yield, nitrogen yield <u>and</u> protein concentration to be negatively correlated with blackgrass head numbers. Similarly in pot experiments, increased wild oat populations have reduced both protein content and yield (Wimschneider *et al.*, 1990).

Similar inconsistencies are evident from herbicide experiments. Penny and Jenkyn (1975), for example, found that an application of dichlorprop plus MCPA increased grain protein content but only when nitrogen was applied as Nitro-chalk prills rather than as a liquid, and only when a fungicide was also applied. In earlier experiments a similar herbicide treatment had little effect (Penny and Freeman, 1974). In more recent studies in India, a range of herbicides increased wheat grain protein content in proportion to the degree of weed control achieved (Porwal and

Gupta, 1993). In Spain, though, control of severe infestations of *Avena sterilis* with triallate had no effect (Ponce *et al.*, 1990). It might be expected that the impact of herbicides in low nitrogen situations will be reduced because the competitive ability of certain weeds relative to wheat are reduced in these circumstances. This certainly appears to explain herbicide × nitrogen interactions with regards to yield, but similar interactions on grain protein content are much rarer (Ponce *et al.*, 1990).

Hagberg falling number

Green weed material present in the crop during grain filling may delay grain drying, particularly if it causes lodging, which might be expected to reduce Hagberg falling number (Gooding *et al.*, 1987a) as well as any detrimental effects associated with delayed harvest. Recent results (N. Cosser, Royal Agricultural College, Cirencester, 1995, unpublished data) have shown that weeding plots by hand in winter and spring can give small, but statistically significant, improvements in Hagberg falling number in the absence of effects on lodging. Pre-harvest applications of glyphosate might be expected to improve Hagberg falling number due to desiccation of both the weeds and crop. Such applications have, however, had inconsistent or no effects on Hagberg falling number, grain protein content, thousand grain weight, specific weight, germination rate or baking quality (Darwent *et al.*, 1994).

Direct effects of herbicides

There is some evidence that particular herbicides may have direct influences on wheat protein content, not mediated through weed control. In the USA, for example, subherbicidal levels of simazine and terbacil have increased both yield and protein content of wheat simultaneously. Similarly Pandey and Srivastava (1985) found that several herbicides including triallate, metoxuron and methabenzthiazuron increased the nitrate reductase activity in leaves and grain protein content of both bread and durum wheats. Chlortoluron has also been shown to increase grain protein content, associated with increases in gluten forming proteins (Kostowska *et al.*, 1984). The closely related linuron has increased the storage of both starch and protein in grain (Bil *et al.*, 1987). In other experiments, however, herbicides have both increased and decreased grain protein content in different years (Martin *et al.*, 1989) and sometimes protein content increases are only associated with grain yield reductions (Martin *et al.*, 1986, 1990). Others have found grain protein responses to herbicides to be highly dependent on application timing (Abo-Hamed *et al.*, 1993).

Recent trends in herbicide use

Despite the large increase in herbicide usage that occurred in many regions during the 1970s and 1980s (e.g. see Table 1.8). Recent surveys of use in northwestern Europe have shown reductions in the amount of active ingredient applied during the 1990s. This has been due to the impact of technological, economic and political factors. One of the most major advances has been the development of herbicides that are active at much lower dose rates. Of particular note has been the use of sulphonylurea herbicides. This group contains both foliar and residual soil activities and since 1982, seven sulphonylurea active ingredients have been commercialized or are in advanced stages of development for use on wheat and barley for different weed spectra in a range of cropping systems and climates. Research of this group has been intensive and the number of products, and their market importance, is set to continue in growth (Brown *et al.*, 1995). It has been estimated that the use of sulphonylurea reduces chemical application rates to treated fields by 95–99% relative to commonly used alternative products. The corresponding reduction in chemical manufacturing waste and packaging material ensuing from sulphonylurea adoption could, it has been argued, result in significant environmental benefits (Brown *et al.*, 1995). Other herbicides which have been increasingly used, which are also active at relatively low application rates, are fluroxypyr and fenoxaprop-ethyl (Fig. 7.14). This shift, although having certain environmental benefits, does not necessarily reduce impact on wildlife in the field as increased activity might substitute for reduced dose rates (Cooke and Burn, 1995). As well as the move to more active chemicals, however, application rates of existing compounds have tended to decline (Fig. 7.14), which would be expected to reduce the risks of pollution and effects on non-target organisms.

PLANT GROWTH REGULATORS

There is a large scope for altering the growth and development of wheat plants by applying chemicals that mimic or, more commonly, suppress the production of, or sensitivity to, plant growth hormones. In practical wheat production, however, there have only been two processes that have gained widespread usage and both are principally aimed at reducing plant height. The first technique is the blocking of gibberellin production, for example, by using chlormequat (CCC). Gibberellin is heavily implicated in the division and elongation of cells immediately below the apical meristem. It is in this region that enough divisions occur to provide a large proportion of the total number of cells involved in primary growth of the stem. The application of chlormequat, therefore, suppresses cell division in the subapical region, greatly reducing the internode length. Chlormequat applied at the start of stem extension can, therefore, result in shorter, stronger stems which are less liable to lodge, and thereby allow a greater response to nitrogen fertilizer. The second technique is the application of ethylene liberating compounds, such as ethephon, after flag leaf emergence. Ethylene can shorten stems, possibly because it inhibits the transport of auxin which is implicated in the expansion of cells.

The control of lodging in wheat with chlormequat and/or ethephon has been confirmed in many different production environments (Gruzdev, 1986; McClean, 1987; Knapp and Harms, 1988; Oglezneva and Berkutova, 1987; Oskarsen, 1989; Mohamed *et al.*, 1990). Their effects on yield and quality, however, not only depend on the timing and severity of lodging controlled but possibly many other plant processes which are known to be influenced by gibberellin, auxin and ethylene. In some cases there appears to be a clear link between lodging control and growth regulator effect. For example, in the UK, McClean (1987) found that chlormequat only increased yields and specific weights, and reduced protein contents when lodging had been controlled. In Norway, lodging control by chlormequat also increased Hagberg falling number (Oskarsen, 1989), possibly because pre-harvest sprouting was induced by the damp conditions which would be expected to develop in a lodged crop. Many other findings are, however, difficult to reconcile solely with effects on lodging. The Norwegian work, for example, found chlormequat to reduce thousand grain weight and specific weight in the absence of effects on yield and protein. Also in marked contrast to the work of McClean (1987), chlormequat application in Russia has lead to simultaneous improvements in yield, thousand grain weight <u>and</u> protein contents. Some work in North America has found chlormequat to increase yields (Knapp and Harms, 1988) but in other cases the reductions in straw height have been considered to be excessive, leading to increased foliar disease and reduced thousand grain weights (Caldwell and Starratt, 1987).

The effect of ethephon on yield has also been varied ranging from positive responses when severe lodging has been controlled (Gruzdev, 1986; Oglezneva and Berkutova, 1987), to no effect (Mohamed *et al.*, 1990; Petroczi, 1991), and even negative responses (Knapp and Harms, 1988). Effects on protein content appear to be more consistent. Several eastern European experiments have shown ethephon to increase grain protein contents either when yield has been increased or not affected (Gruzdev, 1986; Oglezneva and Berkutova, 1987; Dziamba and Mikos, 1988; Petroczi, 1991). In other regions, however, the same effect has not been observed (Bodson *et al.*, 1989).

The balance of beneficial versus detrimental effects is, therefore, difficult to predict and probably depends greatly on the precise timing of application and environmental conditions. This uncertainty suggests that

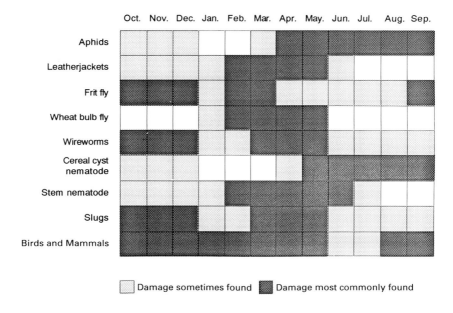

Fig. 7.15. Seasonal occurrence of some cereal pests in the UK (modified from Empson and Gair, 1982).

plant growth regulators should be avoided unless there is a high risk of lodging without their application. In this respect, however, growth regulators can have an indirect, positive effect on quality in permitting higher nitrogen application rates than would otherwise be possible.

PEST CONTROL AND PESTICIDE APPLICATION

There are a multitude of pests that can damage wheat plants and suppress yield and quality throughout the world, including nematodes, molluscs, mites, insects, birds and mammals. The number that cause common and widespread damage to wheat are, however, relatively restricted. Figure 7.15 shows representatives of the most important groups, and the periods during the year in the UK when damage is seen.

Insect Pests

Aphids (order Homoptera, suborder Sternorrhyncha)

Aphids (family Aphididae) are insects which have sucking mouth parts that are designed to pierce through the epidermal layers of vegetation so as to obtain juices from the veins of the plant. Most aphid species over-winter as eggs on a primary host which is usually perennial. In spring, the eggs hatch to produce females which can give birth to live young (parthenogenesis). Early generations may be wingless but later ones are often winged and migrate to secondary hosts, which are often annual, including wheat, where the same type of reproduction can continue. This permits the rapid build up of large colonies under favourable con-ditions. Later, winged members of both sexes migrate back to the per-ennial host where mating occurs and the females lay the overwintering eggs.

Several aphid species can cause damage by transmitting BYDV as described above. Some species may be present in sufficient numbers, however, to cause direct injury from their feeding activities. This may involve stems and foliage being stunted, with spots occurring on the leaves, signifying feeding sites. Aphids on the ear can intercept assimilates, including nitrogen, destined for the grain. Aphids are, therefore, among the most destructive insect pests of wheat throughout the world (Burton et al., 1991).

The species causing the highest levels of direct damage in the UK is the grain aphid (*Sitobion avenae*). This is a very widely distributed pest, also being found in such diverse areas as Egypt and South America following recent accidental introductions (Zuniga, 1991). The grain aphid is a relatively large species (up to 3 mm long) and varies in colour from yellowish green to reddish brown (Empson and Gair, 1982). The eggs are laid on grasses from where winged forms migrate to cereals and grasses in late spring. They are often first seen on the flag leaves of plants at the edge of the field. Under favourable conditions they can soon be found throughout the crop. The aphids move to the ears as they emerge. Early attacks, before anthesis, can reduce grain numbers. As the ears mature, and the grain harden during July, large numbers of the aphids move from the wheat to grasses where they overwinter.

As well as directly reducing yield, and transmitting BYDV, the grain aphid can have significant effects on quality. Lee et al. (1981) found that very high levels of infestation reduced flour extraction rate, impaired flour colour, suppressed formation of HMW glutenins, and reduced SDS-sedimentation volumes. In contrast, however, the aphids had a beneficial effect in reducing α-amylase activity. Reductions in protein quality, as evidenced by the Pelschenke test, have also been reported by others (Rautapaa, 1966). Oakley et al. (1993) found that effects were strongly dependent on the time of infestation, with aphid populations peaking before flowering decreasing yield but having no effects on quality. Inten-sive aphid feeding after flowering, however, reduced thousand grain weight, specific weight and Hagberg falling number, but when yields were substantially reduced, grain protein concentrations were increased.

As quality was only reduced when yield was also severely affected the authors concluded that aphid spray thresholds could be determined by consideration of yield effects alone.

The rose grain aphid (*Metopolophium dirhodum*) is light green with a dark green stripe running along the back (Gair *et al.*, 1987). It overwinters as an egg on roses. It can cause significant direct damage to wheat in Europe, including the UK, and has also been introduced to the Americas (Zuniga, 1991). The winged aphids migrate to cereals and grasses in late spring and their wingless progeny can later be found in the centre of fields. They are usually found on the lower leaves, where they cause premature senescence. Significant direct damage, however, is usually dependent on the aphid reaching the upper leaves. Severe attacks can reduce grain yield by as much as 10%.

Other aphids causing significant levels of direct damage include the green bug (*Schizaphis graminum*), found particularly in Texas, Oklahoma, Kansas and Missouri, USA, but also in areas of South America and North Africa (Zuniga, 1991; Miller, 1994). The Russian grain aphid (*Diuraphilis noxia*) has recently been introduced to South Africa and the Americas where, in the absence of many of its natural predators, has caused serious damage (Burton *et al.*, 1991; Zuniga, 1991). From its detection in the USA in 1986, for example, the Russian wheat aphid caused economic damage in the USA of over US$130 million in 1988 (Burton *et al.*, 1991). Other widely distributed aphids causing significant damage to wheat include the corn leaf aphid (*Rhopalosiphum maidis*), found in North and South America, and North Africa; and the bird cherry aphid (*Rhopalosiphum padi*), found in Europe, America, North Africa and Oceania (Burton *et al.*, 1991; Zuniga, 1991; Miller, 1994).

Non-chemical methods of aphid control include complete turf burial when ploughing grass for autumn wheat to prevent direct transfer from grass to the crop. Also of use is the complete destruction of all volunteer plants. Encouraging natural predation of aphids by, for example, ladybirds, hoverfly larvae, ground beetles and earwigs would also prove beneficial. This might be achieved by restricting the use of broad-spectrum insecticides, avoiding routine spraying, and by maintaining a diversity of flora and habitats in field margins. The development and use of resistant varieties also appears to offer some promise for integrated control programmes. In the USA, for example, considerable research has focused on the development of varieties resistant to green bugs and Russian grain aphid (Burton *et al.*, 1991).

Many farmers, however, rely on the application of aphicides. In many countries, aphids are the most often cited reason for applying insecticide. In England and Wales in 1990, for example, of the 1.53 Mha of land sprayed for named pest problems, 1.39 Mha (over 90%) was for aphid control (Davis *et al.*, 1991). Commonly used chemicals for aphid control included the synthetic pyrethroids cypermethrin (0.61 Mha in England and Wales in 1990), deltamethrin (0.20 Mha) and fenvalerate (0.08 Mha). As a group, the pyrethroids have advantages in giving a rapid knock-down of aphid numbers, are active at relatively low rates (e.g. 5–30 g ha^{-1}) and have relatively low mammalian toxicities. They interfere with the nervous system of the insect and work through both contact action and via ingestion. Some, including cypermethrin, have residual activity and are broad spectrum, controlling a wide range of insect pests. Another large group of insecticides is the organophosphates. Common examples with good activity against aphids include demeton-*s*-methyl (0.10 Mha in England and Wales in 1990) and dimethoate (0.21 Mha). These also have a broad spectrum of activity but are well suited to aphid control being systemic in plant sap. The insect is killed through cholinesterase inhibition. Application rates and mammalian toxicity of this group tend to be higher than the pyrethroids. One of the most selective aphicides, and therefore suitable for use in integrated control programmes, is pirimicarb (0.10 Mha in England and Wales in 1990). This is a carbamate which also kills the insect through cholinesterase inhibition. It has fast acting fumigant and translaminar properties and can also move within the xylem.

Other sucking insects that can cause similar damage to the aphids belong to the order Heteroptera (often grouped as a suborder with Homoptera in Hemiptera). Of particular note with regards effects on quality are the stink bugs (family Pentatomidae). The genera *Aelia* and *Eurygaster* (shield bugs) occur throughout Central and Eastern Europe, the Mediterranean and the Near East and certain species are very damaging to cereals. The eggs are laid in the spring and hatch in the summer. The bugs then thrive and multiply in sunny, warm and dry conditions. They can attack the leaves and stems, but most damage occurs when they invade developing kernels. The grain is inoculated with salivary enzymes so attacks at the dough milky-ripe stages can greatly modify the gluten due to proteolytic activity. Severely affected crops thereby produce doughs which fail to rise when baked. The problem is less severe with varieties which mature earlier as they pass the milky-ripe growth stage before the main proliferation of the insects (Matsoukas and Morrison, 1990). Similarly in New Zealand, wheat bug (*Nysius huttoni*) has also been implicated as causing dramatic effects on wheat quality. Visual symptoms consist of an opaque yellow path on the kernel, often with a dark puncture mark at the centre. Most severe effects are seen with kernels infested at late anthesis as evidenced by grain shrivelling, high α-amylase activity, low thousand grain weight and poor germination capacity (Every, 1990). The proteinase activity from the bug causes a reduction in the SDS-sedimentation test volume, and the almost complete disappearance of the HMW-glutenin subunits from SDS polyacrylamide gel electrophoresis

(SDS–PAGE) gels. This would be expected to have critical effects on dough rheology and baking potential.

True flies (order Diptera)

The true flies constitute one of the largest insect orders comprising over 85,000 species. The adults have two wings and the mouth parts are designed for sucking. Their life cycle consists of egg, larva (i.e. the grub or maggot), pupa and adult. Many adult forms of the Diptera have severe effects on farmers and livestock, particularly through their transmission of diseases. This is exemplified by mosquito transmission of malaria, and tsetse fly transmission of sleeping sickness in humans and nagana in livestock in tropical Africa. Direct damage to wheat crops, however, is principally caused by the larvae feeding off the young stem and leaf tissue.

Leatherjackets

Leatherjackets are the larvae of a number of species of crane fly (suborder Nematocera, family Tipulidae). They occur in many wheat producing regions of the world, particularly in damp areas. The most common species in the UK is *Tipula paludosa*. Other species which can cause damage include *T. oleracea, T. vernalis, Nephrotoma appendiculata* and *N. flavescens*. The females of *T. paludosa* commonly lay eggs in grassland in September so that damage is most common in the first year after a grass ley, particularly in spring cereals when the preceding grass has been ploughed late. Grassy and weedy cereal stubbles can also attract the females so damage to crops in combinable crop rotations is also possible. The larvae hatch about ten days after laying and feed during autumn and winter, becoming much more active in spring (Fig. 7.15). Wheat plants are detached close to soil level and leaf tips that touch the ground may be dragged below the surface. Damaged tissues may also appear to be torn with ragged edges (Empson and Gair, 1982). The larvae are thick skinned, have no legs nor obvious head end and can grow to 40–50 mm in length by June. They can be found near the soil surface, often close to damaged plants or under clods where conditions are damper.

There are many natural killers of leatherjackets including dry weather in autumn, viruses and birds. Non-chemical control includes ploughing grass or grassy stubbles before the middle of August, whereas rolling and nitrogen fertilizer may help the crop to recover. Where this is not sufficient, chemical control may be justified. Gair *et al.* (1987) suggest that pesticides may be warranted when 15 or more leatherjackets are found in ten 30 cm lengths. Commonly used chemicals have included gamma-hexachlorocyclohexane (HCH), an organochlorine insecticide which can be sprayed onto the soil or applied as pelleted baits. Organophosphate alternatives include chlorpyrifos, triazophos and fenitrothion.

Frit fly

Frit fly (suborder Cyclorrhapha, family Chloripidae), also known as chloropid fly, covers a number of small species of fly whose larvae are destructive to cereals. The important pest in the UK from this category is *Oscinella frit*. As with leatherjackets eggs are laid, and maggots develop, predominantly in grassland. The generation of maggots that cause most damage to wheat are those laid between August and October. When grass is ploughed in before a wheat crop, therefore, the larvae move from the decaying turf to attack the cereal seedlings. Maggots may, however, also develop in grassy stubbles and on cereal volunteers, and eggs can be laid directly onto early sown crops. The maggots are active at temperatures above 7.2°C and bore into the leaf sheaths to attack the growing point. This causes the central leaf to turn yellow to give the so-called 'dead heart' symptom. The freshly attacked shoot will contain at least one white, thin, pointed maggot. These are less than 5 mm in length and, after about three weeks of feeding, will change into puparia which are brown, cylindrical and about 3 mm long. Very young plants may be killed by this type of attack but in established plants, tillering can be stimulated which may partly compensate for the loss of the main stem. These new tillers may also be attacked and the extra tillering that occurs gives the crop a grassy appearance. A later generation of maggots may affect the ears of spring wheat, causing the tips to be white and empty, but this type of damage can be caused by many other factors and only rarely is frit fly to blame.

The most effective method of non-chemical control is to extend the period between grass or stubble ploughing and the sowing of the following crop to at least six weeks. Many modern practices such as minimal cultivation and early sowing, however, make this difficult and winter wheat crops at risk may be sprayed with organophosphate insecticides such as chlorpyrifos, omethoate and triazophos before dead heart symptoms have been seen (Empson and Gair, 1982; Gair *et al.*, 1987).

Wheat bulb fly

Wheat bulb fly (*Delia coarctata*, suborder Cyclorrhapha, family Anthomyinae) occurs throughout central and northern Europe and parts of the former USSR. In the UK, eggs are laid in bare soil between mid-July and the end of August. Most eggs appear to be laid in dry soil with a rough surface, shortly after cultivation. The eggs hatch between January and March and the white, legless maggots are attracted to the wheat plants. The maggot bores into the shoot just below ground level and its subsequent feeding kills the central leaf causing dead heart symptoms. The maggots can move from one shoot to the next such that by the end of April each maggot may have entered four or five shoots. This causes the wilting and yellowing of the central leaf in each one. By May the maggots have usually finished feeding and move into the soil where they produce a brown, barrel-shaped puparium. The adults emerge during June. When crops are grown on infested land, they can be devastated. The most susceptible to severe yield loss are winter wheat crops which are backward and poorly tillered at the time of attack. Plants which have not tillered by late winter and early spring can be killed. Recovery depends on the ability of the plants to compensate with extra tillering in late spring and early summer. With spring wheat, only crops sown before mid-March are susceptible. Fields at risk from attack can be identified by pre-cropping egg counts and are most commonly found when wheat is following a summer fallow or a root crop such that rough ground has been free of cover during the egg laying period. Egg laying can be reduced by preparing a fine seedbed soon after primary cultivations, or if following an early harvested crop such as early potatoes or vining peas, a catch crop might be planted. Set-aside land in the UK, bare in the summer, is ideal for egg laying. A non-cereal such as oilseed rape could be grown after set-aside to lower the risk. Sowing winter wheat early may also help because tillering is encouraged such that each plant could 'afford' to loose two tillers in early spring and still be able to recover. The first wheat after potatoes or peas is often a bread-maker, which would benefit from high tillering capacity against wheat bulb fly, e.g. cv. Avalon. Similarly nitrogen applications after the first signs of damage can assist compensatory growth. Chemical control is not completely effective but is the main defence in the UK. Suppression is commonly achieved with either sprays or seed treatments of organophosphorus insecticides including fonofos, chlorfenvinphos and chlorpyrifos (Davis *et al.*, 1991). The recommended treatment threshold in the UK is 2.5 Mha^{-1}, based on egg counts. In the UK the only 'dead heart' stage spray recommended is dimethoate.

Hessian fly

The Hessian fly (*Mayetiola destructor*) is an example of a gall midge (suborder Nematocera, family Cecidomyiidae) which are tiny flies, rarely seen as adults. They are thought to have originated in Asia from where they were transported to Europe and to North America. Distribution of the pest is widespread in the UK but is rarely common enough to cause significant injury (Empson and Gair, 1982). In many other parts of the world, however, particularly in North America, Hessian fly is extremely important. The females lay their eggs on the leaves of wheat from where larvae hatch about a week later. The larvae then proceed down the leaf and under the leaf sheaths. Damage occurs in autumn and spring as larvae, situated between the leaf sheaths and stem, extract juices near to the base of the plant. This often kills small and young tillers, while those that survive may be weakened and liable to lodge. Damaged shoots are also often stunted while the ears can be pale and thin containing shrivelled grain.

Larvae can persist on wheat stubble so deep ploughing soon after harvest and crop rotation can help suppress Hessian fly numbers. Sowing winter or spring crops late will also effect good control such that the adult flies will have emerged and died before the wheat plants appear above the ground. In the central and eastern states of the USA, for example, recommended drilling dates have been governed by predetermined 'fly-free' dates (Martin *et al.*, 1976). Where Hessian fly is not present, autumn drilling can proceed 7–10 days earlier than would otherwise be advisable. Latterly, breeding programmes have identified varieties resistant to Hessian fly, although selection is complicated by the existence of several races of the pest which differentially attack a range of varieties. Some systemic insecticides have also been effective (Empson and Gair, 1982).

Orange wheat blossom midge

Recent outbreaks of the orange wheat blossom midge (*Sitodiplosis mosellana*) in the UK have caused significant reductions in quality. Orange or reddish maggots are often found singly in each attacked floret for about a month after flowering. Attacked grain are typically deformed with blackened, dented regions which give reduced thousand grain weights. Additionally, attack often results in premature sprouting with the attendant increase in α-amylase activity and reduced Hagberg falling numbers. This appears to be due to the loosening of the pericarp which allows easier ingress of water and hence sprouting in poor weather. Direct increases in α-amylase activity from insect origin do not appear to be important in reducing Hagberg falling number as these disappear before grain maturity (Lunn *et al.*, 1995). Proteolytic enzymes of insect origin are, however, important in reducing protein quality.

Other Dipteran larvae causing damage to wheat in the UK include the saddle gall midge (*Haplodiplosis marginata*) which causes a saddle-shaped depression in the stem beneath the sheath of one of the upper leaves. If the 'saddle' rots, the ear may be lost. Bibionid flies (*Bibio* spp. and *Dilophus febrilis*) can cause damage in wheat following grass as the larvae eat out seeds, severing lateral roots and sometimes chewing through the stem below ground. Larvae of some flies feed between the upper and lower surfaces of the leaves. These so-called leaf miners (e.g. *Hydrellia griseola* and *Agromyza nigrella*) leave pale, air filled blisters on the leaf lamina. The subsequent loss of leaf tissue has reduced yields by up to 8% in exceptional years. The gout fly (*Clorops pumilionis*) causes damage when the larvae burrow down to the centre of the shoot. Plants remain stunted and the stem bases become swollen or gouty. The shoots, although initially looking very green, eventually senesce and rot. The larvae of *Opomyza florum* often attack early sown wheat and causes symptoms similar to wheat bulb fly as the larvae burrow into the stem base and kill the central shoot. More detailed descriptions of these more minor pests, their symptoms and control is given in Empson and Gair (1982) and Gair *et al.* (1987).

Beetles and weevils (order Coleoptera)

The Coleoptera comprise the largest order of insects with approximately 250,000 known species. Several are termed as *beneficials* in that they prey on crop pests. Important examples in this respect are members of the Carabidae (ground beetles) and Staphylinidae (rove beetles) which feed on larvae, soft-bodied adult insects and insect eggs. In addition both larvae and adults of most of the Coccinellidae (ladybirds) feed on aphids. Other members of the Coleoptera, however, can be extremely damaging to crop plants and some are significant pests of wheat.

Wireworms

Wireworms are the larvae of several species of click beetles (suborder Polyphaga, family Elateridae). In the UK, troublesome species include *Agriotes lineatus*, *A. obscurus*, *A. sputator*, *Athous haemorhoidalis* and *Cternicera* spp. which naturally inhabit grassland. Wireworms are so called because of their long, slender, cylindrical shape. Eggs are laid in summer and hatch to reveal small pale larvae. The larvae then feed for four to five years, developing a bright yellowish-brown colour with a dark brown head and three pairs of short legs (Gair *et al.*, 1987). Wheat plants are damaged by the wireworm biting through the plants below soil level. This will usually kill young plants while leaves of larger plants turn yellow. Larvae may work along a drill length such that several successive plants are damaged, and the larva responsible will be found close to the next healthy one. Damage is most likely in the first four years following long-term grass on heavy soils.

In the USA, the Great Basin wireworm (*Otenicera pruinina* var. *noxia*) has caused considerable damage to wheat in areas of Washington, Oregon and Idaho where rainfall is less than 360 mm per year. The larvae can live for three to ten years and cause damage in similar ways to that described above.

Non-chemical methods of control include clean fallowing such that the larvae are starved. Damage can also be limited by drilling into a firm seedbed with adequate supplies of phosphate. Rolling and nitrogen applications can help plants recover from partial damage, and higher seed rates may compensate for the removal of some plants by the pest.

In the UK, levels of wireworm damage have declined over the past 20 years. This is partly because of the widespread use of insecticide seed treatment, i.e. dressing the seed with gamma-HCH can limit the damage from moderate to low infestations. Such applications are not always necessary but their relatively low cost often encourages farmers to use them as an insurance against possible attack.

Order Hymenoptera

The Hymenoptera contain sawflies, ants, bees, wasps and parasitic wasps. The adults have two pairs of transparent wings and the mouth parts are designed for biting or biting and sucking. There are only a few members of this group that cause damage to cereals and all are relatively minor pests in the UK. In the USA, however, significant damage is caused by wheat jointworm (*Harmolita tritici*) where it is second in importance only to the Hessian fly. The adult lays an egg in the stem and the resultant grub feeds on stem juices. This activity causes warts to appear above the node. The stem is subsequently weakened and easily lodges. The closely related wheat strawworm (*Harmolita grandis*) is also a serious pest in certain areas of the USA and causes bulb-like structures to develop on tillers at the point of infestation. Each tiller infected in spring is killed.

The wheat stem sawfly (*Cephus cinctus*) occurs throughout major growing wheat areas in Russia, Europe and North America. The name sawfly derives from the egg laying apparatus of the female (the ovipositor) which is armed with saw-like teeth for cutting slits into plant tissue. In the case of wheat stem sawfly, the female slits the stem to lay an egg and the larvae feed downward. When a larva reaches just above soil level it chews a groove or ring in the stem such that the plant lodges. Other sawflies that can attack wheat in both the USA and Europe are the black grain stem sawfly (*Cephus tabidus*), and the European wheat stem saw fly (*C. pygmaeus*).

Nematodes (Eelworms)

Nematode plant parasites are long, slender, transparent, non-segmented roundworms without limbs. They are too small to be seen with the naked eye but are visible under low magnification and are eel shaped. They develop from eggs and pass through larval stages (often four) before adulthood during a period of weeks or months depending on conditions and species.

The most important nematode pest of wheat worldwide is probably the cereal cyst nematode (*Heterodera avenae*) which occurs in New Zealand, Australia, Peru, the USA, Canada, throughout Europe, the former USSR, countries surrounding the Mediterranean, Japan and India. The most susceptible crops are those sown in spring on light, well-drained soils. Feeding by large numbers of larvae cause roots to become short, but highly branched and fibrous. Above ground, the plants are stunted and become yellow, sometimes developing a pink or purple discoloration. Some of the larvae develop into females which swell up to form a sack containing hundreds of eggs. The female dies but the sack persists to form a lemon-shaped cyst about 0.5 mm in diameter. Cysts can lie dormant in the soil for several years or larvae may emerge in spring from eggs that have overwintered in a cyst. The juveniles enter the roots and feed near to the vascular tissue. The females erupt through the root as they swell up. Reductions in yield by cereal cyst nematode can be related to the number of cysts in the soil. Some experiments have recorded average yield losses of 188 kg ha^{-1} for every ten eggs per gram of soil (Gair *et al.*, 1987).

Non-chemical control involves reducing the frequency of susceptible crops in the rotation, which includes the small grained cereals and, to a lesser extent, the ryegrasses. Grass weeds such as spring wild oat can also help increase nematode populations so these should be controlled. The crop can be made more tolerant of the pest if it is provided with adequate nutrition and moisture. Excessive nitrogen applications can, however, favour nematode multiplication. There are sources of cyst nematode resistance in some varieties and these have been exploited in some problematic areas such as Australia. Different pathotypes of the nematode exist which react differentially with different varieties.

Levels of cereal cyst nematode in areas of northern Europe, including the UK, have been in decline. The reasons for this are not clear but it might partly be explained by the presence of nematophagous fungi such as *Nematophthora gynophila* and *Verticillium chlamydosporium* which develop under rotations with a high frequency of cereals (Wiese, 1987).

In areas where the most damaging pathotypes occur such as Australia, the use of nematicides has produced economic yield responses in cyst eelworm invested soil but this practice is not widely adopted (Wiese, 1987).

Other nematodes attacking the roots include a number of species causing root-knots, i.e. root swellings or galls (*Meloidogyne* spp.). Damage has been observed in many areas of Europe, and North and South America. These are most visible in spring or summer and comprise expanded root cortex cells and the swollen bodies of female nematodes with large numbers of eggs. In the UK, root-knot nematode (*M. naasi*) is common in Wales and southwest England and can cause the death of seedlings. On more mature plants the root damage leads to stunting, premature senescence of the older leaves, reduced tillering, and reduced grain size or number. The most effective control method is to rotate wheat with non-hosts such as oats and potatoes. Some varieties of wheat differ in their ability to support nematode reproduction such that the build up of numbers under successive wheat crops may vary. Winter wheat varieties appear to tolerate attacks better than spring wheats (Empson and Gair, 1982; Gair *et al.*, 1987; Wiese, 1987).

Lesions and cavities in roots can be a consequence of attack by root lesion nematodes (*Pratylenchus* spp.). These can cause stunting and yield suppression directly, or may provide entry points for fungal and bacterial rots and diseases. Damage in the UK appears to be mostly due to *P. fallax* whilst in areas of North and Central America and Australia the most damaging causal organisms are *P. neglectus* and *P. thornei*. In Europe, damage has been particularly associated with soil acidity problems (Gair *et al.*, 1987).

One of the longest recognized diseases of wheat has been seed-gall, first characterized in England in 1743. The causal nematode (*Anguina tritici*), contrasts with those above in that yield loss is principally though damage to the ears. Larvae move in water films from the soil to the meristematic tissues, where they invade the developing florets. At ear emergence the grain sites are replaced by galls about 3 mm in diameter containing large numbers of nematode eggs and larvae. The disease has spread with wheat seed throughout the world but has been reduced to negligible levels in many areas by effective seed cleaning. Wiese (1987) reports, however, that significant levels may still occur, in eastern Asia, India, the former Yugoslavia and southeastern Europe. A two year break from wheat or rye will reduce soil borne larvae to non-damaging levels.

Slugs

Slugs are unsegmented soft-bodied animals which excrete slime. As a result they can substantially modify their body shape to move and squeeze easily through small holes and crevices in soil. They are classified as molluscs of the class Gastropoda, subclass Pulmonata (Jones and Jones, 1973), which contains various species of slugs and snails of agricultural importance. They are most numerous on heavier soils in wet, mild, regions than drier habitats. In the UK they are more common in the west, than the east and north, feeding on a wide host range of crop plants including, in particular, potatoes, root crops, peas and clover as well as cereals. Nine species of slugs are significant in the UK in causing crop damage (Jones and Jones, 1973).

Field slug

Deroceras reticulatum (Attwood, 1985) the grey field slug, or netted slug, is most common in damaging cereals. Late sown winter wheat in the UK, following oilseed rape, peas, beans, maize or grass, on clay loams or silty clay loams is most at risk, particularly after a wet summer (Gair *et al.*, 1987). The damage is greatest when the crop is direct drilled, rather deep, into loose cobbley seedbeds. The seeds are stressed by the deeper drilling into a rapidly cooling wet substrate, and the openness encourages slug activity in particular down slots created by the drill. Late October to November is a susceptible period for slug damage, especially under conditions described. Spring wheat is less susceptible, and light textured soils and fen peats are not so prone to slug activity.

The damage can be of three basic types. Grains can be hollowed out by the rasping, circular saw, type mouth parts of slugs, removing the germ and preventing germination. Alternatively, the young shoot and leaves can be severed, killing the seedling. In these cases gaps are created in crops. It feeds mainly at night when the humidity is higher and temperatures lower, and is much affected by relative humidity at the soil surface.

The colour of the field slug varies from pale grey when young, to a mottled grey with red or yellow tinge when adult (Jones and Jones, 1973). The underlying 'foot' is cream, with a darker central zone. When extended it reaches 3–4 cm, and is hemispherical when contracted. In northern Europe, including the UK, it takes 12–15 months to complete the life cycle and may breed throughout the year in favourable weather. Egg laying peaks are March to May and September to October, and about 300 eggs are reportedly laid in batches of ten to 30 in soil crevices (Jones and Jones, 1973). The field slug is most active in the UK in June and September to November. Being cold blooded its activity slows down considerably in late autumn and winter, and it may migrate deeper into the soil at these times to avoid surface frosts. Periods of drought also limit activity and may result, if prolonged, in a significant population decline.

Application of organic manures to the soil encourages slugs by providing a food substrate and maintaining a higher moisture status in their habitat. The ploughing in of straw has a similar effect, and more slug activity has been observed in wet autumns in the UK since the ban on straw burning.

Population estimates are often highly inaccurate, unless careful soil analysis is undertaken. Trapping slugs under wet sacking or tiles with poison baits to estimate numbers can give only a partial picture of populations near the surface, and is less satisfactory than soil or turf flotation techniques. However, Wibberley (1989) recommends baited tiles every 50 m across the field, and six slugs per tile as a treatment threshold. The distribution of traps and bait across a field can have an important influence on monitoring accuracy (Hunter and Symunds, 1970).

Cultural control of slugs has a greater chance of success in reducing populations than chemical use, and no genetic difference can be exploited between wheat varieties for high risk situations. Soil inversion from ploughing can expose many slugs, facilitating bird predation and death from dehydration. Populations of the three main slug species were decreased to a third or a quarter by rotating soils to a fine tilth in a comparison of cultivation techniques (Hunter, 1967). Rolling seedbeds after drilling can be particularly effective in limiting slug activity, which is prevented in firm conditions. Metaldehyde baits on bran, beet pulp or in pellets are commonly used as a molluscicide stomach poison (Attwood, 1985). They can cause water loss or the inability of exposed slugs to move, increasing their exposure to drying and enemies. Dehydration may fail to occur, however, in damp conditions. Methiocarb is generally more effective and less susceptible to prevailing moisture for toxicity. The effectiveness of such baits can be reduced by dampness and long exposure, and many pellets also contain a fungicide to reduce mould development. Chemical control of slugs is not very effective on its own in reducing high populations.

Various natural enemies eat slugs. These include rooks, starlings, gulls, blackbirds, thrushes and ducks. Toads, hedgehogs, moles, shrews, staphylinid and carabid beetles are also predators. Slugs are also found infested with flukes, nematodes, protozoa and fungi. Some of these organisms are being explored as potential biological control agents. In spite of all of these natural enemies the slug population does not seem to be much moderated by predators and parasites and the number of slugs can be considerable under favourable conditions with corresponding potential for high levels of crop damage.

Birds and Mammals

Significant levels of crop loss can occur through the feeding of birds and mammals. Many birds, for instance, will reduce plant populations by eating seed. Grain on the soil surface are commonly taken by pigeons and doves (*Columba* spp.) while buried seed can be taken by rooks and jackdaws (*Corvus* spp.), starlings (*Sturnus vulgaris*) and pheasants (*Phasianus colchicus*). Such feeding can cause lengths of missing and displaced plants adjacent to unaffected rows. When autumn sown seedlings have emerged they may be severed by the feeding of winter flocks of small birds such as skylarks (*Alauda arvensis)* and finches (family Fringillidae) or geese (*Anser* spp. and *Bransa* spp.) as evidenced by a diagonal or V-shaped cut. Rooks, jackdaws and starlings may also pull up the seedling in order to reach the seed. Geese may continue to graze cereals as the crop matures in spring and summer. During grain maturation the kernels may be taken by small birds such as sparrows (family Passeridae) which perch on the heads. Heavier birds such as rooks and pigeons may physically lodge the crop in order to get at the ripening grain. These and many other species will also take seed from crops lodged by other processes (Empson and Gair, 1982).

Bird damage can be reduced by scaring methods ranging from the traditional scarecrow to modern noise scarers. These measures are only partially successful, and invariably temporary, as the birds become conditioned to them. Shooting birds to reinforce 'bang' scarers may improve longevity but large-scale shooting is not often effective at reducing long-term pressures from birds. Mammals such as rats (*Rattus* spp.) and field mice (*Apodemus* spp.) may dig up seed, but most damage is through shoots being grazed in autumn, winter and spring by many species including rabbits (*Oryctolagus cuniculus*), hares (*Lepus* spp.), field mice, voles (*Microtus* spp.) and deer (family Cervidae). Severe grazing of wheat delays maturity, may increase or reduce tiller number depending on timing, and reduces final crop height and competitivity. Shooting, trapping and gassing have all been used to reduce mammal damage. Rabbits and larger mammals may be excluded by specialized fencing. Rabbits have also been controlled biologically with myxomatosis, a fatal, infectious, viral disease. The disease was isolated from South American rabbits and introduced into western Europe and Australia in the 1950s. Since then rabbit populations have almost recovered to pre-myxoma levels in a number of areas. Other viruses have, therefore, been tested as potential biological control agents (Anderson, 1995).

8

Postharvest Management of Grain

CROP LOSS

Most cereals are stored for various periods after harvest, during which time considerable grain loss and quality deterioration can occur. Christensen and Kaufman (1969) quite rightly stress that grain is both 'exceedingly durable and highly perishable'. Worldwide losses of grain after harvest have been estimated as high as 50%, from a combination of microbial deterioration, insect attack, and rodent and bird predation (Adams, 1977; Salunkhe *et al.*, 1985; Pomeranz, 1987). Estimates vary widely, and are markedly influenced by the nature of the crop, type of storage, climatic influences, and regional considerations. Christensen and Kaufman (1969) quote a Food and Agriculture Organization (FAO) survey figure of 30% annual grain loss after harvest in India, parts of Africa and some South American countries. Losses of between 2 and 52% of wheat grain were estimated in various developing countries in a study by the National Academy of Sciences (1978) of the USA. Postharvest loss of grain due to fungal contamination alone is estimated to be of the order of 10–30% (Chelkowski, 1991). Dunkel (1992) has reported annual wheat losses in underground storage pits in Morocco of 25% as a result of fungal attack, insect damage and grain weight loss. Annual grain loss through rodent consumption has been estimated between 2 and 9% in Bangladesh, the Philippines and Brazil (Salunkhe *et al.*, 1985). Although storage technology is improving, and insecticides are increasingly available in a number of developing countries, there are still problems in effectively protecting grain from insect attack in stores in the tropics, and substantial grain losses still occur on farms in parts of Asia and Africa. Economic losses of stored grain in temperate, and generally more prosperous, areas is lower than warmer regions. Detection of pests and close monitoring of temperature often leads to remedial measures being taken before an escalation of damage. The means of doing this in some developing countries may be neither affordable nor available (Wilkin and Rowlands, 1988). Nonetheless even in temperate climates laboratory experiments suggest that certain individual weevils and mites could reduce grain weight by 10% over a six week period if left unchecked (Wilkin and Rowlands, 1988). Of arguably greater significance in developed countries is the effect of poor storage on quality (Harein, 1982). For selling to the intervention agencies of the Economic Union (EU), for example, a complete freedom from live pests is specified as a minimum requirement. Despite this, a survey of English farms in 1987 found 10% to have live insect pests and 72% to have mites (Wilkin and Rowlands, 1988). It has been suggested that insects cause more problems in France, Germany and the USA than in the UK, but in Canada the cold temperatures during winter greatly restricts damage (Wilkin and Rowlands, 1988).

HARVEST CONSIDERATIONS

Wheat grain can start to deteriorate in the field prior to harvest. Preharvest rainfall can, for example, stimulate grain sprouting, encourage grain diseases, weather and leach grain, increase soil contamination, and increase grain moisture content such that intensive drying is required to achieve reliable storage (Yarham, 1980; Barnes, 1989; Lockhart, 1989; Gooding *et al.*, 1996). Delayed harvesting has reduced grain specific weight, protein content and Hagberg falling number (Hayward, 1987). To limit this deterioration wheat crops in the UK for quality outlets, such as milling and seed, are often harvested first, in preference to wheat destined for animal feeds.

Grain is subjected to profound changes at harvest. The condition of the wheat crop and method of harvesting, together with prevailing weather influences, are all critical for grain quality. If ears are too dry, and they are combine threshed at an inappropriately high speed, a large proportion of the grains can shatter or crack. The speed of the harvester can, therefore, influence grain loss over a range of harvesting conditions and correct combine settings are crucial (Nash, 1985; Barnes, 1989).

The type of grain damage incurred at harvesting is influenced by such factors as species, variety, moisture content and grain shape. Longer grains incur greater damage than short grains; moist grain can undergo so-called bruising and skinning, and dry grain can scratch and break (Nash, 1985). Damaged grain is more readily attacked by fungi, mites and insects than sound grain. Slight, often not visible, grain damage is sufficient to encourage grain colonization and later store infestation by mites (Wilkin and Rowlands, 1988).

Harvesting redistributes fungal inoculum in grain and introduces further contamination. Storage fungi are few and at low levels generally on grain in temperate regions before harvest, but an increase in *Penicillium* species inoculum by more than 250 times through contamination with residues and soil in the combine from earlier harvests are reported (Lacey, 1990). The prevalence of weeds also interacts with combine harvester performance, to reduce harvest yield, increase moisture and contaminate grain batches. The timing of harvest can be very significant, as it relates in particular to grain moisture. Harvesting wheat too early in South China to accommodate land preparation and the sowing of rice, in a triple crop system, has compounded fungal deterioration due to the storage of too moist grain – which could not be better dried due to the lack of sun at harvest (Dunkel, 1992).

GRAIN DRYING

The development of combine harvesters has provided the capacity to gather large quantities of grain in a short time period. This contrasts with the time consuming and less efficient stooking of sheaves in the field following cutting. This traditional practice, however, did provide a period of pre-drying in the field before grain storage but there was a great risk of quality deterioration in wet climates. In cool temperate regions grain moisture in the field at harvest is rarely less than 15%, and many crops are harvested around 20% moisture content. The need for pre-storage drying is substantially less, therefore, in warmer climates, such as Australia where wheat is often harvested at 10–12% moisture. Considerable drying occurs in the standing crop. Grain is not completely enclosed in the ear and equilibrates moisture readily with the surrounding air. If grain is left too long in the field at low moisture contents, however, individual grains can be shed resulting in losses amounting to several tonnes per hectare in the UK (Lockhart, 1989).

In hot climates, grain may be left exposed to the sun after harvesting to achieve adequate drying. In Bangladesh, for example, the need for low technology solutions has led to the practice of drying wheat to about 10% moisture content (mc) by leaving it in the sun for at least six or seven consecutive days (Ahmed, 1985). The results can, however, be variable and substantial grain losses may occur from birds and predators (Salunkhe *et al.*, 1985).

In more temperate areas, the benefits of drying grain and keeping it cool for long term storage have been recognized for many years. Nash (1985) describes a system of natural ventilation for grain stores developed by the Romans, and ventilation systems driven by bellows for aeration of grain in France in the mid-eighteenth century. All drying systems rely on air of suitable relative humidity and temperature being passed through the crop (McLean, 1987). Drying systems can, however, be broadly categorized into two groups depending on the air temperature used, i.e. near ambient or bulk driers, or high temperature driers.

Near Ambient or Bulk Driers

Many modern systems still rely on drying grain over a long period with forced ventilation while the grain is stored in bulk in bins or on floors. Large quantities of grain are sometimes dried at ambient air temperatures but the drying power is substantially improved by heating the circulating air by 2–7°C or by the use of dehumidifiers (McLean, 1987). With these bulk-storage driers, using temperatures at or near the ambient, the drying times can be several weeks and care must be taken to prevent microbial deterioration (Nellist and Bruce, 1987). However, these techniques allow crops to be dried in deep layers so that they can be dried in store. Handling costs are thereby reduced.

In bulk driers the desired power output of fans will depend on the required ventilation rate and the resistance to air flow through the crop (Nash, 1985). When using ambient air, the rate of drying is related directly to the airflow, the air temperature and relative humidity (Loewer *et al.*, 1994). McLean (1989) argues that for average UK conditions ventilating with 0.05 m^3 s^{-1} for 24 h will reduce moisture content by approximately 0.5%. Computer simulation models showed how increasing relative humidity and reducing temperatures below 18°C increased the required ventilation rate due to reduced moisture carrying capacity of the air. These models suggested, however, that at temperatures above 18°C increased ventilation rates were required because a greater rate of grain deterioration due to mould occurred in the warmer conditions. Once the required ventilation rates have been decided, consideration of resistance to airflow is necessary. This is reduced by large seed sizes, is directly proportional to crop depth, and increases with increasing air velocity and dust and weed seed contamination. The method of filling the drier can also have a large impact upon the bulk density of grain and hence resistance to airflow. High bulk density increases storage capacity but will increase the requirements of the ventilation system. Walking on the grain, possibly for inspection, can greatly increase bulk density (McLean, 1989). Other important factors include duct size and the design of floors and air entry points. It is essential that the system is well maintained and blockages alleviated quickly.

High Temperature Driers

McLean (1989) defines high temperature driers as those in which if the grain remained in contact with the air until its equilibrium moisture had been reached, severe overdrying would result. Hence drying only continues until the desired moisture content has been reached. This is followed by cooling before storage. Nash (1985), McLean (1989) and Loewer *et al.* (1994) describe a number of rapid grain drying practices including continuous flow systems, batch drying, and storage driers. Batch and continuous driers typically use air temperatures between 40 and 120°C. These temperatures can dry grain sufficiently after only 4–0.5 h, which is too short for significant deterioration by microorganisms, but has to be carefully controlled to avoid thermal damage (Nellist and Bruce, 1987). Drying moist grain too quickly at high temperatures, for example, can cause considerable structural stresses in grain leading to brittleness and cracking problems. However, levels and types of heat damage tolerable for different markets, and therefore safe drying conditions, vary considerably with the intended end-use. The palatability of grain for ruminants can actually be improved if it has been slightly roasted (Nellist and Bruce, 1987). For non-ruminants, however, it is important that the protein is not denatured such that it becomes unavailable to the livestock. This can, for example, be related to lysine availability which depends on both drying temperature and duration approximately according to equation 8.1, where $T_{10\%}$ is the temperature in °C at which lysine availability is reduced by 10% and h is the duration in hours (Nellist and Bruce, 1987).

$$T_{10\%} = 126 - 7h \qquad [8.1]$$

For non-ruminants, a 10% reduction in availability of lysine has been considered acceptable (Nellist and Bruce, 1987), so as long as 100°C is not exceeded for more than 3 h, risk of heat damage is still small (Table 8.1).

For milling and seed purposes, temperature of drying can be more critical. Both germination and loaf volumes show large reductions when drying air temperature is 80°C compared with 70°C, particularly for wetter grain (Fig. 8.1). Nellist and Bruce (1987) concluded from laboratory studies, drier tests and other experiments that drying wheat from 20 to 15% mc with simple batch or cross-flow continuous driers would not harm germination or baking potential as long as drying air temperatures

Table 8.1. Critical temperatures for grain drying with heated air (Nash, 1985).

Grain for milling	Maximum of 65°C
Grain for seed	55–65°C
Grain for feed	Up to 100°C

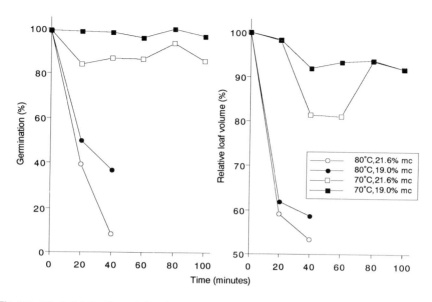

Fig. 8.1. Effect of drying time, drying air temperature and initial moisture content (mc) on germination and baking potential of wheat (after Wassermann *et al.*, 1983).

did not exceed 66°C. This is consistent with critical temperatures for drying grain with heated air as reviewed by Nash (1985) (Table 8.1).

Grain which has been artificially dried with hot air must be cooled to help prevent problems in storage. Continuous flow driers often have a cooling section where air at ambient temperature is forced through the grain (McLean, 1989). In batch and bulk systems, grain is often cooled using low volume aeration, i.e. the passing of small volumes of ambient air through grain to reduce temperature to prevent deterioration. The airflow necessary to do this may only be 5% of that required to dry grain in bulk and can be particularly effective when cool night-time air is utilized (Nellist, 1986). This technique can also help reduce temperatures of crops harvested in hot conditions.

GRAIN CLEANING

The loose chaff is mainly ejected with straw and small grain during combine harvesting as a result of being blown over the sieves. Some chaff can remain, however, together with other crop debris and weed contaminants. Further cleaning operations prior to storage require the use of various mechanical and aspiration machines. As standards for quality grain are raised the requirements for pre-cleaning of grain become more important.

In the past much of the cleaning of grain was left to grain merchants, following short-term farm storage. Grain contamination is now carefully assessed and the requirements for cleaning can influence the price in trading with merchants. Rigorous additional cleaning takes place at mills prior to grain preparation for milling into flours. A significant amount of stones and other debris is often removed at this late stage. The most sophisticated techniques are reserved for that grain which is saved for seed (Clarke, 1985). This often commands the highest price and a wide variety of machines may be used to remove inert matter, weed seeds and the low germination fraction. For example, McLean (1989) refers to wild oat separators in which contaminated seed is passed onto a rotating velvet-like surface. The awns of the wild oat seed are caught on the surface, whereas wheat seed readily roll off it. Indented cylinders can also be used for more thorough cleaning than can be achieved with sieves and fans alone. Grain is fed into inclined rotating cylinders with indented sides. Short seeds remain in the indentations long enough during rotation for them to fall into a collecting tray. Longer seeds and material falls from the indentations sooner and travel along the cylinder to be discharged at the lower end (McLean, 1989).

Effective cleaning of grain for longer term storage is essential. Contamination can have a number of adverse effects on storage potential – including the blocking of intergranular spaces which can reduce airflow and lead to areas of high relative humidity. Chaff, straw and weed seeds can also be a source of mould growth, and encourage the breeding of mites and insects leading to 'hot spot' development.

As well as removing impurities, grain cleaning can change other quality characteristics. Removing small shrivelled grain with further sieving and blowing will increase thousand grain weight and can also increase grain specific weight (McLean, 1987). For specialist markets, many other cleaning techniques can also be employed, either as the grain goes in or comes out of storage. A potentially valuable technique is the use of specific gravity separators. These involve passing grain through aerated beds designed to classify particles of differing density. They have been particularly useful in removing impurities such as ergot from seed when more normal cleaning procedures have been inadequate (McLean, 1987). More latterly, however, it has been observed that removing low density grain can also be used to increase specific weight, which sometimes leads to increased Hagberg falling number. This is possibly partly due to specific weight and Hagberg falling number both being reduced by lodging. If this is the case, gravity separation could selectively remove grain that originated from lodged ears, and by implication, grain with a greater chance of having a higher α-amylase activity (McLean, 1987).

The economics of grain cleaning are not simple. The cost of cleaning itself depends greatly on the capital invested and throughput of grain.

McLean (1987), for example, accounted for the charge to depreciation, repairs, interest, labour, and power in estimating the cost of grain cleaning (Fig. 8.2). It was shown that the break-even point for cleaning grain was when the value of cleaned grain (y, £ t^{-1}) related to the value of uncleaned grain (x, £ t^{-1}), the cost of cleaning (c, £ t^{-1}), the reduction in weight of saleable grain (p, %), and the return obtained from screenings (s, £ t^{-1} of grain cleaned) as described in equation 8.2:

$$y = \frac{(x + c)\ 100 - s}{100 - p} \qquad [8.2]$$

Values of y that exceed this break-even value represent a positive contribution. Sometimes the value of the screenings can be as high as the uncleaned grain. For example, using gravity separation to increase Hagberg

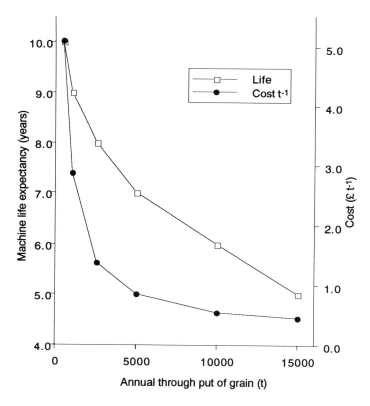

Fig. 8.2. The relationship between annual use of grain cleaning equipment, life expectancy and the cost of grain cleaning per tonne of grain. Based on a capital cost of £15,000; cleaner capacity of 20 t h^{-1}; interest at 12%; labour at £2 h^{-1}; and electricity at £1 h^{-1} (after McLean, 1987).

falling number for bread-making may leave screenings that are still suitable for livestock feed.

GRAIN STORAGE

Moisture Content and Storage Potential

Most grain is stored by achieving and maintaining sufficiently low moisture content. The dehydration of cereal seed during maturation is an adaptation for species survival, having evolved as the reproductive strategy, and this loss of moisture during the drying down process provides the basis for long-term grain storage. This feature alone of cereals has contributed to their success historically as major food crops (Wibberley, 1989). Moisture content of grain has a major influence on the suitability of crops for harvest and their storage potential. Wheat grain containing 12.6% water has been stored experimentally for up to 16 years, during which time the water content only increased by 0.7% (Pixton and Hill, 1967; Pixton *et al.*, 1975). Christensen and Sauer (1982) state that cereal seed with moisture contents of 13.5% and below can be stored for any length of time without damage from storage fungi. This figure is taken as the upper figure for traded wheat. At moisture content above 13.5%, storage risk is also determined by temperature and time.

Brooker *et al.* (1981) propose a number of critical moisture contents of wheat grain relating to harvest and storage requirements (Table 8.2). Drying grain to 12–13% mc is considered unacceptable to cereal farmers in western Europe, despite the clear benefits to storage potential. This is mainly due to drying costs and reported difficulties of processing too dry grain. A moisture content of 14% is considered adequate for bulk storage in cooler climates, and as high as 17–18% for short periods of storage (Nash, 1985). The risks of maintaining a higher moisture level in the grain are less, however, in these cool conditions than they would be in warmer

Table 8.2. Moisture content of wheat grain during harvest and for safe storage.

	% Moisture content
Maximum at harvest	38
Optimum at harvest for minimum grain loss	18–20
Usual moisture range when harvested	9–17
Requirements for safe storage	
1 year	13–14
5 or more years	11–12

regions. Longer term storage at these higher moistures carries a greater risk. Grain damaged at harvest has a lower storage potential, and should be dried more quickly and to a lower moisture content than is normal to improve its storage life (Nash, 1985).

Wheat stored at 12% moisture content is in equilibrium with interseed air at 60% relative humidity if the temperature remains above 27°C. But when this air rises to the surface of a batch of grain and cools to 15.6°C its relative humidity approaches saturation, and the wheat near the surface begins to absorb moisture from the air. If the air cools further it passes the saturation point and free water begins to be deposited on the grain. In cold weather in some grain stores water drips from roof supports back into the grain (Bailey, 1982).

Estimates of storage potential based on grain moisture content are not particularly dependable, in part because of the variability in moisture in a grain sample. Moisture content of grain at harvest can vary considerably within a field, and during a harvesting day (Bailey, 1982).

In terms of storage potential, the absolute moisture content of grain is considered less important than the relative humidity of intergranular air with which it is in equilibrium (Pomeranz, 1982). The water availability in grain is considered best indicated by stating equilibrium relative humidity (ERH percentages) or as a water activity (aw) in decimals. ERH is the relative humidity of the intergranular atmosphere in equilibrium with water in the grain, and aw values are the ratio of vapour pressure of water over grain to that over pure water at the same temperature and pressure. ERH and aw are numerically the same but expressed differently. Grain is hygroscopic, losing or gaining water from the air or material with which it is in contact towards an equilibrium state characterized by ERH or aw values.

Millers prefer wheat for flour production to be stored at less than 14% moisture content to reduce the risk of 'taints' developing. Maximum storage duration with respect to germination, and therefore crop seed quality, is also heavily dependent on moisture content but in interaction with temperature (Fig. 8.3).

Storage Less Reliant on Drying

More moist grain, around 22% moisture content, is sometimes stored for short to medium term periods at low temperatures (e.g. 5°C), but some storage fungi can develop under these conditions which can limit preservation. A more highly developed system is practised in some countries such as France, Belgium and Germany, involving the rapid artificial chilling of grain of moderate moisture (16–18°C) using powerful refrigeration units. There are apparent advantages to this in that it can take less energy

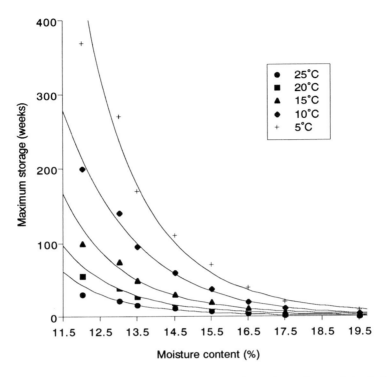

Fig. 8.3. The effect of temperature and moisture content on the maximum storage of wheat with respect to germination (plotted from data by Kreyger, 1963).

to chill grain compared with the energy required to dry it. The required ventilation flows and ducts are also much reduced. McLean (1989) points out, however, that chilled grain is less stable than dry grain and, as grain is usually marketed at 15% mc or below, some drying may still be required for grain moving off the farm. In addition field fungi which do not usually persist in grain stores are more likely to do so at low temperatures under moist conditions. Some fungi can continue to develop at very low temperatures, increasing the need for expensive rechilling. Mites can also continue to develop at temperatures as low as 3°C. In contrast to fungi and mites, damage from insects is largely curtailed at temperatures below 15°C and this reduction is one of the greatest benefits of grain chilling.

There are reports of moist grain stored cool making satisfactory flour, and giving higher loaf volumes on baking (Nash, 1985). There are strong concerns, however, that 'taints' can be transmitted from moist grain to flour and bread during processing and drier storage is much preferred. However, grain is often stored on farms in cooler climates at a higher moisture for animal feed over shorter periods, often following low volume aeration with ambient air (McLean, 1989).

Grain can also be stored at high moisture contents if oxygen levels are depleted in airtight silos. The oxygen is used up by microbial and, to a lesser extent, seed respiration and as a result the carbon dioxide and, to a small degree, humidity levels increase (Nash, 1985). In the UK this techniques has been used to store grain at 18–24% moisture content. This is most appropriate for grain destined for livestock feed. Germination percentage declines in such circumstances. Hermetically sealed grain silos are reported to have been used for storing grain above 16% moisture, which has later been used for bread-making following mixing with drier-stored grain. However, bread made from grain stored at 22% mc in a sealed store had poor texture and taints (Oxley and Hyde, 1955).

Controlled atmosphere storage may also be utilized in different ways. Gas injection systems, involving carbon dioxide or nitrogen, are utilized in some countries with warm climates to control insect activity.

MICROFLORA CONTAMINATION

Colonization of ripening grain begins soon after ear emergence, firstly by bacteria, then yeasts, followed by filamentous fungi. The greatest increase in filamentous fungi occurs during ripening as tissues senesce, in particular in cool and damp weather (Lacey, 1990). However, the microclimate around growing seeds is considered sufficiently humid most of the time to facilitate tissue invasion by some fungi. Mycelium of *Alternaria* is reportedly common under the pericarps of wheat even when grown in dry areas.

More than 150 species of yeasts and filamentous fungi have been reported as colonizers of cereal grain (Christensen and Sauer, 1982). Fungi are the most tolerant of low water availability in a grain environment. Bacteria and yeasts are less of a problem in most storage conditions as they often require free water for development (Wilkin and Rowlands, 1988). The interrelationships of microorganisms occurring on grain before harvest and during storage with environmental factors and between species have been most effectively studied by principal component analysis (Abramson, 1991), allowing large bodies of data to be analysed to discern significant relationships in the system. The principal components provide a structural picture of the grain system and, as well as indicating fungal succession and grain environment changes, can also include such significant grain quality criteria as mycotoxin development, the appearance of fungal odours, seed germination, and changes in grain moisture content in stores. Abramson (1991) undertook a pooled analysis of variables measured in spring wheat grain stored for different periods at 15

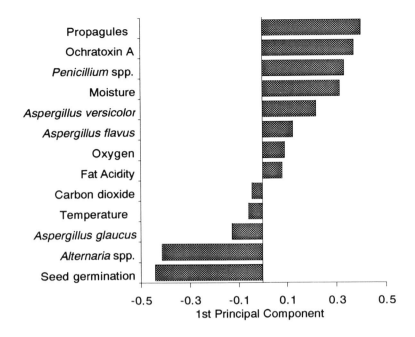

Fig. 8.4. Principal components analysis of variables in prairie spring wheat stored at 15 and 19% initial moisture content for 60 weeks in a farm granary (from Abramson, 1991).

and 19% moisture. The first principal component was associated positively with mycotoxin Ochratoxin A, *Penicillium* spp. and moisture content, and negatively associated with seed germination and *Alternaria* spp. (Fig. 8.4). This type of approach can, therefore, also help to reveal competition between species for niches on colonized grain, such as that reported by Magan and Lacey (1984), who found less abundant *Alternaria* colonization of grain with a high incidence of *Fusarium culmorum*.

Fungi affecting grain have been classified into two ecological groups as field fungi and storage fungi (Christensen and Sauer, 1982). So-called 'field' fungi can invade the seeds while the plants are still growing. However, the incidence of contamination by true field fungi (Fig. 8.5) during storage commonly declines. Some field fungi, however, carry over into storage, blurring a clear distinction between the two proposed groups. Examples of these fungi are presented in Table 8.3. The microflora occurring on grain vary greatly with climate and location.

The ecological niche and succession of microorganisms on grain in store is governed principally by moisture content, the temperature of grain (Fig. 8.6; Table 8.4), and to a lesser degree by the gaseous composition of the intergranular atmosphere. As water becomes increasingly

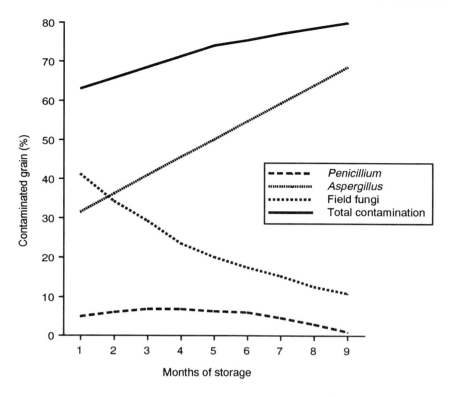

Fig. 8.5. Changes in the prevalence of fungal contaminants sampled from 20 Polish grain stores (redrawn from Chelkowski, 1991).

available, more fungi are able to grow. Grain stored under different conditions develop their own particular microflora that can characterize the storage environment. Christensen and Sauer (1982) report, for example, that *Aspergillus halophilicus* will invade wheat embryos only if the grain is held for some months at grain moisture contents of 13.8–14.3%. At higher moisture it is replaced by other fungi.

FUNGAL DAMAGE

Grain can be adversely affected by fungi in a number of ways that can substantially reduce its quality and market value. These include decreases in germination due to invasion of the embryo (Christensen and Sauer, 1982), discoloration, heating, mustiness, increase of fatty acids (rancidity), production of mycotoxins, and loss of weight (Christensen and Kaufmann, 1969; Wilkin and Rowlands, 1988). Unfortunately, all these detrimental

Table 8.3. Examples of field and storage fungi colonizing cereal grain (after Christensen and Sauer, 1982; Lacey, 1990).

Field fungi	Storage fungi
Temperate climates	
Cladosporium cladosporioides	*Aspergillus* and
Alternaria alternata	*Penicillium* species
Verticillium lecanii	which include
Epicoccum purpurascens	*A. flavus*
Fusarium species which include	*A. restrictus*
F. culmorum	*A. glaucus* group
F. graminearum	*A. fumigatus*
F. avenaceum	*P. aurantiogriseum*
F. tricinctum	*P. ruglosum*
F. poae	*P. corylophilum*
Helminthosphorium	*Fusarium*
	Rhizopus
	Mucor
Warm climates	
In addition to the above species	
Curvularia	*Nigrospora*
Cochliobolus	
Phoma sorghina	
Penicillium chrysogenum	
Penicillium funiculosum	
Penicillium oxalicum	
Aspergillus flavus	

effects can occur without the fungi necessarily being visible to the naked eye, and therefore substantial deterioration can occur before detection. Fungi, with insect activity and respiration contributions from the grain substrate, can induce temperature increases in store leading to 'hot spot' development and overheating. Deterioration of grain can be rapid under damp conditions and once initiated creates a conducive environment for further fungal colonization and spoilage.

The fungi responsible for damage in store are principally *Aspergillus* spp. and *Penicillium* spp. (Wilkin and Rowlands, 1988). A decrease in total carbohydrates in the endosperm of wheat grain and increases in reducing sugars have been reported following colonization, in particular, by *A. flavus* fungi in store and associated respiration increases. More free fatty acids, due to extracellular fungal lipase activities, have also been reported. This reaction in infected grain occurs at an early stage of fungal colonization,

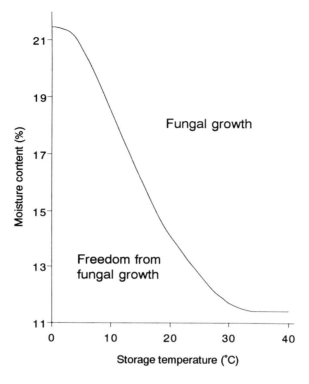

Fig. 8.6. The relationship between moisture content and storage temperature on fungal growth (redrawn from Nash, 1985).

Table 8.4. Effects of water activity and spontaneous heating on the colonization of cereal grains (Lacey, 1990).

Water activity (wa)	Maximum temperature (°C)	Predominant fungi
< 0.60	Ambient	–
0.75	Ambient	*A. restrictus*
0.90	25	*A. versicolor*
		P. aurantiogriseum
		P. citrinum
		P. funiculosum
		P. hordei
		P. janthinellum
		P. variabile

and is accelerated with increasing moisture availability. Unsaturated fatty acids are most rapidly metabolized, and their degradation contributes to the production of volatiles which contribute to grain off-odours (Wasowicz, 1991). Total protein content, as measured by nitrogen, is reported to remain relatively constant during storage (Bothast, 1978). Fungal colonization, however, can increase protein concentration because respiration reduces the amount of carbohydrate present (Wasowicz, 1991). Invasion of the embryo resulting in grain death has been attributed in particular to *A. restrictus* (Christensen and Sauer, 1982).

Mycotoxins can be produced as a result of grain contamination by *Aspergillus*, *Penicillium* and *Fusarium* species (see Chapter 2). In temperate regions toxin-producing *Penicillium* spp. seem to predominate, with toxigenic *Aspergillus* spp. being more common in semitropical climates. Aflatoxins, which are best known, can be produced, for example, by the *A. flavus* group in wheat grain at moisture contents of 18.3–18.5% and optimum field temperatures of 27–30°C, but the production of these toxins most commonly occurs in maize. Wheat is not reported to be a high-risk crop in terms of aflatoxin production in the USA. Studies in the prairies of Canada of mycotoxin formation in stored medium-protein wheat revealed it was a low risk substrate for ochratoxin A production, unless the grain had a high moisture content (> 19%) at harvest (Abramson, 1991).

A map has been drawn up in Canada indicating regions of particular susceptibility to toxin-producing fungi. This indicated that ochratoxin A was produced in particular by *A. ochraceus* and *P. viridicatum* in store. *Pencillium* toxins were mainly found in grain from the semi-arid prairies, while *Fusarium* mycotoxins occurred in the more humid eastern and maritime provinces (Abramson, 1991).

Insect and Mite Damage

Considerable damage to grain in store can be caused by insects and mites. It has been estimated in the USA that between 1951 and 1960 annual storage losses from insects averaged approximately $472,000,000 (Cotton and Wilbur, 1982). They are a hazard in several ways, e.g. grain and grain dust is eaten, and in the process insects contribute to rises in grain temperature and moisture. Through feeding activities, insects and mites contaminate grain samples with faeces, webbing, and body fragments. Odours are produced from the resultant metabolic products and their activities also encourage microbiological deterioration. Insects and mites constitute, therefore, a major sanitation and quality control problem. A number of biochemical changes (Fig. 8.7) have been reported in wheat grain infested by insects in store (Salunkhe *et al.*, 1985).

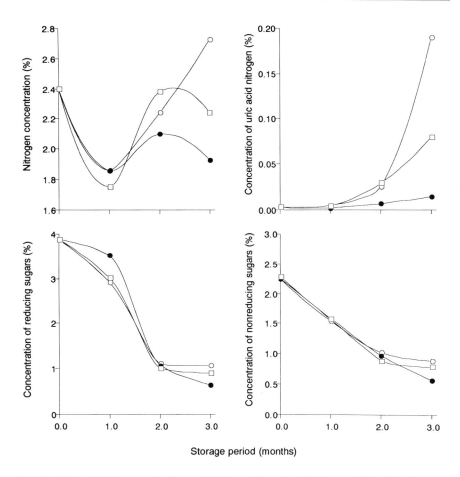

Fig. 8.7. Biochemical changes in wheat grains due to infestation with rice weevil (○, *Sitophilus oryzae*), khapra beetle (●, *Trogoderma granarium*), and red flour beetle (□, *Tribolium castaneum*).

Specific reports of insect and mite damage in grain stores illustrate the magnitude of the potential problem. For example, a loss in weight of wheat grain from direct feeding by larval rice weevil of 50% has been reported over a 9.5 day period (Cotton and Wilbur, 1982). Flour beetles are also reported to secrete pungent liquid containing ethylquinone, toluquinone and methylquinone from glands into infested grain, and large populations turn the flour pink.

It has been estimated that several hundred different species of insects are associated in one way or another with stored grain (Cotton and Wilbur, 1982). Only a few species, however, cause serious damage to wheat seed in good condition. Cotton and Wilbur (1982) group insects in stored grain as either major pests which are well adapted to grain ecosystems; a larger group of minor pests which approach the status of major pests under certain circumstances, often those of poor grain storage, or incidental pests which often accidentally contaminate grain samples; and a group of 'parasites and predators' of grain infesting pests equally regarded through their presence as pests. Field infestations of cereals by 'storage insects' are rare, but newly harvested grain may become infested from many diverse sources. Which insects are of consequence is determined partly by geographical region, prevailing climates, the farming system, and type of crop. However, the wide dispersal of insect contaminants in grain transported amongst countries has resulted in several insects appearing in stores in a wide range of locations. In the UK for example, the most common damaging insect pests of stored grain are the saw-toothed grain beetle (*Oryzaephilus surinamensis*), the rust-red grain beetle (*Crptolestes ferrugineus*), the rust-red flour beetle (*Tribolium castaneum*), the grain weevil (*Sitophilus granarius*) and the rice weevil (*Sitophilus oryzae*). Most of these are of tropical origin so, although they may be able to survive at low temperatures, a completion of their life cycle usually requires temperatures above 18°C, with optima between 27 and 35°C (Wilkin and Rowlands, 1988). Severe damage to grain can be expected from insects at temperatures above 21°C, up to 35°C, at which temperature most insects are killed. Grain moisture is important, but less so than for microbial deterioration. Greater grain moisture increases insect number and activity up to a point, before microorganisms take over. It is often assumed, therefore, that grain is safe from insect attack if temperatures can be kept below 15°C (McLean, 1989). This is relatively easy to achieve in cooler climates such as the UK but insects pose a considerably greater threat to successful storage in warmer climates. In South and Southeast Asia, Singh (1985) and Clements (1988) name the rice weevil, the rust-red flour beetle, the lesser grain borer (*Rhizopertha dominica*), the angoumois grain moth (*Sitotroga cereallea*) and the khapra beetle (*Trogoderma granarium*) as being particular problems. Clements (1988) also mentioned the ant (*Monomorium pharanois*) which was capable of attacking very dry seed.

In cool damp weather, such as that in the UK normally at harvest, mite infestation of grain is more important than insect problems. A survey of UK farms in the 1970s found a 90% infestation by mites with over 90 species found. *Acarus siro*, the flour mite and *Glycyphagus destructor* were, however, by far the most important. As with fungi, mites are small (< 0.5 mm), and difficult to see. They are only readily detected when numbers reach 50,000 kg^{-1}, by which time damage would have occurred in the form of taints and reduced germination (Wilkin and Rowlands, 1988). Mites can also induce allergenic responses in humans and have adverse effects on livestock (Wilkin and Rowlands, 1988). Mites are a potential hazard over a range of storage conditions. Different species can tolerate low temperatures (below freezing) to the higher temperatures prevailing in some grain stores (of 10–16°C). However, mites are particularly sensitive to low humidities and low grain moisture contents. Grain is said to be at risk from mite damage at moisture contents greater than 12%, and temperatures greater than 2°C (McLean, 1989).

Controlling fungal and insect storage problems

Control of biological agents that cause damage in stores is varied. From the above, it can be seen that control of temperature and moisture are the principal non-chemical methods. Sanitation is a further line of defence to include the thorough cleaning of harvesting equipment and grain stores between harvests. Additional techniques, however, are needed when the grain and store climate cannot be adjusted to the desired levels. Where insecticides are prohibitively expensive or unavailable other non-chemical methods can be adopted. In Southeast Asia farmers mix sand or wood ash with grain so as to scratch the cuticle of the insect and thereby desiccate the pest (Clements, 1988). The use of natural products such as neem (*Azadirachta indica*) oil to coat grain for protection against insects has been recommended in India. In such areas, increasing temperatures of the grain to over 40°C by sun drying will also force insects to leave (Clements, 1988).

In temperate areas the use of insecticides in grain stores has been increasing. Despite the technology devoted to drying and cooling grain it has been argued that it is impossible to guarantee pest-free grain unless pesticides are used (Rowlands *et al.*, 1989). Although total eradication is not necessary to reduce significant weight reductions, as stated above many quality outlets and intervention agencies have either very low or zero thresholds of live pests. In the UK the number of farms using insecticide in grain stores nearly doubled between the mid-1960s and mid-1980s, such that by 1984–85 76% of farms in the major cereal growing areas with more than 75 ha of cereals had adopted the practice (Rowlands *et al.*, 1989). Chemical use may be by fumigation of, or application directly on to, the storage areas and machinery or the grain itself (Harein, 1982).

Fumigation refers to the method of killing insects and mites by the application of a gas, often as an emergency measure to control severe infestation. It has been argued that because they diffuse away after use that, unlike other pesticide applications, they do not leave residues in the grain after use. Because of this they have no residual activity. The most commonly used fumigant for use in grain worldwide is phosphine (PH_3) (Chakrabarti, 1994), with the use of other potential fumigants such as carbon tetrachloride, ethylene dichloride and methyl bromide being banned or heavily restricted due to safety and/or environmental problems (Rowlands *et al.*, 1989). Phosphine is a gas and is liberated from either aluminium phosphide or magnesium phosphide on exposure to moist air. Various formulations have differing release characteristics which are detailed by Chakrabarti (1994). Efficacy depends on even distribution through the grain and requires a reasonable degree of airtightness.

Organophosphates, in particular, have been applied directly to grain in the UK, including etrimfos, fenitrothion, malathion, methacriphos and pirimiphos methyl. Usually they have a contact action and residual activity. They need to be applied at rates which result in limits set for maximum residue levels (MRLs) not being exceeded (Rowlands *et al.*, 1989), i.e. commonly 4–10 g t^{-1}. They are usually applied as a spray or dust as the grain moves on a conveyor. A comparison of different application methods is given by Miller and Bins (1994). An increasing problem for such insecticide treatments is the development of greater resistance in target pests (Davies, 1992). Because of the long distance transport of grain there is the possibility of resistant strains being distributed throughout the world relatively rapidly. Rowlands *et al.* (1989) report that 31% of 74 strains of saw-toothed grain beetle from UK farms where there was a high infestation were resistant to malathion. There is also an increasing consciousness of chemical residues on grain. Pre-storage drying of cereal grain, together with systems for the effective control of temperature and humidity in store, is the best approach to limiting microbial deterioration, and insect/mite infestations. A number of studies have attempted to develop more integrated approaches to pest control with greater emphasis on grain cooling at specific times to reduce costs, and the possible restriction of pesticide applications to the surface layers of grain in store (e.g. Armitage *et al.*, 1992). Other approaches include the development of pheromones to interfere with the development of specific species with little or no mammalian toxicity (Rowlands *et al.*, 1989). The enclosed nature of grain stores also renders them possibly more suitable to biological control measures such as the introduction of pest predators.

9

Wheat Vegetation as Forage

Despite the pre-eminence of wheat as a grain crop, the significant biomass generated during crop growth makes available substantial amounts of vegetation, either as straw or green foliage. This is often seen as of low value in prime wheat growing areas, but its efficient utilization may also be important to meet the future needs of an increasing world population and for resource conservation (Wedin and Hoveland, 1987). Wheat vegetation will rarely be the major source of nutrients for livestock, however, as most systems are heavily dependent on pasture species of grass and legume, often supplemented by high energy producing crops such as maize, *Brassica* spp. and/or beet (*Beta vulgaris*) and high protein crops such as clover (*Trifolium* spp.), lucerne (*Medicago sativa*) and pulses.

Wheat vegetation can be used in dual purpose systems where the wheat is grazed or cut in spring and then left for a grain harvest in summer. Alternatively, the wheat may be harvested as a whole crop in summer to zero graze or conserve as silage. After a grain harvest the straw can be fed to livestock and/or the stubble may be grazed.

FORAGE QUALITY CRITERIA

Forage quality encompasses the ability of wheat to provide livestock, particularly ruminants, with the nutrients required to maintain body weight. In the case of growing animals, nutrients are also required to form muscle, adipose tissue and bone, and by the lactating animal to synthesize fat protein and lactose. There will be specific requirements for amino-acids, glucose, fatty acids, minerals and vitamins which depend on the stage of development, size, type and level of production of the animal (Gill *et al.*, 1989). Given that wheat vegetation will usually be only a minor part of an animal's intake, certain components will derive from other sources.

The feeding value of wheat, therefore, is usually expressed in terms of the amount of energy and, to a lesser extent, the amount of protein, made available to a particular class of animal. The provision of energy will depend on factors such as palatability and availability, which will govern the intake of the vegetation, and also the digestibility of the forage. As the wheat plant matures, the digestibility generally declines. This is because the proportion of cell wall constituents such as hemicellulose, cellulose and, in particular, lignin increases. This increase is particularly associated with the elongation and strengthening of the stem required to support the developing ear. The highest digestibility, therefore, occurs before the start of stem elongation but at this time the total forage yield is low (Fig. 9.1). Apparent digestibilities can be calculated by determining the amount eaten compared with the amount expelled in faeces. For example, the dry matter (DM) digestibility coefficient can be expressed as in equation 9.1:

$$\text{DM digestibility (\%)} = \frac{\text{DM consumed} - \text{DM in faeces} \times 100}{\text{DM consumed}} \qquad [9.1]$$

Similar equations can be substituted to calculate the digestibility of other feed components such as crude protein (DCP) and organic matter digestibility (OMD). Because energy availability is more closely associated with organic matter than DM, a much more commonly used coefficient is the amount of digestible organic matter (DOM) contained within the dry matter, i.e. the *D*-value as calculated in equation 9.2.

$$D\text{-value} = \frac{\text{OM consumed} - \text{OM in faeces} \times 100}{\text{DM consumed}} \qquad [9.2]$$

The decline in *D*-value as the wheat crop matures is clearly shown in Fig. 9.1. This provides a measure that relates to the amount of energy digested, and can also be related to the metabolizable energy (ME) provided by a particular wheat forage.

Assessing *D*- and ME values directly is laborious and time consuming. Several procedures have, therefore, been developed which attempt to assess digestibility with laboratory techniques. As cellulose and lignin fibres are associated with poor digestibility, many procedures are aimed at quantifying the amount and indigestibility of fibre present. Common procedures include the neutral-detergent fibre (NDF) analysis. This quantifies the residue left after boiling with neutral solutions of sodium lauryl sulphate and ethylenediaminetetra-acetic acid (EDTA). The residue comprises principally lignin, cellulose and hemicellulose and is, therefore, associated with the amount of plant cell wall material present. Acid detergent fibre (ADF) is the residue left after refluxing with 0.5 M sulphuric acid and acetyltrimethylammonium bromide which relates to the crude lignin and cellulose

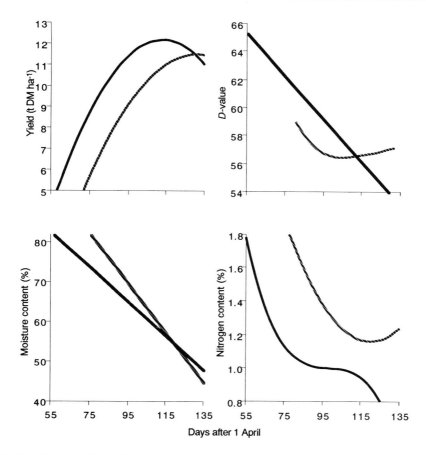

Fig. 9.1. Yield, digestibility (*D*-value), moisture content and nitrogen concentration of vegetation cut 30 mm above soil level during crop maturation of winter wheat (solid line, anthesis 29 June, 90 days after 1 April) and spring wheat (hatched line, anthesis 8 July, 99 days after 1 April) (redrawn after Corrall *et al.*, 1977).

content together with silica. In the UK the ADF procedure is modified by using a longer period of boiling and a higher concentration of acid. The resultant residue is known as the modified acid-detergent fibre (MADF). ADF and MADF correlate well with forage digestibility.

GRAZING WHEAT FOR DUAL PURPOSE PRODUCTION

Wheat is grazed by livestock in many parts of the world including southern USA, Australia, Argentina, Brazil, South Africa and parts of Europe (Aase and Siddoway, 1975; Dunphy *et al.*, 1982; Sharrow and Motazedian,

1987). In the Southern Great Plains, for example, it has been estimated that 4 million cattle graze wheat which will ultimately be harvested for grain (Horn *et al.*, 1977). It is, however, seldom practised now in the UK, although it has been common in the past towards the west on mixed livestock farms, throughout Wales, the south west and West Midlands (Holliday, 1956).

Vigorous spring growth of wheat can provide early grazing (Dann, 1968) at a time of the year when there is often a shortage and may satisfy demand for emergency grazing (Holliday, 1956). Not only is the wheat of value to livestock, but there can be potential benefits to the crop itself. Grazing will, for example, reduce lodging (Dann, 1968) by reducing the length of the lower internodes (Justus and Thurman, 1955) and it has occasionally been used to check excessive spring growth. When lodging is controlled, grazing can increase wheat grain yields as long as there is good soil fertility and a favourable climate (Dann, 1968). It may also be that by removing frost damaged tissue, defoliation can reduce the infection loci of diseases, although there is no evidence of this resulting in yield increases (Dann, 1968). Others have suggested that yield might also be improved by increasing grain numbers (Sharrow and Motazedian, 1987). Grazing can also help to suppress weeds. For example, N.D. Cosser (Royal Agricultural College, Cirencester, 1995, unpublished results) found sheep grazing to reduce seed return by red dead nettle (*Lamium purpureum*). In contrast, however, the competitiveness of the wheat can be reduced such that blackgrass heads increase (Cosser *et al.*, 1995).

More commonly, defoliation and spring grazing have resulted in grain yield reductions (Holliday, 1956; Sharrow, 1990). These reports are mostly associated with lower ear populations and reduced numbers of grain per ear (Sharrow and Motazedian, 1987; Summers, 1990), although thousand grain weight can also sometimes be reduced (Dann, 1968; Summers, 1990). Grazing after the elevation of the apical meristem at the start of stem extension is particularly prone to reducing grain yield with the timing of grazing being much more important than grazing severity. Muldoon (1985) presents a predictive model suggesting that yield reductions are likely if grazing occurs within 150 days of harvest. Defoliation as late as 120 days before harvest can, however, increase yield provided apical meristems are not elevated (Sharrow and Motazedian, 1987). The effects of spring defoliation on wheat grain yield also interact strongly with variety. Gooding *et al.* (1993a) report that cutting wheat to 8 cm twice in April had no significant effect on the yield of Maris Widgeon, an older variety lacking dwarfing genes. The yields of several newer and shorter varieties, however, were significantly reduced by the same treatment. Yield following cutting related to the amount of light intercepted by the crop during the season (Fig. 9.2). Varieties which are better able to tiller and produce a canopy after cutting may, therefore, be better adapted to

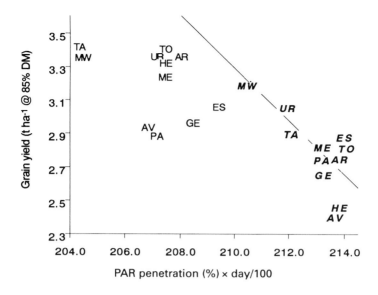

Fig. 9.2. Relationship between penetration of photosynthetically active radiation (PAR) through the wheat canopy and grain yield of non-mown (not significant) and mown ($b = -0.159$, $r^2 = 0.58$) (shown in italics) wheat varieties. SED (72 d.f.) for comparing varieties within mowing treatments are 0.26 t ha^{-1} for yield and 1.34% × day/100 for PAR. AV = Avalon, AR = Aristocrat, ES = Estica, GE = Genesis, HE = Hereward, MW = Maris Widgeon, ME = Mercia, PA = Pastiche, TA = Talon, TO = Torfrida, UR = Urban.

spring grazing. These varieties are also often those which produce greater quantities of forage for grazing by spring.

With careful attention to the wheat growth stage and variety, therefore, useful amounts of forage can be grazed without significant reductions in grain yield. The amount and quality of forage available, however, depends greatly on climate, sowing date, variety and time of grazing. Figures 9.3 and 9.4 and Table 9.1 summarize results for two winter wheat varieties contrasting in height, sown on two dates and grazed with 42 × 50 kg sheep ha^{-1} between the 16th and 19th of March on soil of the Sherborne series in Gloucestershire, UK. Measurements from the ungrazed plots showed that sowing in September dramatically increased the above ground dry matter available in March, particularly of Maris Widgeon, but proportional differences declined by the end of April (Fig. 9.3). The nitrogen content of the forage was also influenced by variety and sowing date but differences varied greatly with sampling date (Fig. 9.4). When the quantity and quality of the forage was determined by comparing plots which had not been grazed with those that had been, immediately after removal of sheep (Table 9.1), it was evident that the

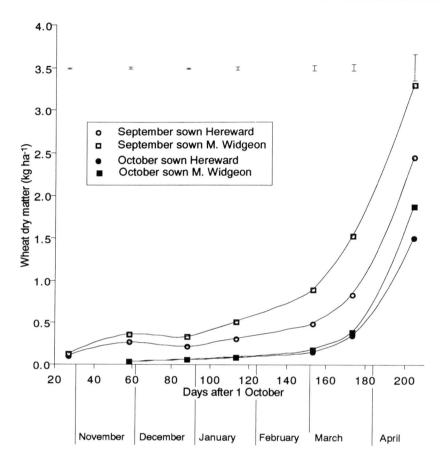

Fig. 9.3. The effect of sowing date on the above ground dry matter produced during winter and spring in the UK by contrasting varieties of winter wheat (from Cosser *et al.*, 1994a). Error bars represent standard errors of the difference between varieties sown on the same date.

October sown plots had provided negligible amounts of forage. In contrast, plots sown a month earlier were highly productive with Maris Widgeon providing nearly 1 t DM of forage with high metabolizable energy and nitrogen content. This was available at a time in the year when grazing from grass pastures is often limited. In this experiment grazing did not significantly affect yield or quality of the wheat grain, although dry matter harvest index was improved. Using a typical figure of 0.33p MJ^{-1}, the harvesting of forage with spring grazing increased the total return from the Maris Widgeon by £35 ha^{-1}. To realize this extra return, however, requires arable fields to be adequately fenced and the availability of appropriate stock early in the season. It is also easier on lighter and/or

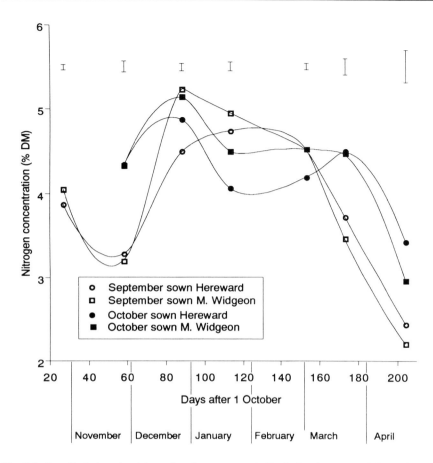

Fig. 9.4. The effect of sowing date on the nitrogen concentration in the above ground dry matter produced during winter and spring in the UK by contrasting varieties of winter wheat (from Cosser *et al.*, 1994a). Error bars represent standard errors of the difference between varieties sown on the same date.

well-drained soils due to the reduced risks of excessive poaching at a time of the year when soil moisture contents can be high.

WHOLE CROP WHEAT PRODUCTION

Whole crop wheat production differs from the dual purpose systems described above in having no separate forage and grain harvest. It usually involves the harvest of entire above ground wheat after ear emergence to produce feed for livestock, typically after a period of conservation. This

Table 9.1. Yield and quality of forage grazed by sheep between the 16th and 19th March from contrasting winter wheat varieties sown on two dates in the UK (from Cosser *et al.*, 1994b).

Sowing date	Variety	Yield and quality of forage removed by sheep				
		DM (t ha^{-1})	ME (MJ kg^{-1})	Mg (%)	P (%)	N (%)
17 September	Hereward	0.41	11.5	0.088	0.27	3.7
17 September	M. Widgeon	0.89	11.4	0.055	0.31	3.8
19 October	Hereward	0.15	12.5	0.094	0.33	5.2
19 October	M. Widgeon	0.19	12.4	0.066	0.25	4.9
SED (within sow dates)		0.111	0.65	0.0222	0.088	0.45

type of wheat forage shares similarities with forage maize which is more productive in many environments. However, there are geographical and climatic limitations to the spread of maize. Wheat can be produced in a different range of environments and may provide an alternative in situations unsuitable for maize (Hill and Leaver, 1990; Leaver and Hill, 1990; Newman, 1992; Bastiman and Pullar, 1993). Compared with other small grained cereals, wheat has produced lower total dry matter yields than barley in semi-arid environments (Hadjichristodoulou, 1976) and triticale in the UK (Tetlow, 1992). There is, however, a large degree of overlap in the performance of varieties of different species (Hadjichristodoulou, 1976) and yields of different genotypes interact with harvest date and environmental conditions (Hadjichristodoulou, 1976; Tetlow, 1992; Weller *et al.*, 1995). Reviews or summaries of research programmes include Anon. (1988) where the suitability of whole crop cereals and entire legume plants is discussed in relation to use in former Czechoslovakia. Grimm (1987) compares the relative merits of whole crop cereals and forage maize in Germany. Sansoucy (1981) reviews whole crop cereals for cattle feeding, and Tetlow (1992) provides an overview of whole crop cereal research conducted at the Institute for Grassland and Animal Production, UK from 1980 to 1990.

The increased use of whole crop wheat has been facilitated and encouraged by a number of factors. The first of these is the development of improved ensiling techniques. Traditional methods of conservation by ensilage have relied on bacterial fermentation of sugars to produce acid, which lowers the pH. To be most effective, this method requires the wheat to be harvested at 30–35% dry matter. Unfortunately, only 80–90% of maximum crop yield has typically been attained by this stage (Tetlow, 1992). Other methods of ensilage have, therefore, been adopted which use additives such as urea, aqueous ammonia, anhydrous ammonia or sodium hydroxide (Bolsen *et al.*, 1983; Davis and Greenhalgh, 1983; Kulik *et al.*, 1983; Chung, 1986; Tetlow and Mason, 1987; Tetlow *et al.*, 1987; Pettigrew *et al.*, 1988; Almeida *et al.*, 1989; Tetlow, 1990). These additives facilitate preservation at relatively high pH. The urea, for example, is hydrolysed to ammonia and reduced to ammonium. These so-called alkaline preservation techniques overcome some of the variability, aerobic instability at the silage face, and low digestibility otherwise recognized for whole crop cereal silage (Woolford *et al.*, 1982; Kristenson, 1991; Tetlow, 1992; Bastiman and Pullar, 1993). Such treatments also allow harvesting at more mature, higher yielding growth stages with higher dry matter contents (45–60%) than would be the case with more traditional ensilage techniques which rely on acid production (Hill and Leaver, 1990; Table 9.2). These additives can also increase digestibility as indicated by fibre analyses (Bolsen *et al.*, 1983; Chung, 1986; Tetlow and Mason, 1987; Pettigrew *et al.*, 1988), ration intake (Tetlow *et al.*, 1987; Deschard *et al.*, 1987; 1988; Anon., 1988; Hill and Leaver, 1990, 1991b; Phipps *et al.*, 1990; Tetlow, 1990) and, in the cases of urea and ammonia, increase nitrogen content (Almeida *et al.*, 1989; Chung, 1986; Deschard *et al.*, 1987, 1988; Hill and Leaver, 1991a; Pettigrew *et al.*, 1988). In the UK the use of urea has been particularly favoured (Table 9.2) because it is relatively safe and easy to use compared with ammonia and sodium hydroxide, it enhances aerobic stability (Tetlow, 1990) and, despite some palatability problems, does not result in the high urine output associated with the use of large quantities of sodium hydroxide.

A major advantage for having the ability to conserve wheat effectively is that it provides an alternative stock feed if grass and other forage

Table 9.2. Average analyses of whole crop wheat and wheat straw products for ruminants in UK surveys (compiled from Ministry of Agriculture Fisheries and Food, 1992 and Weller, 1992).

	Whole crop winter wheat		Winter wheat straw		
	Fermented	Urea-preserved	Untreated	Ammonia treated	Sodium hydroxide treated
pH	3.9	7.6	–	–	–
Dry matter (%)	36.5	56.1	87.2	87.3	84.2
Crude protein (% DM)	9.5	19.4	3.9	6.7	3.6
MADF (% DM)	29.7	28.9	–	–	–
ADF (% DM)	–	–	49.9	54.5	49.4
NDF (% DM)	–	45.6	80.6	77.9	68.9
D-value	66.6	67.4	42.9	50.8	57.0
ME (MJ kg^{-1} DM)	11.2	10.6	6.0	7.5	8.6

production is insufficient (Weller, 1991; Walker, 1992), thus reducing the risk of having to purchase more expensive foodstuffs. This is particularly relevant in years and areas where grass yields are limited by rainfall, when the cereals may perform relatively better (Hadjichristodoulou, 1976; Corrall *et al.*, 1977; Deschard, 1983). Alternatively, if other forages are in good supply the option to harvest the cereal grains conventionally can still be taken, allowing flexibility in management (Kristenson, 1991). This could have increasing importance as rainfall becomes less predictable in many areas where it has previously underpinned high grass yields, possibly as a result of climate change.

Reduced grain prices relative to the cost of grass silage makes the harvesting of cereals for forage more attractive (Doyle *et al.*, 1988b; Castejon and Leaver, 1991). Preliminary financial modelling has suggested that profits can be increased on UK dairy farms by replacing or integrating grass with whole crop cereal (Doyle *et al.*, 1990). Producers report contrasting results on whether whole crop wheat is cheaper to produce than grass (Weller, 1991; Kristenson, 1991; Phipps *et al.*, 1991; Walker, 1992; Weller, 1992), but initial estimates suggest the cost of producing utilizable metabolizable energy with whole crop wheat is at least comparable with that of grass (Harvey, 1990). Whole crop wheat may also reduce alternative costs of grain drying and straw disposal (Doyle *et al.*, 1986, 1988b; Grimm, 1987).

Removing a cereal crop from land prior to grain maturity may result in a number of rotational and other agronomic advantages. In the UK, the timely sowing of rape or grassland leys may be easier following whole crop wheat which will often be 3–6 weeks earlier than a conventional grain harvest (Ingram, 1990; Weller, 1991; Newman, 1992), whilst in some situations whole crop cereal may allow two crops to be grown in one year. This has been possible mostly in warmer European or Mediterranean climates (Ashbell *et al.*, 1984, 1985; Anon., 1988; Brosh *et al.*, 1989, 1990; Baxter *et al.*, 1980; Grimm, 1987; Holzer *et al.*, 1985) but has also been achieved in the UK. Removing the cereal from the land earlier can also increase the chances of successful establishment and later productivity of undersown pasture species (Schwarz *et al.*, 1990).

Of particular environmental concern, the harvesting of whole crop wheat at relatively high dry matter content, in common with maize, greatly reduces the risk of effluent pollution compared with grass silages which are commonly harvested at much lower dry matter contents (Edwards *et al.*, 1968; Sansoucy, 1981; Weller, 1991). In other respects, however, some whole crop wheat systems may carry considerable environmental cost. Gooding and Alliston (1993) attempted to evaluate several environmental aspects of whole crop wheat in comparison with conventional wheat (i.e. harvesting grain and straw separately at full maturity). The results from two whole crop wheat scenarios are shown in Table 9.3.

Table 9.3. Summary of expected environmental benefits (+) and disadvantages (−) of whole crop wheat with efficient harvesting (A) and typical harvesting (B) relative to conventional grain and grass silage production. (=, +/=, −/=, no expected difference, marginal or speculative benefit and marginal or speculative disadvantage, respectively) (from Gooding and Alliston, 1993).

Environmental aspect	Benefit relative to conventional wheat		Benefit relative to grass silage production	
	A	B	A	B
Nitrogen				
Total inorganic N requirement	−	−	−	−
Nitrate leaching	+/=	+/=	−	−
N-volatilization	−	−	−	−
N efficiency (DCP/N appl.)	+	−	+	−/=
Energy				
Total support energy	−	−	−	−
Energy efficiency (UME/support)	=	−	+/=	−/=
Effluent and pesticide pollution				
Silage effluent	=	=	+	+
Pesticide pollution	=	=	−	−

The systems were based on wheat harvested at 50% DM and preserved with the addition of urea at 4% DM (Tetlow, 1992). This type of production is gaining favour with UK farmers and was estimated to have been adopted by 200 UK farms in 1991 (Newman, 1992). It utilizes, in particular, the recent innovations in whole crop conservation (Weller, 1992). The two wheat scenarios presented in Table 9.3 show estimates for an 'average' whole wheat crop under good, but not atypical, farm conditions (A) and what might occur in poorer conditions as indicated by poor experimental and on farm results (B). The only clear benefit that wheat whole crop conserved with urea has is with respect to effluent production, while in comparison with grass cut for silage, nitrate leaching, N-volatilization, support energy requirements and pesticide pollution might all be expected to be worse. This is partly due to the high amounts of urea nitrogen needed to conserve the whole crop wheat. An average UK wheat crop, in terms of nitrogen used to fertilize the crop and crop yield, requires 190 kg N ha^{-1} to grow the wheat and a further 210 kg N ha^{-1} to conserve it. The efficiency of nitrogen applications in terms of DCP kg^{-1} N might be better for whole crop wheat, but this is particularly dependent on efficient harvest with little grain loss.

It is evident that the net environmental effect of whole crop wheat is dependent on the urea applications required to preserve the crop successfully. There is, therefore, a need to determine whether 4% of DM N as urea is necessary. Some recent recommendations have suggested that 2.5–3.5% would be adequate, but the success of these reduced rates depends on the accurate and even distribution of urea which, given the large quantities involved, presents difficulties with commonly used machinery (Newman *et al.*, 1992). In any case, it is predicted that rates of less than 1% DM would be needed for comparability with grass and conventional cereals grown with average N application rates. It is concluded, therefore, that for whole crop wheat to have a net benefit for the environment, methods not involving preservation with urea are most suitable (Bastiman and Pullar, 1993; Gooding and Alliston, 1993).

Most of the agronomy of whole crop wheat is similar to that of wheat grown principally for grain production, not least because the producer would often like to harvest the grain if other sources of forage have been sufficient. From a quality point of view also, much of the energy and protein provided by a whole crop wheat feed derives from the grain. Tetlow (1992) argues that the application of growth regulators to shorten the stems is important due to the poor nutritional quality of the straw relative to the grain and, presumably, the same argument could be made for using semi-dwarf varieties. Weller *et al.* (1995) demonstrate the importance of selecting varieties with the highest grain harvest index when high quality forage is required by showing that over 70% of the variation in MADF could be explained by a negative relationship with grain content. With some systems there may be a case for increasing nitrogen application rates by up to 10% compared with grain production because lodging is less of a risk for the earlier harvest, while late fungicide sprays, particularly after ear emergence, are less likely to give a worthwhile response (Tetlow, 1992).

The point at which the crop is cut can have considerable impact upon yield and quality. Tetlow (1992) suggests that for alkali-treated crops (e.g. by using urea or sodium hydroxide) the optimum harvest time is between 45 and 55% DM content. Crops containing more moisture than this may result in poor quality silage due to clostridial fermentation and the production of butyric acid. Drier crops may not have enough moisture to hydrolyse the urea so that excessive amounts of urea enter the rumen and cause ammonia toxicity (Tetlow, 1992). Judging when the crop has reached optimum moisture content can, however, be difficult. Tetlow (1992) suggests that in the UK, the crop is at about 50% dry matter when the grain is at the medium to hard dough stage; the crop is neither uniformly green or yellow; and there is about 2.5 cm of green either side of the second node from the base of the plant. In a survey of UK wheat crops, Harvey (1992) concluded that crop colour was a good indicator of dry matter content observing that 50–60% DM could be inferred by a predominantly yellow field with traces of green. Unfortunately different varieties reach their maximum yields at different moisture contents. Weller (1992), for example, found that cv. Norman reached maximum yield at a dry matter content of 35.3%, while maximum yield of Beaver was not attained until dry matter content had reached 73.3%. It may, therefore, be possible to identify and select varieties which reach their maximum yields at dry matter contents most suitable for conservation.

The quality and yield of whole crop wheat is also dependent on efficient harvesting. Given that many farmers will also have equipment for harvesting and ensiling grass silage, and that wheat whole crop is often only harvested in cases of shortfall, investment in specialized machinery is not usually economic. Redman and Knight (1992) suggest two main alternatives, that of harvesting directly from the standing crop with a forage harvester with a wide header, or harvesting from a pre-mown swath. The choice will depend on equipment available but careful attention should be paid to reducing grain losses which can typically reach 10% and thereby reduce yield and, more dramatically, quality. Common heights of cut range between 7.5 and 15 cm above ground level (Weller, 1992). Weller *et al.* (1995), however, found significant improvements in quality as the cutting height was raised to 40 cm as the proportion of grain to straw increased. Although this decreased yield, they considered that increasing the height of cut would be beneficial on dairy farms where the available forages were only of modest quality and a high quality forage was desired. The material should be chopped to 50–60 mm, which Redman and Knight (1992) consider to be within the capability of standard chopping mechanisms of forage harvesters. The material is usually less dense than grass and, therefore, requires greater transport capacity. The greatest difficulty is probably in the application of large quantities of urea as these greatly exceed the capacity of many additive applicators used in grass silage making.

STRAW UTILIZATION

Wheat straw differs from the forage products mentioned so far in that virtually all the carbohydrate present is in the form of structural polysaccharides, rather than as water-soluble carbohydrates and starch. By the time of the grain harvest, the cells in the straw are dead, devoid of the cell contents which would otherwise enhance nutritional value, and are extensively lignified. Lignification, in particular, reduces the breakdown of cell walls by gut microorganisms in ruminants, and hence digestibility is poor. Straw is, therefore, often considered as being of marginal value having a low *D*-value and providing only modest amounts of metabolizable energy (Table 9.2), at best only being capable of maintenance levels of nutrition. Nonetheless, the large quantities that are produced warrant consideration of how its utilization can be improved. Dry matter yields of straw, allowing for losses of non-grain material in the field, are probably about 60% of the grain yield in modern varieties, and even larger for older varieties which have a lower grain harvest index (Sundstol, 1988). This amounts to a total world production of about 350 Mt year^{-1}.

Cereal straw has been fed in small quantities, as physical roughage mainly, for very many years. Inclusion of straw in feed on mixed farms provides some scope for cheaper diets, and is particularly useful as a buffer feed in a poor forage year. Compared with other small grained cereals, wheat is generally inferior, often having high fibre contents and D-values ranging between 30 and 53 (Table 9.2). By comparison, typical ranges for barley and oat straw are 33–56, and 38–59 respectively. Wheat varieties do, however, differ with respect to straw quality (Kernan *et al.*, 1984; Capper, 1988). It has been suggested that feeding value may be inversely related to straw height (Sundstol, 1988). In contrast, however, semi-dwarf varieties have sometimes given greater values for *in vitro* measures of digestibilities than older, taller varieties (White *et al.*, 1981; Schulthess *et al.*, 1995). This does not appear to be related to total fibre contents but rather to the greater fibre digestibility. Shorter varieties can also have a greater leaf to stem ratio and leaves, even after complete senescence, are easier to digest than straw (Schulthess *et al.*, 1995). Environmental effects can also be significant, with lower temperatures sometimes being associated with reduced lignification (Schulthess *et al.*, 1995). The height of the stubble left in the field will also be of consequence as lignification declines up the stem.

Larger effects on straw digestibility can be obtained by treatment after harvest. One of the earliest techniques was to treat the straw with sodium hydroxide. A modern development of this procedure is to soak the straw in 1.5% NaOH solution for 0.5 to 1 h and then store it without rinsing for 4–6 days (Sundstol, 1988). Treating wheat straw in this way solubilizes the lignin and increases degradability depending on the amount of NaOH applied but, typically, wheat straw D-values and ME are increased by 30–40% (Table 9.2). Wheat straws are also commonly upgraded by application of either anhydrous or aqueous ammonia. Ammonia treatment increases digestibility, but not to the same extent as sodium hydroxide (Table 9.2). However, ammonia increases the non-protein N content (represented by an increased CP in Table 9.2) which may be used by the rumen microorganisms for protein synthesis. Additionally, ammonia has a fungicidal effect which reduces the need for the straw to be dry when stacked. Chemical treatment of straw also kills large numbers of weed seeds, thereby reducing their distribution around the farm.

10

Ethanol, Starch and Gluten Production

ETHANOL PRODUCTION

Cereal grains and potatoes have been extensively used for many years for the production of pure alcohol and bioethanol for the drinks industry. Wheat, with maize and barley, has also been exploited for bioethanol production for industrial purposes and fuel use in the USA, Finland, Austria and Sweden (Batchelor *et al.*, 1993). In Austria and Sweden ethanol fermentation from grain has been considered an appropriate method of utilizing surplus wheat, with the technology being well developed, and dictated by economics. The process is either undertaken in a dedicated industrial plant for ethanol, with the by-product of distillers' dried grains for animal feed, or in an integrated starch plant where bioethanol is one of a number of subsequent derivatives. In the USA, a number of starch plants combine bioethanol production with development of high fructose sugars for soft drinks.

Some 450,000 t of wheat per year is utilized in the UK for grain alcohol production, as the basis for vodka, gin and whisky. Most distillers are in Scotland and much of the wheat (about 370,000 t) is reported to be home-grown (Taylor *et al.*, 1993). Wheat replaced imported maize in 1981 with a change of European Community regulation, and lower wheat costs.

Distillers in Scotland initially selected wheat as a raw material based on the same quality specifications as animal feed wheat (Taylor *et al.*, 1993). Wheat varieties grown for human food and bread-making outlets are not considered best for ethanol, having been bred for a higher protein content and hard endosperm (Batchelor *et al.*, 1993). It was found that hard grained wheats gave a lower spirit yield than soft grained varieties, and the distilling industry in Scotland introduced a hardness value of less than 10 on a calibrated near-infrared reflectometer as a specification. Although laboratory studies did not confirm this relationship (Taylor and Roscrow, 1990), commercial processing problems due to grain hardness have resulted in the adoption of a hardness measure. Basically, it proved difficult commercially to breakdown starch from hard endosperm varieties. Rifkin (1989) confirmed that soft endosperm textured varieties were preferable, with some hard wheats yielding reasonable amounts of ethanol when the total nitrogen content was low.

A relationship was found between thousand grain weight and alcohol per tonne of grain, presumably due to a higher proportion of starch in the larger grains. Specific weight measures were useful in indicating well-filled grain generally, but there was no clear relationship between high specific weights and high spirit yield. Other workers specify a minimum specific weight as being desirable (Entwistle and Kinloch, 1989). Grain nitrogen percentage has been found to be important in determining alcohol yield. Variety trials in Scotland indicated that there was a 5.8 l t^{-1} decrease in alcohol yield for 0.1% increase in grain nitrogen (Taylor and Roscrow, 1990).

Ethanol has been used as a petrol additive and a petrol substitute. Batchelor *et al.* (1993) concluded that bioethanol has greater possibilities as an oxygenate additive in fuel, than as a complete fuel *per se*. Increasing interest in renewable energy sources and a change in the economics of alternative fuels could improve the prospects for using wheat in this way. In the meantime, the use of bioethanol in the production of cosmetics and toiletries is considered a better prospect than fuel use (Batchelor *et al.*, 1993). Co-production of ethanol with starch, and other products, is thought to offer the best scope for greater industrial utilization of wheat. Currently, bioethanol alone from wheat does not have an overall cost advantage over other sources of potential fuel ethanol. However, if political policies for encouraging industrial cropping on set-aside land are supported, the economics may change, providing new opportunities for wheat. Public perception of countryside and managed land use through agricultural production of industrial cropping, rather than set-aside options, could influence political opinion. Public impressions of renewable natural products, such as bioethanol from wheat rather than synthetic alcohols, could also make crop sourcing more attractive in the market place.

STARCH AND GLUTEN EXTRACTION

Globally, wheat is used industrially to produce 25 million tonnes (Mt) of starch. In the 12 EU countries, 2.6 Mt is utilized industrially mainly for starch and gluten production (Anon., 1995c). Starch is a mixture of two polymers, amylose and amylopectine, which derive from the linkage of large numbers of glucose molecules. With amylose the glucose linkage is end to end, and with amylopectin there are branch linkages. The macromolecular structure is reported to give starch unique properties in relation to water and temperature such as solubility, viscosity, gelling and adhesion. It is also easily hydrolysed to simpler molecules such as glucose, from which a wide variety of derivatives can be produced. This simple process to produce a multiplicity of products from plant-derived starch has been termed 'green chemistry' (Leygue, 1993). In the UK about 75% of the starch manufactured is hydrolysed to give a range of glucose or maltose rich syrups (Table 10.1) used in the food industry for confectionery and by the brewing industry in fermentation (Jones, 1987a). Surplus food crop production in the EU and the encouragement of industrial crops, rather than increasing set-aside of land, will increase the output of grain for industrial processing. The present uses of starch in the EU are summarized in Fig. 10.1. It is estimated that production of wheat starch could grow in the EU to 5.6 Mt by 1998 (Anon., 1995c), much of which will be exported. However, policy decisions in the EU will continue to affect production. For example, a recent move to reduce potato surplus in The Netherlands has resulted in a strong incentive to increase wheat starch production, and EU tariffs on imports have boosted the domestic starch price. The fact that wheat production in the EU is generally more efficient than maize also encourages wheat grain extraction, and some of the wet maize processing facilities are being re-equipped with new plants for wheat.

The interchangeability of wheat, maize and potatoes as raw materials for sourcing starch, all of which have approximately 75% starch on a dry weight basis, is influenced by their availability, cost, market pressures and policy. The starch extraction industry, however, cannot readily exploit different source materials due to the different processes required, although many wheat starch producers are also involved with maize extraction. Because of the high cost of entry and technology required, the sector is dominated by multinational groups, with small company involvements

Table 10.1. Production and consumption of starch and glucose in the UK (from Jones, 1987a).

	Tonnes per annum	
	Production	Consumption
Unmodified starch	70,000	180,000
Modified starches	70,000	90,000
Solid dextrose/glucose	30,000	50,000
Glucose syrups	400,000	440,000
Isoglucose syrups	30,000	50,000

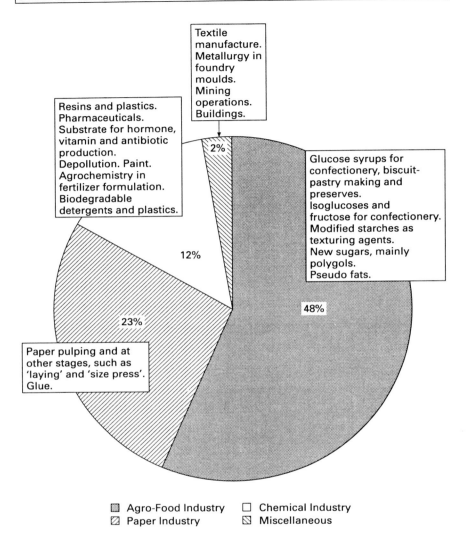

Fig. 10.1. Products derived from starch in the EU in 1990–91 (after Leygue, 1993).

having to be scaled up to a cost-effective size to remain competitive. Currently in the EU 4.25 Mt of starch is produced from maize, 2.3 Mt from wheat, and 1.3 Mt from potatoes (Bodjuniak and Sturgess, 1995). In the UK roughly 75% of starch is extracted from mainly imported maize (Fig. 10.2). Worldwide, corn starch dominates the industry.

Vital wheat gluten (VWG) is a key co-product of the starch extraction process. On a weight basis it is more valuable than starch, and part of

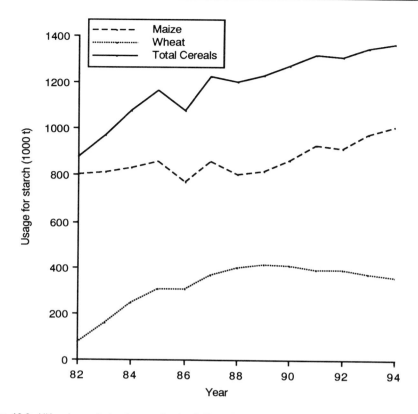

Fig. 10.2. UK maize and wheat usage for starch (from data presented in Bodjuniak and Sturgess, 1995).

the profitability of the starch extraction process is linked to VWG. The growing demand from baking and other food companies for gluten is driving the extraction process worldwide. To achieve 1 t of gluten, 6–7 t of starch are produced and both need to be sold to ensure profitability. In the EU the domestic demand for starch is only increasing at 1.5–2% annually, requiring large quantities to be exported. Currently approximately 30% is exported to the USA (Bodjuniak and Sturgess, 1995).

Historically the high price of VWG provided the basis for a thriving wheat gluten industry. During the last ten years, however, the price of gluten has continued to drop and the gluten industry has become, with time, more of a starch extraction industry. It is estimated that more than 600 products are made from extracted starch and its derivatives (Leygue, 1993).

With improvements in production and processing technology the differences between corn, wheat and potato starches is becoming less distinct. At one time wheat starch separated into two fractions: type A whose purity was similar to corn starch, and secondary type B which contained impurities. Mature starch granules of 15–20 μm average diameter made up 75–85% of wheat starch, and constituted type A. The residue, type B, which used to make up 15–25% of the extract, contained small and fragmented granules with soluble protein, fibre and pentosan impurities (Jones, 1987a). Traditionally type B was utilized mainly as low grade syrups and animal feed (Leygue, 1993). Extracted wheat starch is now a great deal more homogeneous and competitive.

Reliable and high levels of wheat production in the EU have underpinned the growing wheat starch industry and displaced maize imports from North America. France is currently the biggest industrial user of wheat at 700,000 t, followed by The Netherlands at 500,000 t, the UK and Germany at approximately 460,000 t, and Belgium and Luxembourg at 390,000 t (Anon., 1995c).

Healthy diets are benefiting from the availability of non-fattening starch products involving new sugars (polygols) and pseudo-fats, and this demand is likely to grow.

Fuel ethanol could also derive from a starch substrate in future (Leygue, 1993) given economic encouragement. Ethanol from wheat is already obtained from starch for the drinks industry, and more could be produced in integrated starch plants.

The biotechnology industries increasingly utilize glucose from starch for antibiotic and ascorbic acid production, and it is considered that this use will continue to grow (Jones, 1987a).

The profitability of starch and gluten manufacture is influenced not only by raw material costs, competitive products, product quality, and processing plant operational effectiveness, but also by waste treatment costs. Considerable effluent, with a strong chemical oxygen demand, is produced during the extraction process which needs to be treated before disposal, often by anaerobic digestion (Jones, 1987a).

Class II bread-making wheats are recommended in particular for efficient starch and gluten extraction. The variety Avalon was reportedly the most popular wheat selected by starch manufacturers in the UK in the late 1980s (Jones, 1987a), which is now classified as a class I wheat for bread-making purposes. The quality criteria on intake of wheat for starch are reportedly broadly similar to those for milling for bread making. These checks include variety testing, moisture, protein quantity and quality tests, specific weight, and the yield of gluten on washing (Jones, 1987a). High α-amylase or low Hagberg falling numbers are particularly unwelcome as these cause excessive starch digestion during processing and a sugary effluent which would be expensive to treat.

References

Aase, J.K. and Siddoway, F.H. (1975) Regrowth of spring-clipped winter wheat in the northern Great Plains of the United States. *Canadian Journal of Plant Science* 55, 631–633.

Abdel, H.T., Bassiouni, A.A., El-Hyatemy, Y.Y. and Shafik, I. (1987) Evaluation of fungicides for the control of yellow stripe rust of wheat. *Egyptian Journal of Phytopathology* 19, 85–96.

Abo-Hamed, S.A., Mansour, F.A. and Al-Desuquy, H.S. (1993) Growth and protein content of wheat kernels treated with three growth regulators. *Qatar University Science Journal* 13, 53–56.

Abramson, D. (1991) Development of moulds, mycotoxins and odours in moist cereals during storage. In: Chelkowski, J. (ed.) *Cereal Grain: Mycotoxins, Fungi and Quality in Drying and Storage.* Elsevier, Amsterdam, pp. 119–148.

Abrol, Y.P., Uprety, D.C., Ram, A. and Tikoo, S. (1971a) Phenol colour reaction as an indicator of chapati quality in wheat. *Sabrao Newsletter* 3, 17–21.

Abrol, Y.P., Uprety, D.C., Ahuja, V.P. and Naik, M.S. (1971b) Soil fertilizer levels and protein quality of wheat grains. *Australian Journal of Agricultural Research* 22, 195–200.

Adams, J.M. (1977) A review of the literature concerning losses in stored cereals and pulses. *Tropical Science* 19, 1.

Adams, D., Bacon, R. and Marx, D. (1988) *1987 Arkansas Wheat Test Weights: Fact Sheet 2036.* Cooperative Extension Service, University of Arkansas, USA.

Agcaoili, M.C. and Rosegrant, M.W. (1994) *World Supply and Demand Projections for Cereals, 2020.* International Food Policy Research Institute, Washington, USA.

Ageeb, O.A.A. (1994) Agronomic aspects of wheat production in Sudan. In: Saunders, D.A. and Hettel, G.P. (eds) *Wheat in Heat-Stressed Environments: Irrigated, Dry Areas and Rice–Wheat Farming Systems.* CIMMYT, Mexico D.F., pp. 67–74.

Aggarwal, P.K. and Sinha, S.K. (1987) Performance of wheat and triticale varieties in a variable soil water environment. IV. Yield components and their action with grain yield. *Field Crops Research* 17, 45–53.

Agricultural Development and Advisory Service (1994) *Agricultural Strategy: An Independent Outlook for the Agricultural Industry.* ADAS Publications, Cambridge.

Ahmad, S., Ahmad, N., Ahmad, R. and Hamid, M. (1989) Effect of high temperature stress on wheat reproductive growth. *Journal of Agricultural Research Lahore* 27, 307–313.

Ahmadi-Esfahani, F.Z. and Stanmore, R.G. (1994) Values of Australian wheat flour characteristics. *Agribusiness* 10, 529–536.

Ahmed, S.M. (1985) Wheat seed production, storage and distribution in Bangladesh. In: *Wheats for More Tropical Environments, A Proceedings of an International Symposium*. CIMMYT, Mexico D.F., pp. 291–296.

Alauddin, M. and Tisdell, C.A. (1991) Welfare consequences of Green Revolution technology. *Development and Change* 22, 497–517.

Alexander, M. (1977) *Soil Microbiology*, 2nd edn. John Wiley and Sons, New York.

Ali, F.M. (1984) Socioeconomic and agroeconomic implications of growing wheat in Sudan. In: *Wheats for More Tropical Environments, A Proceedings of an International Symposium*. CIMMYT, Mexico D.F., pp. 332–338.

Ali, M. (1993) Wheat/chickpea intercropping under late-sown conditions. *Journal of Agricultural Science, Cambridge* 121, 141–144.

Ali, N.M. (1995) Performance of high-protein mutant lines of *Triticum aestivum* (L.) under semi-arid conditions of Syria. *Field Crops Research* 41, 101–108.

Al-Mustafa, W.A., El-Shall, A.A., Abdallah, A.E. and Modaihsh, A.S. (1995) Response of wheat to sewage sludge applied under two different moisture regimes. *Experimental Agriculture* 31, 355–359.

Allan, R.E. (1986) Agronomic comparisons among wheat lines nearly isogenic for three reduced-height genes. *Crop Science* 26, 707–710.

Alldrick, A.J. (1991) *The Nature, Sources, Importance and Uses of Cereal Dietary Fibre*. HGCA Research Review No. 22. Home-Grown Cereals Authority, London.

Allen, P.J. and Goddard, D.R. (1938) A respiratory study of powdery mildew of wheat. *American Journal of Botany* 25, 613–621.

Almeida, J.A.A., Ribeiro, J.M. and Ferriera, A.M. (1989) Conserving and upgrading the nutritive value of whole-crop silage with urea as an ammonia precursor. In: *Proceedings of the International Symposium on the Constraints and Possibilities of Ruminant Production in the Dry Subtropics*. Centre for Agricultural Publishing and Documentation, Wageningen, p. 251.

Alshallash, K.S. and Drennan, D.S.H. (1993) Competition between spring wheat and *Lolium multiflorum* (Italian rye-grass). In: *Brighton Crop Protection Conference, Weeds – 1993*. British Crop Protection Council, Thornton Heath, UK, pp. 113–117.

Altschul, A.M. (1965) *Proteins: Their Chemistry and Politics*. Chapman and Hall, London.

American Association of Cereal Chemists (1995) *Wheat End-Uses Around the World*. American Association of Cereal Chemists, St Paul, Minneapolis.

Anderson, I. (1995) Deadly rabbit virus out of control. *New Scientist* 148, 7.

Anderson, J.A., Sorrels, M.E. and Tanksley, S.D. (1993) RFLP analysis of genomic regions associated with resistance to preharvest sprouting in wheat. *Crop Science* 33, 453–459.

Anon. (1970) *Recommended Varieties of Cereals, Farmers leaflet No. 8 1970*. National Institute of Agricultural Botany, Cambridge.

Anon. (1974–1994) *The Cereal Harvest, 1974–1993*. Home Grown Cereals Authority, London.

Anon. (1984–1988) *Cereal Statistics*. Home-grown Cereals Authority, London.

Anon. (1985) *Nitrogen and Winter Wheat Quality.* Ministry of Agriculture, Fisheries and Food, Cambridge.

Anon. (1986) *Grain Quality Guide.* Schering Agriculture, Nottingham.

Anon. (1988) Harvesting, preservation and use of whole cereal and legume plants (whole plant silage system – WPS) (Review). *Vedeckotechnicky Rozvoj v Zemedelstvi* No. 8.

Anon. (1992a) *World Grain Statistics 1992.* International Wheat Council, London.

Anon. (1992b) *The Influence of Agriculture on the Quality of Natural Waters in England and Wales: Water Quality Series No. 6.* The National Rivers Authority, Bristol.

Anon. (1992c) Intervention in the UK 1992/93; requirements for wheat, barley, and rye offers. *HGCA Marketing Note: Supplement to Weekly Bulletin* 27 (4), 1–4.

Anon. (1994) Intervention in the UK 1994/95. *Marketing Note, Supplement to the Weekly Bulletin, Home-Grown Cereals Authority* 29, (7).

Anon. (1995a) *Cereal Exports – What the Buyers Want.* British Cereal Exports, London.

Anon. (1995b) *Cereal Variety Handbook, NIAB Recommended Lists of Cereals 1995.* National Institute of Agricultural Botany, Cambridge.

Anon. (1995c) Duelling over gluten. *World Grain* 13 (9), 6–12.

Archer, J. (1985) *Crop Nutrition and Fertiliser Use.* Farming Press, Ipswich.

Armitage, D.M., Wilkin, D.R., Fleming, D.A. and Cogan, P.A. (1992) *Integrated Pest Control Strategy for Stored Grain – Surface Pesticide Treatments of Aerated Commercial and Farm Stores to Control Insects and Mites. Project Report No. 57.* Home-Grown Cereals Authority, London.

Ascher, J.S., Graham, R.D., Elliot, D.E., Scott, J.M. and Jessop, R.S. (1994) Agronomic value of seed with high nutrient content. In: Saunders, D.A. and Hettel, G.P. (eds) *Wheat in Heat-Stressed Environments: Irrigated, Dry Areas and Rice–Wheat Farming Systems.* CIMMYT, Mexico D.F., pp. 297–307.

Ashbell, G., Theune, H.H. and Sklan, D. (1984) Effect of formic acid and urea phosphate–calcium propionate on amino acids in wheat silage. *Journal of Agricultural and Food Chemistry* 32, 849–852.

Ashbell, G., Theune, H.H: and Sklan, D. (1985) Ensiling whole wheat at various maturation stages: Changes in nutritive ingredients during maturation and ensiling and upon aerobic exposure. *Journal of Agricultural and Food Chemistry* 33, 1–4.

Astbury, J.M. and Kettlewell, P.S. (1990) Optimising the management of nitrogen-containing fluid fertilizers for breadmaking quality of intensively-managed wheat in the United Kingdom: field experiments in 1989. In: Smith, J.J. (ed.) *Advances in Fluid Fertilizer Agronomic and Application Management Technology.* Fluid Fertilizer Foundation, Manchester, Missouri, pp. 201–217.

Astbury, J.M. and Kettlewell, P.S. (1992) *Late-season Husbandry and Hagberg Falling Number of Wheat.* Home-Grown Cereals Authority, London.

Attwood, P.J. (1985) *Crop Protection Handbook – Cereals.* BCPC Publications, Croydon.

Aulakh, M.S. and Rennie, D.A. (1986) Nitrogen transformations with special reference to gaseous N losses from zero-tilled soils of Saskatchewan, Canada. *Soil and Tillage Research* 7, 157–171.

Austin, R.B., Bingham, J., Blackwell, R.D., Evans, L.T., Ford, M.A., Morgan, C.L. and Taylor, M. (1980) Genetic improvements in winter wheat yields since

1900 and associated physiological changes. *Journal of Agricultural Science, Cambridge* 94, 675–689.

Axford, D.W.E., McDermott, E.E. and Redman, D.G. (1979) Note on the sodium dodecyl sulphate test of bread-making quality: comparison with Pelshenke and Zeleny tests. *Cereal Chemistry* 56, 582–584.

Ayers, G.S., Wert, V.F. and Ries, S.K. (1976) The relationship of protein fractions and individual proteins to seedling vigour in wheat. *Annals of Botany* 40, 563–570.

Bagrintseva, V.N. and Khodzhaeva, N.A. (1992) Effect of non-litter manure and fertilizers on yield and quality of winter wheat grain in a crop rotation on chestnut soil. *Agrokhimiya* No. 7, 105–110.

Bailey, J.E. (1982) Whole Grain Storage. In: Christensen, C.M. (ed.) *Storage of Cereal Grain and their Products*. American Association of Cereal Chemists, St Paul, Minnesota, pp. 53–116.

Bakhsh, A., Khattak, J.K. and Bhatti, A.U. (1986) Comparative effect of potassium chloride and potassium sulfate on the yield and protein content of wheat in three different rotations. *Plant and Soil* 96, 273–277.

Balyan, R.S., Malik, R.K., Panwar, R.S and Singh, S. (1991) Competitive ability of winter wheat cultivars with wild oat (*Avena ludoviciana*). *Weed Science* 39, 154–158.

Barber, J.S. and Jessop, R.S. (1987) Factors affecting yield and quality in irrigated wheat. *Journal of Agricultural Science, Cambridge* 109, 19–26.

Barnes, P.J. (1989) Wheat in milling and baking. In: Palmer, G.H. (ed.) *Cereal Science and Technology*. Aberdeen University Press, Aberdeen, pp. 367–412.

Bastiman, B. and Pullar, D. (1993) Yields, in-silo losses and feed quality of fermented and urea-treated whole-crop wheat. In: Hopkins A. and Younie D. (eds) *Forward with Grass into Europe, Occasional Symposium of the British Grassland Society, No. 27*. British Grassland Society, Reading, pp. 157–159.

Batchelor, S., Booth, E.J., Walker, K.G. and Cook, P. (1993) *The Potential for Bioethanol Production from Wheat in the UK. H-GCA Research Review No. 29*. Home-Grown Cereals Authority, London.

Bateman, G.L., Hornby, D. and Gutteridge, R.J. (1990) Effects of take-all on some aspects of grain quality of winter wheat. In: *Aspects of Applied Biology 25, Cereal Quality II*. Association of Applied Biologists, Warwick, pp. 339–348.

Batten, G.D. and Khan, M.A. (1987) Effect of time of sowing on grain yield, and nutrient uptake of wheats with contrasting phenology. *Australian Journal of Experimental Agriculture* 27, 881–887.

Baxter, H.D., Montgomery, M.J. and Owen, J.R. (1980) Digestibility and feeding value of corn silage fed with boot stage wheat silage and alfalfa silage. *Journal of Dairy Science* 63, 225.

Bayles, R.A. (1977) Poorly filled grain in the cereal crop I. The assessment of poor grain filling. *Journal of the National Institute of Agricultural Botany* 14, 232–240.

Bayles, R.A., Evers, A.D. and Thorne, G.N. (1978) The relationship of grain shrivelling to the milling and baking quality of three winter wheat cultivars grown with different rates of nitrogen fertilizer. *Journal of Agricultural Science, Cambridge* 90, 445–446

Bechtel, D.B., Kaleikau, L.A., Gaines, R.L. and Seitz, L.M. (1985) The effects of *Fusarium graminearum* infection on wheat kernels. *Cereal Chemistry* 62, 191–197.

Becker, D., Brettschneider, R. and Loerz, H. (1994) Fertile transgenic wheat from microprojectile bombardment of scutellar tissue. *Plant Journal* 5, 299–307.

Bell, B.M. (1985) A rapid method of dietary fibre estimation in wheat products. *Journal of the Science of Food and Agriculture* 36, 815–821.

Beninati, N.F. and Busch, R.H. (1992) Grain protein inheritance and nitrogen up take and redistribution in a spring wheat cross. *Crop Science* 32, 1471–1475.

Benzian, B. and Lane, P.W. (1986) Protein concentration of grain in relation to some weather and soil factors during 17 years of English winter-wheat experiments. *Journal of the Science of Food and Agriculture* 37, 435–444.

Bhardwaj, R.B.L., Jain, N.K., Wright, B.C., Sharma, K.C., Gill, G.S. and Krantz, B.A. (1975) *The Agronomy of Dwarf Wheats.* The Indian Council of Agricultural Research, New Delhi.

Bhatt, G.M. and Derera, N.F. (1980) Potential use of Kenya 321-type dormancy in a wheat breeding programme aimed at evolving varieties tolerant to pre-harvest sprouting. *Cereal Research Communications* 8, 291–295.

Bil, K.Y., Fomina, I.R., Gerts, S.M., Komarova, V.P., Makarov, A.D. and Biryukov, S.V. (1987) Effect of linuron on metabolization of exogenous sucrose by winter wheat plants of different genotypes. *Soviet Agricultural Sciences* No. 3, 4–8.

Biscoe, P.V. and Gallagher, J.N. (1978) A physiological analysis of cereal yield I. Production of dry matter. *Agricultural Progress* 53, 34–50.

Blackman, J.A. and Payne, P.I. (1987) Grain quality. In: Lupton, F.G.H. (ed.) *Wheat Breeding, Its Scientific Basis.* Chapman and Hall, London, pp. 455–485.

Blaxter, K.L. (1975) The energetics of British agriculture. *Journal of the Science of Food and Agriculture* 26, 1055–1064.

Blumenthal, C.S., Bekes, F., Batey, I.L., Wrigley, C.W., Moss, H.J., Mares, D.J. and Barlow, E.W.R. (1991) Interpretation of grain quality results from wheat variety trials with reference to high temperature stress. *Australian Journal of Agricultural Research* 42, 325–334.

Blumenthal, C.S., Barlow, E.W.R. and Wrigley, C.W. (1993) Growth environment and wheat quality: the effect of heat stress on dough properties and gluten proteins. *Journal of Cereal Science* 18, 3–21.

Bodjuniak, R. and Sturgess, I. (1995) Cereals and oilseeds products. In: Strak, J. and Morgan, W. (eds) *The UK Food and Drink Industry.* EuroPA and Associates, Northborough, pp. 77–115.

Bodson, B., Haquenne, W. and Maddens, K. (1989) Influence of the application of growth regulators on the quality of winter wheat. *Mededelingen van de Faculteit Landbouwwetenschappen, Rijksuniversiteit Gent* 54, 409–417.

Boggini, G., Tusa, P. and Pogna, N.E. (1995) Bread making quality of durum wheat genotypes with some novel glutenin subunit compositions. *Journal of Cereal Science* 22, 105–113.

Bolsen, K.K., Tetlow, R.M. and Wilson, R.F. (1983) The effect of calcium and sodium hydroxides and of sodium acrylate on the fermentation and digestibility *in vitro* of ensiled whole-crop wheat and barley harvested at different stages of maturity. *Animal Feed Science and Technology* 9, 37–47.

Booth, M.R. and Melvin, M.A. (1979) Factors responsible for the poor bread-making quality of high yielding European wheat. *Journal of the Science of Food and Agriculture* 30, 1057–1064.

Borlaug, N.E. and Dowsell, C.R. (1993) Fertilizer: To nourish infertile soil that feeds a fertile population that crowds a fragile world. *Proceedings of the 3rd Annual Conference, International Agribusiness Management Association (IAMA), May 22–25*, IAMA, San Francisco, California.

Borlaug, N.E., Ortega, J., Narvaez, I., Garcia, A. and Rodriguez, R. (1964) *Hybrid Wheat In Perspective*. Crop Quality Council, Minneapolis, Minnesota.

Bosshart, U., Kickuth, R. and Vogtmann, H. (1986) The influence of conventional agricultural methods on nitrate leaching (study of a natural object). In: Vogtmann, H., Boechncke, E. and Fricke, J. (eds) *The Importance of Biological Agriculture in a World of Diminishing Resources*. Verlagsgruppe, Witzenhausen.

Bothast, R.J. (1978) Grain proteins during storage. In: *Postharvest Biology and Biotechnology*. Food and Nutrition Press Inc., Westport, pp. 210–243.

Boucqué, C.V. and Fiems, L.O. (1988) Vegetable by-products of agroindustrial origin. *Livestock Production Science* 19, 97–135.

Bowerman, P., Rule, J.S. and Freer, J.B.S. (1993) Effect of cultivations and soil type on the seed emergence of barren brome, meadow brome and winter barley. In: *Brighton Crop Protection Conference, Weeds – 1993*. British Crop Protection Council, Thornton Heath, UK, pp. 317–322.

Boyacioglu, D. and Hettiarachchy, N.S. (1995) Changes in some biochemical components of wheat grain that was infected with *Fusarium graminearum*. *Journal of Cereal Science* 21, 57–62.

Braken, A.F. and Bailey, C.H. (1928) Effect of delayed harvesting on the quality of wheat. *Cereal Chemistry* 5, 128.

Branlard, G. and Dardevet, M. (1994) A null Gli-D1 allele with a positive effect on bread wheat quality. *Journal of Cereal Science* 20, 235–244.

Braunack, M.V. and Dexter, A.R. (1990) Soil aggregation in the seed-bed: a review. *Soil Tillage Research* 14, 281–298.

Bremner, P.M. (1972) The accumulation of dry matter and nitrogen by grains in different positions of the wheat ear as influenced by shading and defoliation. *Australian Journal of Biological Sciences* 25, 657–681.

Brennan, A.F. (1989) Effect of superphosphate and superphosphate plus flutriafol on yield and take-all of wheat. *Australian Journal of Experimental Agriculture* 29, 247–252.

Brennan, J.P. (1990) *Valuing the Breeding Characteristics of Wheat. Agricultural Economics Bulletin 7*. NSW Agriculture and Fisheries Wagga Wagga, Australia.

Brennan, C.S., Sulaiman, B.D., Schofield, J.D. and Vaughan, J.G. (1993) The immunolocation of friabilin and its association with wheat endosperm texture. In: *Aspects of Applied Biology 36, Cereal Quality III*. Association of Applied Biologists, Warwick, pp. 69–74.

British Agrochemicals Association (1993) Focus on wheat (*Triticum aestivum*). *Grapevine* No. 37. BAA, Peterborough, UK.

Brooker, D.B., Bakker-Arkema, F.W. and Hall, C.W. (1981) *Drying Cereal Grains*. Avi Publishing Company, London.

Brosh, A., Holzer, Z., Bar-Tsur, A., Levy, D., Ilan, D. and Kali, J. (1989) High wheat and maize silage diets for growth and fattening young cattle. *Animal Feed Science and Technology* 26, 287–298.

Brosh, A., Holzer, Z. and Levy, D. (1990) The effect of source of nitrogen used for supplementation of high wheat silage diets for cattle. *Animal Production* 51, 109–114.

Brown, H.M., Lichtner, F.T., Hutchison, J.M. and Saladini, J.A. (1995) The impact of sulfonylurea herbicides in cereal crops. In: *Brighton Crop Protection Conference, Weeds – 1995*. British Crop Protection Council, Thornton Heath, UK, pp. 1143–1152.

Brun, L. (1982) Combined variety and nitrogen and fertiliser trial with cereals, 1975–1979. *Forskning og Forsok i Landbruket* 33, 133–142.

Bryson, R.J., Sylvester-Bradley, R., Scott, R.K. and Paveley, N.D. (1995) Reconciling the effects of yellow rust on yield of winter wheat through measurements of green leaf area and radiation interception. In: *Aspects of Applied Biology 42, Physiological Responses of Plants to Pathogens*. Association of Applied Biologists, Warwick, pp. 9–18.

Buchanan, B.B., Hutcheson, S.W., Magyarosy, A.C. and Metalbini, P. (1981) Photosynthesis in healthy and diseased plants. In: Ayers, P.G. (ed.) *Effects of Disease on the Physiology of the Growing Plant*. Cambridge University Press, Cambridge, pp. 14–28.

Bulmer, R.H.C., Eleftherohorinos, I., Snaydon, R.W. and Drennan, D.S.H. (1985) The effect of metoxuron and chlorsulfuron on competition between *Bromus sterilis* (sterile brome) and winter wheat. In: *Aspects of Applied Biology 9, The Biology and Control of Weeds in Cereals*. Association of Applied Biologists, Warwick, pp. 75–80.

Bunting, A.H. and Drennan, D.S.H. (1966) Some aspects of the morphology and physiology of cereals in the vegetative phase. In: Milthorpe, F.L. and Ivins, J.D. (eds) *The Growth of Cereals and Grasses*. Butterworths, London, pp. 20–38.

Burleigh, J.R., Roelfs, A.P. and Eversmeyer, M.G. (1972) Estimating damage to wheat caused by *Puccinia recondita tritici*. *Phytopathology* 62, 944–946.

Burton, R.L., Porter, D.R., Baker, C.A., Webster, J.A., Burd, J.D. and Puterka, G.J. (1991) Development of aphid-resistant wheat germplasm. In: Saunders, D.A. (ed.) *Wheat for the Nontraditional, Warm Areas*. CIMMYT, Mexico D.F., pp. 203–213.

Butorina, E.P., Yagodin, B.A. and Feofanov, S.N. (1991) Effect of a late foliar application of urea and molybdenum on winter wheat grain yield and quality. *Agrokhimiya* No. 4, 17–20.

Byerlee, D. (1990) *Technical Change, Productivity and Sustainability in Irrigated Cropping Systems of South Asia: Emerging Issues in the Post-Green Revolution Era*. CIMMYT Economics Working Paper 90/06. CIMMYT, Mexico, D.F.

Byerlee, D. and Moya, P. (1993) *Impacts of International Wheat Breeding Research in the Developing World, 1966–1990*. CIMMYT, Mexico, D.F.

Byerlee, D. and Siddiq, A. (1994) Has the Green Revolution been sustained? The quantitative impact of the seed-fertiliser revolution in Pakistan revisited. *World Development* 22, 1346–1361.

Byers, M., Franklin, J. and Smith, S.J. (1987) The nitrogen and sulphur nutrition of wheat and its effect on the composition and baking quality of the grain. In: *Aspects of Applied Biology, 15, Cereal Quality*. Association of Applied Biologists, Warwick, pp. 337–344.

Cahill, M.J. (1988) Diflufenican and isoproturon: A case history of its development in winter cereals in Ireland 1983–1987. In: *Aspects of Applied Biology 18, Weed Control in Cereals and the Impact of Legislation on Pesticide Application*. Association of Applied Biologists, Warwick, pp. 247–264.

Caldwell, C.D. and Starratt, C.E. (1987) Response of Max spring wheat to management inputs. *Canadian Journal of Plant Science* 67, 645–652.

Camblin, P.H. and Gall, P.H. (1987) Cultural techniques and cereal quality – effect of foliar sulphur. In: *Aspects of Applied Biology, 15, Cereal Quality.* Association of Applied Biologists, Warwick, pp. 345–358.

Campbell, C.A. and Read, D.W.L. (1968) Influence of air temperature, light intensity and soil moisture on the growth, yield and some growth analysis characteristics of Chinook wheat grown in the growth chamber. *Canadian Journal of Plant Science* 48, 299–311.

Campbell, C.A., Pelton, W.L. and Neilson, K.F. (1969) Influence of solar radiation and soil moisture on growth and yield of Chinook wheat. *Canadian Journal of Plant Science* 49, 685–699.

Campbell, C.A., Davidson, H.R. and Winkleman, G.E. (1981) Effect of nitrogen, temperature, growth stage and duration of moisture stress on yield components and protein content of Manitou spring wheat. *Canadian Journal of Plant Science* 61, 549–563.

Campbell, C.A., Selles, F., Zetner, R.P., McLoed, J.G. and Dyck, F.B. (1991) Effect of seeding date, rate and depth on winter wheat grown on conventional fallow in S.W. Saskatchewan. *Canadian Journal of Plant Science* 71, 51–61.

Cannell, R.Q., Ellis, F.B., Christian, D.G., Graham, J.P. and Douglas, J.T. (1980) The growth and yield of winter cereals after direct drilling, shallow cultivation and ploughing on non-calcareous clay soils, 1974–8. *Journal of Agricultural Science, Cambridge* 94, 345–359.

Capper, B.S. (1988) Genetic variation in the feeding value of cereal straw. *Animal Feed Science and Technology* 21, 127–140.

Carr, P.M., Jacobsen, J.S., Carlson, G.R. and Nielson, G.A. (1992) Influence of soil and N fertilizer on performance of barley and spring wheat cultivars. *Canadian Journal of Plant Science* 72, 651–661.

Carter, E.P. and Young, G.Y. (1945) Effect of moisture content, temperature, length of storage on the development of sick wheat in sealed containers. *Cereal Chemistry* 22, 418–428.

Castejon, M. and Leaver, J.D. (1991) Integration of grazing and whole crop wheat for calves in spring. *Animal Production* 52, 606–607.

Catania, P.J. (1993) Price and quality in futures markets. In: *Aspects of Applied Biology 36, Cereal Quality III.* Association of Applied Biologists, Warwick, pp. 1–8.

Chakrabarti, B. (1994) *Methods of Distributing Phosphine in Bulk Grain. Research Review No. 27.* Home-Grown Cereals Authority, London.

Chamberlain, N., Collins, T.H. and McDermott, E.E. (1982) The influence of α-amylase on loaf properties in the UK. *Proceedings of the 7th World Cereal and Bread Congress.* International Association for Cereal Science and Technology, Vienna, pp. 841–845.

Chandhanamutta, P. (1985) Thailand winter cereals programme. In: *Wheats for More Tropical Environments.* CIMMYT, Mexico D.F., pp. 83–85.

Chaney, R.L. and Giordano, P.M. (1977) Microelements as related to plant deficiencies and toxicities. In: *Soils for the Management of Organic Wastes and Wastewaters.* SSSA: Madison, WI, pp. 235–279.

Charles, A.H. (1958) The effect of undersowing on the cereal cover crop. *Field Crop Abstracts* 11, 233–239.

Chaudhri, A.B. and Muller, H.G. (1970) Chapaties and chapati flour. *Milling* 152(11), 22–25.

Chaudri, A.M., McGrath, S.P., Crosland, A.R. and Zhao, F. (1993) Mineral status of British wheat. In: *Aspects of Applied Biology 36, Cereal Quality III.* Association of Applied Biologists, Warwick, pp. 347–353.

Chelkowski, J. (1991) Fungal pathogens influencing cereal seed quality at harvest. In: Chelkowksi, J. (ed.) *Cereal Grain: Fungi and Quality in Drying and Storage.* Elsevier, Amsterdam, pp. 53–66.

Christensen, C.M. and Kaufmann, H.H. (1965) Deterioration of stored grain by fungi. *Annual Reveiw of Phytopathology* 3, 69–84.

Christensen, C.C. and Kaufmann, H.H. (1969) *Grain Storage: The Role of Fungi in Quality Loss.* University of Minnesota Press, Minneapolis.

Christensen, N.W. and Mientz, V.W. (1982) Evaluating N fertilizer sources and timing for winter wheat. *Agronomy Journal* 74, 840–844.

Christensen, C.C. and Sauer, D.B. (1982) Microflora. In: Christensen (ed.) *Storage of Cereal Grain and their Products.* American Association of Cereal Chemists, St Pauls, Minnesota, pp. 219–240.

Chung, T.Y. (1986) Effect of ammonia treatment of rye silage on the nutritive value and livestock performance. *Korean Journal of Dairy Science* 8, 1–15.

Church, D.C. (1984) *Livestock Feeds and Feeding,* 2nd edn. O and B Books, Corvallis, USA.

CIMMYT (1993) *1992/93 CIMMYT World Wheat Facts and Trends. The Wheat Breeding Industry in Developing Countries: An Analysis of Investments and Impacts.* CIMMYT, Mexico D.F.

CIMMYT (1995) *CIMMYT in 1994. Modern Maize and Wheat Varieties: Vital to Sustainable Agriculture and Food Security.* CIMMYT, Mexico D.F.

Clare, R.W., Hayward, C.F. and Jordan, V.W.L. (1990) Interaction between fungicides and nitrogen fertiliser application on yield and quality of winter wheat. In: *Aspects of Applied Biology 25, Cereal Quality II.* Association of Applied Biologists, Warwick, pp. 363–374.

Clark, H.E. (1978) Cereal based diets to meet protein requirements of adult man. *World Review of Nutritional Dietetics* 90, 423–427.

Clark, W.S. (1993) Interaction of winter wheat varieties with fungicide programmes and effects on grain quality. In: *Aspects of Applied Biology 36, Cereal Quality III.* Association of Applied Biologists, Warwick, pp. 397–406.

Clark, W.S. (1995) What does resistance mean to farmers? In: Hewitt, H.G., Tyson, D., Hollomon, D.W., Smith, J.M., Davies, W.P. and Dixon K.R. (eds) *A Vital Role for Fungicides in Cereal Production.* BIOS, Oxford, pp. 165–170.

Clark, J.H., Blair, A.M. and Moss, S.R. (1994) The testing and classification of herbicide resistant *Alopecurus myosuroides* (black-grass). In: *Aspects of Applied Biology 37, Sampling to Make Decisions.* Association of Applied Biologists, Warwick, pp. 181–188.

Clarke, B. (1985) *Manual on Crop Seed Processing Equipment.* ELE International Ltd, Hemel Hempstead, UK.

Clarkson, J.D.S. (1981) Relationship between eyespot severity and yield loss in winter wheat. *Plant Pathology* 30, 125–131.

Cleal, R.A.E. (1993) Effect of growth regulators on the grain yield and quality of triticale and wheat grown as a second cereal on light soil. In: *Aspects of Applied*

Biology 35, Cereal Quality III. Association of Applied Biologists, Warwick, pp. 281–286.

Clements, D.J. (1988) Wheat seed storage under tropical conditions. In: Klatt, A.R. (ed.) *Wheat Production Constraints in Tropical Environments.* CIMMYT, Mexico D.F., pp. 360–365.

Conner, R.L. and Kuzyk, A.D. (1988) Effectiveness of fungicides in controlling stripe rust, leaf rust, and blackpoint in soft white spring wheat. *Canadian Journal of Plant Pathology* 10, 321–326.

Cook, R.J. (1987a) The classification of wheat cultivars using a standard reference electrophoresis method. *Journal of the National Institute of Agricultural Botany* 17, 273–281.

Cook, R.J. (1987b) Effect of late season fungicides on winter wheat quality. In: *Aspects of Applied Biology 15, Cereal Quality.* Association of Applied Biologists, Warwick, pp. 417–423.

Cook, R.J. and Hims, M.J. (1990) Influence of disease on wheat Hagberg falling number. In: *Aspects of Applied Biology 25, Cereal Quality II.* Association of Applied Biologists, Warwick, pp. 355–362.

Cook, R.J. and King, J.E. (1984) Loss caused by cereal diseases and the economics of fungicidal control. In: Wood, R.K.S. and Jellis, G.J. (eds) *Plant Diseases: Infection, Damage and Loss.* Blackwell Scientific Publications, Oxford, pp. 237–245.

Cooke, A.S. and Burn, A.J. (1995) The environmental impact of herbicides used in intensive farming systems. In: *Brighton Crop Protection Conference, Weeds – 1995.* British Crop Protection Council, Thornton Heath, UK, pp. 603–612.

Cooksley, J. (1995) Why overseas buyers need Chopin Alveograph results. *Arable Farming* 22 (18), 22–24.

Cornford, C.A. and Black, M. (1985) α–Amylase content of premature unsprouted wheat grains. *Journal of Cereal Science* 3, 295–304.

Corrall, A.J., Heard, A.J., Fenlon, J.S., Terry, C.P. and Lewis, G.C. (1977) *Whole Crop Forages; Relationship between Stage of Growth, Yield and Forage Quality in Small-Grained Cereals and Maize. Technical Report No. 22.* Grassland Research Institute, Hurley.

Cosser, N.D., Thompson, A.J., Gooding, M.J., Davies, W.P. and Froud-Williams, R.J. (1994a) Influences of variety and establishment date on nitrogen accumulation during winter in the shoot system of winter wheat. In: *Aspects of Applied Biology 39, The Impact of Genetic Variation on Sustainable Agriculture.* Association of Applied Biologists, Warwick, pp. 195–200.

Cosser, N.D., Thompson, A.J., Gooding, M.J., Davies, W.P. and Froud-Williams, R.J. (1994b) Effects of sowing date and cultivar on the yield and nutritional quality of wheat for sheep grazing in the early spring. *Proceedings of the 4th Research Conference of the British Grassland Society.* British Grassland Society, Reading.

Cosser, N.D., Gooding, M.J. and Froud-Williams, R.J. (1995) The effect of wheat dwarfing genes on competition against blackgrass. In: *Brighton Crop Protection Conference, Weeds – 1995.* British Crop Protection Council, Thornton Heath, UK, pp. 361–362.

Cosser, N.D., Gooding, M.J., Davies, W.P. and Froud-Williams, R.J. (1996) Effects of wheat dwarfing genes on grain yield and the quality of wheat in competition with *Alopecurus myosuroides. Second International Weed Control Congress,*

Copenhagen. Department of Weed Control and Pesticide Ecology, Flakhebjerg, Denmark, pp. 1089–1094.

Cotton, R.T. and Wilbur, D.A. (1982) Insects. In: Christensen, C.M. (ed.) *Storage of Cereal Grain and their Products*. American Association of Cereal Chemists, St Paul, Minnesota, pp. 281–318.

Cousens, R., Pollard, F. and Denner, R.A.P. (1985) Competition between *Bromus sterilis* and winter cereals. In: *Aspects of Applied Biology 9, The Biology and Control of Weeds in Cereals*. Association of Applied Biologists, Warwick, pp. 67–74.

Cox, M.C., Qualset, C.O. and Rains, D.W. (1986) Genetic variation for nitrogen assimilation and translocation in wheat: III. Nitrogen translocation in relation to grain yield and protein. *Crop Science* 26, 737–740.

Cramer, H.H. (1967) *Plant Protection and World Crop Production*. Bayer Pflanzenschutz, Leverkusen.

Cressey, P.J., Macgibbon, D.G. and Grama, A. (1987) Hexaploid wild emmer wheat derivatives grown under New Zealand conditions 3. Influence of nitrogen fertilisation and stage of grain development on protein composition. *New Zealand Journal of Agricultural Research* 30, 53–58.

Croll, B. (1987) The water authority view. *Proceedings from Water Pollution – the Farming Perspective*. National Agricultural Conference, Stoneleigh, pp. 1–10.

Cromack, H.T.H. and Clark, A.N.S. (1987) Winter wheat and winter barley – the effect of seed rate and sowing date on grain quality. In: *Aspects of Applied Biology 15, Cereal Quality*. Association of Applied Biologists, Warwick, pp. 171–177.

Cunfer, B.M. (1994) Taxonomy and nomenclature of *Septoria* and *Stagnospora* species on cereals. In: Arsenuik, E. Goral, T. and Czembor, P. (eds) *Proceedings of the 4th International Workshop on Septoria of Cereals*. Swiss Federal Research Station for Agronomy, Zurich, pp. 15–19.

Curic, R. (1988) Investigation of the effect of late application of nitrogen on wheat yield and nitrogen accumulation in the grain. In: Jenkinson, D.S. and Smith, K.A. (eds) *Nitrogen Efficiency in Agricultural Soils*. Elsevier, London, pp. 137–144.

Curtis, B.C. (1988) The potential for expanding wheat production in marginal and tropical environments. In: Klatt, A.R. (ed.) *Wheat Production Constraints in Tropical Environments*. CIMMYT, Mexico D.F., pp. 5–11.

Cutforth, H.W., Campbell, C.A., Brandt, S.A., Hunter, J., Judiesch, D., DePauw, R.M. and Clark, J.M. (1990) Development and yield of Canadian western red spring and Canada prairie spring wheats as affected by delayed seeding in the brown and dark brown soil zones of Saskatchewan. *Canadian Journal of Plant Science* 70, 639–660.

Da Silva, D.B. (1991) Effect of sowing depth on irrigated wheat in the Cerrado region. *Pesquisa Agropecuaria Brasileira* 26, 769–773.

Dale, C.J., Unwin, D.S. and Young, T.W. (1987) Malted wheat as a raw material for beer production. In: *Aspects of Applied Biology 15, Cereal Quality*. Association of Applied Biologists, Warwick, pp. 57–62.

Dampney, P.M.R. (1987) The effect of applications of nitrogen during stem extension and grain filling on the quality of wheat grain used for bread-making. In: *Aspects of Applied Biology 15, Cereal Quality*. Association of Applied Biologists, Warwick, pp. 239–248.

Dampney, P.M.R. and Salmon, S. (1990) The effect of rate and timing of late nitrogen applications to breadmaking wheats as ammonium nitrate or foliar urea-N, and the effect of sulphur application I. Effect on yield, grain quality and recovery of nitrogen in the grain. In: *Aspects of Applied Biology 25, Cereal Quality II*. Association of Applied Biologists, Warwick, pp. 229–241.

Dann, P.R. (1968) Effect of clipping on yield of wheat. *Australian Journal of Experimental Agriculture and Animal Husbandry* 8, 731–735.

Darwent, A.L., Kirkland, K.J., Townley-Smith, L., Harker, K.N., Cessna, A.J., Lukow, O.M. and Lefkovitch, L.P. (1994) Effect of preharvest applications of glyphosate on the drying, yield and quality of wheat. *Canadian Journal of Plant Science* 74, 221–230.

David, C.C. and Otsuka, K. (1994) *Modern Rice Technology and Income Distribution in Asia*. Lynne Reiner Publishers, Boulder, Colorado and IRRI, Manila.

Davies, B. (1988) *Reduced Cultivation for Cereals. HGCA Research Review No. 5*. Home-Grown Cereals Authority, London.

Davies, D.B. (1990) Using nitrogen wisely. *Journal of the Royal Agricultural Society of England* 151, 161–224.

Davies, W.P. (1992) Prospects for pest resistance to pesticides. In: *Pest Management and the Environment in 2000*. CAB International, Wallingford, pp. 95–110.

Davies, W.P. (1995) Developing integrated farming systems. *Annual Report of the Royal Agricultural College 1994*. Royal Agricultural College, Cirencester, pp. 17–22.

Davies, D.B. and Cannell, R.Q. (1975) Review of experiments on reduced cultivation and direct drilling in the United Kingdom, 1957–1974. *Outlook on Agriculture* 8, 216–220.

Davies, W.P. and Gooding, M.J. (1995) Developing lower-input wheat production systems. *Agricultural Progress* 70, 95–105.

Davis, M. and Greenhalgh, J.F.D. (1983) Effects of urea, aqueous NH_3, anhydrous NH_3 and NaOH treatment on the intake and digestibility of mature whole-crop barley. *Animal Production* 36, 511.

Davis, R.P., Garthwaite, D.G. and Thomas, M.R. (1991) *Pesticide Usage Survey Report 85, Arable Crops 1990*. MAFF Publications, London.

Dawson, I.A. and Wardlaw, I.F. (1989) The tolerance of wheat to high temperatures during reproductive growth. III. Booting to anthesis. *Australian Journal of Agricultural Research* 40, 965–980.

De Marco, D.G. (1990) Effect of seed weight, and seed phosphorous and nitrogen concentrations on the early growth of wheat seedlings. *Australian Journal of Experimental Agriculture* 30, 545–549.

De Waard, M.A., Banga, M. and Ellis, S.W. (1992) Characterization of the sensitivity of *Erysiphe graminis* f.sp. *tritici* to morpholines. *Pesticide Science* 34, 374–376.

Deschard, G. (1983) Alkali treatment of whole-crop cereal silage. PhD thesis, University of Reading.

Deschard, G., Tetlow, R.M. and Mason, V.C. (1987) Treatment of whole-crop cereals with alkali. 3. Voluntary intake and digestibility studies in sheep given immature wheat ensiled with sodium hydroxide, urea or ammonia. *Animal Feed Science and Technology* 18, 283–293.

Deschard, G., Mason, V.C. and Tetlow, R.M. (1988) Treatment of whole-crop cereals with alkali. 4. Voluntary intake and growth in steers given wheat ensiled with sodium hydroxide, urea or ammonia. *Animal Feed Science and Technology* 19, 55–66.

Devitt, D.A., Stolzy, L.H. and Labanauskas, C.K. (1987) Impact of potassium, sodium, and salinity on the protein and free amino acid content of wheat grain. *Plant and Soil* 103, 101–109.

Dexter, J.E., Matsuo, R.R., Kosmolak, F.G., Leisle, D. and Marchylo, B.A. (1980) The suitability of the SDS-sedimentation test for assessing gluten strength in durum wheat. *Canadian Journal of Plant Science* 60, 25–29.

Dickie, A.J. (1987) UK grain quality requirements for export. In: *Aspects of Applied Biology 15, Cereal Quality.* Association of Applied Biologists, Warwick, pp. 49–56.

Dowdell, R.J. and Crees, R. (1980) The uptake of ^{15}N-labelled fertiliser by winter wheat and its immobilisation in a clay soil after direct drilling or ploughing. *Journal of the Science of Food and Agriculture* 31, 992–996.

Doyle, C.J., Mason, V.C. and Baker, R.D. (1986) Straw disposal and utilization: An economic evaluation of the alternative end-uses for wheat straw in the UK. In: *Animal and Grassland Research Stations Annual Report 1986.* AFRC Institute for Grassland and Animal Production, Hurley, p. 95.

Doyle, A.D., Moore, K.J. and Herridge, D.F. (1988a) The narrow-leafed lupin (*Lupinus angustifolius* L.) as a nitrogen-fixing rotation crop for cereal production. III. Residual effects of lupins on subsequent cereal crops. *Australian Journal of Agricultural Research* 39, 1029–1037.

Doyle, C.J., Mason, V.C. and Baker, R.D. (1988b) Straw disposal and utilization: an economic evaluation of the end-uses for wheat straw in the UK. *Biological Wastes* 23, 39–56.

Doyle, C.J., Mason, V.C. and Baker, R.D. (1990) The economic consequences of integrating whole-crop wheat with grass production on dairy farms in the UK: a computer simulation. *Grass and Forage Science* 45, 179–190.

Draper, S.R. and Stewart, B.A. (1980) Procedures for the comparative assessment of quality in crop varieties. III. Methods used in assessing grain protein content, Hagberg falling number, ease of milling and the baking quality of wheat varieties. *Journal of the National Institute of Agricultural Botany* 15, 194–197.

Draper, S.R., Kingston, I.B. and Holden, M. (1979) Procedures for the comparative assessment of quality in crop varieties. I. Methods of analysis and assessment. *Journal of the National Institute of Agricultural Botany* 15, 1–14.

Dubetz, S. (1977) Effects of high rates of nitrogen on Neepawa wheat grown under irrigation. I. Yield and protein content. *Canadian Journal of Plant Science* 57, 331–336.

Dubin, H.J. (1985) Reflections on foot rots of wheat in warmer, nontraditional wheat-growing climates. In: *Wheats for More Tropical Environments, A Proceedings of an International Symposium.* CIMMYT, Mexico D.F., pp. 182–185.

Dubin, H.J. and van Ginkel, M. (1991) The status of wheat diseases and disease research in warmer areas. In: Saunders, D.A. (ed.) *Wheat for the Nontraditional, Warm Areas.* CIMMYT, Mexico D.F., pp. 125–145.

Dubin, H.J., Nambiar, K.K.M., Burton, R.L., Hobbs, P.R., Ries, E.M., Derpsch, R. and Byerlee, D.R. (1991) Are the objectives of sustainability and disease and insect control incompatible with respect to tillage and cropping systems in wheat. In: Saunders, D.A. (ed.) *Wheat for the Nontraditional, Warm Areas.* CIMMYT, Mexico D.F., pp. 447–448.

Dukes, J. Toma, R.B. and Wirtz, R. (1995) Cross-cultural and nutritional values of bread. *Cereal Foods World* 40, 384–385.

Dunkel, F.V. (1992) The stored grain ecosystem: A global perspective. *Journal of Stored Products Research* 28, 73–87.

Dunphy, D.J., McDaniel, M.E. and Holt, E.C. (1982) Effect of forage on wheat grain yield. *Crop Science* 66, 106–109.

Dyke, G.V. and Stewart, B.A. (1992) Factors affecting the grain yield, milling and bread-making quality of wheat, 1969–72. I. Grain yield, milling quality and flour protein content in relation to variety and nitrogen fertilizer. *Plant Varieties and Seeds* 5, 115–128.

Dziamba, S. and Mikos, M. (1988) Effect of Flordimex T (Camposan) and of fertilizer application on the protein content of triticale, wheat and rye grain and the lysine content in the protein. *Roczniki Nauk Rolniczych, A Produkcja Roslinna* 107 (2), 9–21.

Echeverria, H.E., Navarro, C.A. and Andrade, F.H. (1992) Nitrogen nutrition of wheat following different crops. *Journal of Agricultural Science, Cambridge* 118, 157–163.

Edwards, C.A. (1975) Effects of direct drilling on the soil fauna. *Outlook on Agriculture* 8, 243–244.

Edwards, R.A., Donaldson, E. and MacGregor, A.W. (1968) Ensilage of whole-crop barley. I. Effects of variety and stage of growth. *Journal of the Science of Food and Agriculture* 19, 656–660.

El Titi, A., Boller, E.F. and Gendrier, J.P. (1993) Integrated production – principles and technical guidelines. *IOBC/WPRS Bulletin* 16, pt. 1.

Ellen, J. (1990) Effects of nitrogen and plant density on growth, yield and chemical composition of two winter wheat (*Triticum aestivum* L.) cultivars. *Journal of Agronomy and Crop Science* 164, 174–183.

Ellis, S.A. (1996) The pathology, control and effect on breadmaking quality of blackpoint on wheat. PhD thesis, the University of Bristol.

Ellis, S.A. and Gooding, M.J. (1995) Blackpoint – the latest stain on quality. *Arable Farming* (February 14th), 34–41.

Ellis, S.A., Gooding, M.J., Davies, W.P. and Fenwick, R. (1994) Blackpoint infection of spring wheat cultivars. *Tests of Agrochemicals and Cultivars* 15, *Annals of Applied Biology* 124 (Supplement), 130–131.

Ellis, S.A., Gooding, M.J. and Thompson, A.J. (1996) Factors influencing the relative susceptibility of wheat cultivars (*Triticum aestivum* L.) to blackpoint. *Crop Protection* 15, 69–76.

El-Yamani, M. and Hill, J.H. (1991) Crop loss assessment and germplasm screening for resistance to barley yellow dwarf virus in west-central Morocco. *Phytopathologia Mediterranea* 30, 93–102.

Empson, D.W. and Gair, R. (1982) *Cereal Pests*. Her Majesty's Stationery Office, London.

Engels, A.J.G. (1995) Integrated control of powdery mildew in cereals and its impact upon fungicide use. In: Hewitt, H.G., Tyson, D., Hollomon, D.W., Smith, J.M., Davies, W.P. and Dixon K.R. (eds) *A Vital Role for Fungicides in Cereal Production*. BIOS, Oxford, pp. 117–128.

Entwistle, G. and Kinloch, M. (1989) *The Production and Utilization of Wheat in Scotland, Report No. 14*. Scottish Agricultural Colleges, Edinburgh.

Evans, L.T. (1993) *Crop Evolution, Adaptation and Yield.* Cambridge University Press, Cambridge.

Evans, L.T., Wardlaw, I.F. and Fischer, R.A. (1975) Wheat. In: Evans, L.T. (ed.) *Crop Physiology.* Cambridge University Press, London, pp. 101–150.

Evers, A.D. and Flintham, J. (1994) *Improvement of Resistance to Sprouting in Cereals through Exploitation of a Novel Grain Component and Advanced Breeding Techniques.* Project No 36/3/91. Home-Grown Cereals Authority, London.

Evers, A.D., Cox, R.I., Shaheedullah, M.Z. and Withey, R.P. (1990) Predicting milling extraction rate by image analysis of wheat grains. In: *Aspects of Applied Biology 25, Cereal Quality II.* Association of Applied Biologists, Warwick, pp. 417–426.

Evers, A.D., Flintham, J. and Kotecha, K. (1995) α-Amylase and grain size in wheat. *Journal of Cereal Science* 21, 1–3.

Every, D. (1990) Wheat-bug damage in New Zealand wheats: the feeding mechanism of *Nysius huttoni* and its effect on the morphological and physiological development of wheat. *Journal of the Science of Food and Agriculture* 50, 297–309.

Eyal, Z. (1972) Effect of *Septoria* leaf blotch on the yield of spring wheat in Israel. *Plant Disease Reporter* 56, 983–986.

Faridi, H. and Faubion, J.M. (1995) *Wheat End–Uses Around the World.* American Association of Cereal Chemists, St Paul, Minnesota, USA.

Farrand, E.A. (1972) Potential milling and baking value of home grown wheat. *Journal of the National Institute of Agricultural Botany* 12, 464–470.

Farrar, J.F. (1995) Just another sink? Sources of assimilate for foliar pathogens. In: *Aspects of Applied Biology 42, Physiological Responses of Plants to Pathogens.* Association of Applied Biologists, Warwick, pp. 81–89.

Fatyga, J. (1991) Effect of sowing date and nitrogen fertilizer application on yield and quality of spring wheat grain. *Roczniki Nauk Rolniczych. Seria A, Produkcja Roslinna* 109, 71–84.

Fawell, J. (1992) Pesticides in water – How big a problem? In: *Proceedings from Pesticides in Water.* National Agricultural Conference, Stoneleigh, pp. 1–5.

Fearne, A. and Warren, R. (1993) Cereal production in the European Community. *British Food Journal* 95 (2), 25–29.

Fenwick, R. (1993) Cereal variety quality evaluation in a changing agricultural environment. In: *Aspects of Applied Biology 36, Cereal Quality III.* Association of Applied Biologists, Warwick, pp. 169–178.

Fielder, A. (1988) *Interactions between Variety and Sowing Date for Winter Wheat and Winter Barley, HGCA Research Review No. 6.* Home-Grown Cereals Authority, London.

Fifield, C.C. (1945) Quality characteristics of wheat varieties grown in the western United States. *USDA Technical Bulletin* 87, 1–35.

Finney, M.E. (1979) The influence of infection by *Erysiphe graminis* D.C. on the senescence of the first leaf of barley. *Physiological Plant Pathology* 14, 31–36.

Finney, K.F. and Fryer, H.C. (1958) Effect on loaf volume of high temperatures during the fruiting period of wheat. *Agronomy Journal* 50, 28–34.

Finney, K.F., Meyer, J.W., Smith, F.W. and Fryer, H.C. (1957) Effect of foliar spraying on Pawnee wheat with urea solutions on yield, protein content, and protein quality. *Agronomy Journal* 49, 341–347.

Firbank, L.G., Mortimer, A.M. and Putwain, P.D. (1985) *Bromus sterilis* in winter wheat: a test of a predictive population model. In: *Aspects of Applied Biology 9, The Biology and Control of Weeds in Cereals*. Association of Applied Biologists, Warwick, pp. 59–66.

Fischer, R.A. (1981) Developments in wheat agronomy. In: Evans, L.T. and Peacock, W.J. (eds) *Wheat Science – Today and Tomorrow*. Cambridge University Press, London, pp. 249–269.

Fischer, R.A. (1985) Physiological limitations to producing wheat in semi-tropical and tropical environments and possible selection criteria. In: *Wheats for More Tropical Environments, A Proceedings of an International Symposium*. CIMMYT, Mexico D.F., pp. 209–230.

Fischer, R.A. and Byerlee, D.R. (1991) Trends of wheat production in warmer areas: Major issues and economic considerations. In: Saunders, D.A. (ed.) *Wheat for the Nontraditional, Warm Areas*. CIMMYT, Mexico D.F., pp. 3–27.

Fischer, R.A. and Varughese, G. (1994) Developments in CIMMYT wheat program in 1993. *Annual Wheat Newsletter, Colorado State University* 40, 161–164.

Fischer, R.A., Lindt, J.H. and Glave, A. (1977) Irrigation of dwarf wheats in the Yaqui valley of Mexico. *Experimental Agriculture* 13, 353–367.

Fleming, G.A. and Delaney, J. (1961) Copper and nitrogen in the nutrition of wheat on cutaway peat. *Irish Journal of Agricultural Research* 1, 81–82.

Flintham, J.E. and Gale, M.D. (1988) Genetics of pre-harvest sprouting and associated traits in wheat: review. *Plant Varieties and Seeds* 1, 87–97.

Ford, M. (1987) Quality requirements for milling and baking. In: *Aspects of Applied Biology 15, Cereal Quality*. Association of Applied Biologists, Warwick, pp. 10–17.

Foulkes, M.J., Scott, R.K. and Sylvester-Bradley, R. (1994) Suitabilities of UK winter wheat (*Triticum aestivum* L.) varieties to soil and husbandry conditions. *Plant Varieties and Seeds* 7, 161–181.

Francois, L.E., Maas, E.V., Donovan, T.J. and Youngs, V.L. (1986) Effect of salinity on grain yield and quality, vegetative growth, and germination of semi-dwarf and durum wheat. *Agronomy Journal* 78, 1053–1058.

Froud-Williams, R.J. (1983) The influence of straw disposal and cultivation regime on the population dynamics of *Bromus sterilis*. *Annals of Applied Biology* 103, 139–148.

Froud-Williams, R.J. (1985) The biology of cleavers (*Galium aparine*). In: *Aspects of Applied Biology 9, The Biology and Control of Weeds in Cereals*. Association of Applied Biologists, Warwick, pp. 189–195.

Froud-Williams, R.J., Pollard, F. and Richardson, W.G. (1980) Barren brome: a threat to winter cereals? *Report of the Weed Research Organization 1978–1979*, pp. 43–51.

Fuller, P. and Stewart, B.A. (1968) The effects of nitrogenous fertilizer on the milling and baking quality of Maris Widgeon wheat. *Bulletin of the Flour Milling and Baking Research Association*, 201–204.

Fullington, J.G. and Nityagopal, A. (1986) Effect of rust infection on the protein components of wheat. *Phytochemistry* 25, 1289–1292.

Fullington, J.G, Miskelly, D.M., Wrigley, C.W. and Kasarda, D.D. (1987) Quality related endosperm proteins in sulfur-deficient and normal wheat grain. *Journal of Cereal Science* 5, 233–246.

Gair, R., Jenkins, J.E.E. and Lester, E. (1987) *Cereal Pests and Diseases.* Farming Press, Ipswich.

Gale, M.D. and Youssefian, S. (1985) Dwarfing genes in wheat. In: Russell, G.E. (ed.) *Progress in Plant Breeding.* Butterworth, London, pp. 1–35.

Gale, M.D., Flintham, J.E. and Arthur, E.D. (1983) α-Amylase production in the late-stages of grain development – an early sprouting damage risk period? In: Kruger, J.E. and LaBerge, D.E. (eds) *3rd International Symposium on Pre-harvest Sprouting in Cereals.* Westview Press, Boulder, Colorado, pp. 29–35.

Gallagher, J.N. and Biscoe, P.V. (1978) A physiological analysis of cereal yield I. Partitioning of dry matter. *Agricultural Progress* 53, 51–70.

Gallagher, E.J., Doyle, A. and Dilworth, D. (1987) Effect of management practices on cereal yield and quality. In: *Aspects of Applied Biology 15, Cereal Quality.* Association of Applied Biologists, Warwick, pp. 151–160.

Gareis, M. and Ceynowa, J. (1994) Influence of the fungicide Matador (tebuconazole/triadimenol) on mycotoxin production by *Fusarium culmorum. Zeitschrift fur Lebensmittel Untersuchung und Forschung* 198, 244–248.

Garstang, J.R., Clark, W.S. and Dampney, P.M.R. (1994) *Production Methods for Cereals within the Reformed CAP, Research Review No. 26.* Home-Grown Cereals Authority, London.

Gedye, D.J. and Joyce, D. (1978) Controlling crop inputs to improve cereal yields. In: *Developments in the Business and Practice of Cereal Seed Trading and Technology.* Gavin Press, London, pp. 88–127.

Ghaderi, A. and Everson, E.H. (1971) Genotype–environment studies of test weight and its components in soft winter wheat. *Crop Science* 11, 617–620.

Gilchrist, L.I. (1985) CIMMYT methods for screening wheat for *Helminthosporium sativum* resistance. In: *Wheats for More Tropical Environments, A Proceedings of an International Symposium.* CIMMYT, Mexico D.F., pp. 149–151.

Gill, M., Beever, D.E. and Osbourn, D.F. (1989) The feeding value of grass and grass products. In: Holmes, W. (ed.) *Grass, Its Production and Utilization.* Blackwell Scientific Publications, Oxford, pp. 89–129.

Givens, D.I. (1994) *Effect of Nitrogen Fertiliser on the Protein Quality of Wheat for Ruminants. Project Report No. 86.* Home-Grown Cereals Authority, London.

Gold, C.M., Duffus, C.M. and Russell, G. (1990) Environmental effects on α-amylase activity during grain development of winter wheat cultivars. In: *Aspects of Applied Biology 25, Cereal Quality II.* Association of Applied Biologists, Warwick, pp. 159–162.

Gooding, M.J. (1988) Interactions between late-season foliar applications of urea and fungicide on foliar disease, yield and breadmaking quality of winter wheat. PhD thesis (CNAA), Harper Adams Agricultural College.

Gooding, M.J. and Alliston, J.C. (1993) Environmental aspects of whole crop cereal production. In: Pain, B.F., Gooding, M.J., Chadd, S.A. and Alliston, J.C. (eds) *Maize, Whole Crop Cereals and the Environment.* Ministry of Agriculture, Fisheries and Food, London.

Gooding, M.J. and Davies, W.P. (1992) Foliar urea fertilization of cereals: A review. *Fertilizer Research* 32, 209–222.

Gooding, M.J., Davies, W.P., Kettlewell, P.S. and Hocking, T.J. (1986a) The influence of late-season fungicide application and disease on the Hagberg falling number of grain from breadmaking varieties of winter wheat. *Cereal Research Communications* 14, 245–249.

Gooding, M.J., Kettlewell, P.S., Davies, W.P. and Hocking, T.J. (1986b) Effects of spring nitrogen fertilizer on the Hagberg falling number of grain from breadmaking varieties of winter wheat. *Journal of Agricultural Science, Cambridge* 107, 475–477.

Gooding, M.J., Kettlewell, P.S., Davies, W.P. and Hocking, T.J. (1987a) Do fungicides reduce Hagberg falling number of wheat by maintaining grain moisture content during grain development? In: *Aspects of Applied Biology 15, Cereal Quality*. Association of Applied Biologists, Warwick, pp. 413–416.

Gooding, M.J., Kettlewell, P.S., Davies, W.P., Hocking, T.J. and Salmon, S.E. (1987b) Interactions between late-season foliar urea and fungicide applications on the breadmaking quality of winter wheat. In: *Aspects of Applied Biology 15, Cereal Quality*. Association of Applied Biologists, Warwick, pp. 385–394.

Gooding, M.J., Kettlewell, P.S. and Davies, W.P. (1988) Disease suppression by late season urea sprays on winter wheat and interaction with fungicide. *Journal of Fertilizer Issues* 5, 19–23.

Gooding, M.J., Kettlewell, P.S. and Hocking, T.J. (1991) Effects of urea alone or with fungicide on the yield and breadmaking quality of wheat when sprayed at flag leaf and ear emergence. *Journal of Agricultural Science, Cambridge* 117, 149–155.

Gooding, M.J., Thompson, A.J. and Davies, W.P. (1993a) Interception of photosynthetically active radiation competitive ability and yield of organically grown wheat varieties. In: *Aspects of Applied Biology 34, Physiology of Varieties*. Association of Applied Biologists, Warwick, pp. 355–362.

Gooding, M.J., Davies, W.P., Thompson, A.J. and Smith, S.P. (1993b) The challenge of achieving breadmaking quality in organic and low input wheat in the UK – A review. In: *Aspects of Applied Biology 36, Cereal Quality III*. Association of Applied Biologists, Warwick, pp. 189–198.

Gooding, M.J., Thompson, A.J., Collingborn, F.M.B., Smith, S.P. and Davies, W.P. (1993c) Blackpoint on wheat grain: Influences of cultivar, management and season on symptom severity. In: *Aspects of Applied Biology 36, Cereal Quality III*. Association of Applied Biologists, Warwick, pp. 391–396.

Gooding, M.J., Smith, S.P., Davies, W.P. and Kettlewell, P.S. (1994) Effects of late-season applications of propiconazole and tridemorph on disease, senescence, grain development and the breadmaking quality of winter wheat. *Crop Protection* 13, 362–370.

Gooding, M.J., Smith, G.P., Davies, W.P. and Kettlewell, P.S. (1997) The use of residual maximum likelihood to model grain quality characters of wheat with variety, climatic and nitrogen fertiliser effects. *Journal of Agricultural Science, Cambridge* (in press).

Gorbacheva, A.E., Lapko, P.G. and Dzyubinskii, N.F. (1989) Effectiveness of different fertilizer systems for field crops. *Vestnik-Sel'skokhozyaistvennoi Nauki, Moscow* No. 1, 85–89.

Graham, R.D. (1991) Breeding wheats for tolerance to micro-nutrient deficient soils: Present status and priorities. In: Saunders, D.A. (ed.) *Wheat for the Nontraditional, Warm Areas*. CIMMYT, Mexico D.F., pp. 315–332.

Grama, A., Porter, N.G. and Wright, D.S.C. (1987) Hexaploid wild emmer wheat derivatives grown under New Zealand conditions 2. Effect of foliar urea sprays on plant and grain nitrogen and baking quality. *New Zealand Journal of Agricultural Research* 30, 45–51.

Gray, R.C. (1977) Foliar fertilisation with primary nutrients during the reproductive stage of plant growth. *Proceedings of the Fertiliser Society, London.* No. 164. Fertiliser Society, London.

Graybosch, R.A., Peterson, C.J., Baenziger, P.S. and Shelton, D.R. (1995) Environmental modification of hard red winter wheat flour protein composition. *Journal of Cereal Science* 22, 45–51.

Gregory, P.H. (1973) *Microbiology of the Atmosphere.* Leonard Hill, London.

Grifanov, V.K. and Davydov, A.M. (1972) Effect of molybdenum and urea in inducing the protein content of the grain of spring wheat. *Agrokhimiya* No. 10, 137–140.

Griffith, D.R., Mannering, J.V. and Box, J.E. (1986) Soil and moisture management with reduced tillage. In: Sprague, M.A. and Triplett, G.B. (ed.) *Notillage and Surface-tillage Agriculture.* John Wiley and Sons, New York, pp. 1–18.

Grimm, K. (1987) Whole crop silage – WCS – as an alternative in forage production. *Bayerische Landwirtschaftliches Jahrbuch* 64, 45–57.

Grundy, A.C., Froud-Williams, R.J. and Boatman, N.D. (1992) The effects of nitrogen rate on weed occurrence in a spring barley crop. In: *Aspects of Applied Biology 30, Nitrate and Farming Systems.* Association of Applied Biologists, Warwick, pp. 377–380.

Gruzdev, L.G. (1986) Effectiveness of mixtures of growth regulators when winter wheat and spring barley plants are treated with them. *Soviet Agricultural Sciences* 3, 37–41.

Gupta, R.B., MacRitchie, F. and Shepherd, K.W. (1989) The relative effect of allelic variation in LMW and HMW glutenin subunits on dough properties in the progeny of two bread wheats. *Theoretical and Applied Genetics* 77, 57–64.

Hadjichristodoulou, A. (1976) Effect of genotype and rainfall on yield and quality of forage barley and wheat varieties in a semi-arid region. *Journal of Agricultural Science, Cambridge* 87, 489–497.

Hagras, A.M. (1985) Response of wheat to nitrogen, phosphorus and potassium fertilization. *Annals of Agricultural Science, Moshtohor* 23, 1023–1035.

Hall, N. (1981) Adjustment in the Australian wheat growing industry. *Quarterly Review of Rural Economy* 3, 163–165.

Hall, D.O. and Rao, K.K. (1994) *Photosynthesis*, 5th edn. Cambridge University Press, Cambridge.

Halverson, J. and Zeleny, L. (1988) Criteria of wheat quality. In: Pomeranz, Y. (ed.) *Wheat Chemistry and Technology.* American Association of Cereal Chemists, Minnesota, pp. 15–45.

Hambridge, K.M. (1981) Zinc deficiency in man: its origins and effects. *Philosophical Transactions of the Royal Society, London* B 294, 129–144.

Haneklaus, S. and Schnug, E. (1992) Baking quality and sulphur content of wheat II. Evaluation of the relative importance of genetics and environment including sulphur fertilization. *Sulphur in Agriculture* 16, 35–38.

Haneklaus, S., Evans, E. and Schnug, E. (1992) Baking quality and sulphur content of wheat I. Influence of grain sulphur and protein concentrations on loaf volume. *Sulphur in Agriculture* 16, 31–34.

Hanson, E.W. and Christensen, J.J. (1953) The blackpoint disease of wheat in the United States. *University of Minnesota Technical Bulletin* No. 206. University of Minnesota, Minnesota.

Hanson, H., Borlaug, N.E. and Anderson, R.G. (1982) *Wheat in the Third World.* Westview Press, Colorado.

Harein, P.K. (1982) Chemical control alternatives for stored-grain insects. In: Christensen, C.M. (ed.) *Storage of Cereal Grain and their Products.* American Association of Cereal Chemists, St Paul, Minnesota. pp. 319–362.

Harlan, J.R. (1981) The early history of wheat: Earliest traces to the sack of Rome. In: Evans, L.T. and Peacock, W.J. (eds) *Wheat Science – Today and Tomorrow.* Cambridge University Press, Cambridge.

Hart, B.J., Manley, W.J., Limb, T.M. and Davies, W.P. (1994) Habitat creation in large fields for natural pest regulation. In: *Field Margins: Integrating Agriculture and Conservation. BCPC Monograph 58.* BCPC Publications, Croydon, pp. 319–322.

Harvey, J.J. (1985) Control of *Poa* spp. in winter cereals – is it worthwhile? In: *Aspects of Applied Biology 9, The Biology and Control of Weeds in Cereals.* Association of Applied Biologists, Warwick, pp. 117–128.

Harvey, J.J. (1990) The cost of utilisable metabolisable energy from forage crops. In: Pollot, G.E. (ed.) *Milk and Meat from Forage Crops, Occasional Symposium No. 24.* The British Grassland Society, Reading, pp. 33–40.

Harvey, J.J. (1992) Assessing whole-crop cereal maturity in the field. In: Stark, B.A. and Wilkinson, J.M. (eds) *Whole-Crop Cereals,* 2nd edn. Chalcombe Publications, Canterbury, pp. 39–49.

Hayward, C.F. (1987) The effect of date of harvest on the grain quality of winter wheat at Terrington E.H.F. 1982–85. In: *Aspects of Applied Biology 15, Cereal Quality.* Association of Applied Biologists, Warwick, pp. 181–187.

Hayward, C.F. (1990a) The effect of sowing date on grain quality of winter wheat. In: *Aspects of Applied Biology 25, Cereal Quality II.* Association of Applied Biologists, Warwick, pp. 163–170.

Hayward, C.F. (1990b) The effect on grain yield and quality of sowing cereal cultivars out of season at Terrington Experimental Husbandry Farm. In: *Aspects of Applied Biology 25, Cereal Quality II.* Association of Applied Biologists, Warwick, pp. 171–176.

Hayward, C.F., Froment, M.A. and Harrison, R. (1993a) The effect of spring applied animal slurries on cereal grain yield and quality. In: *Aspects of Applied Biology 36, Cereal Quality III.* Association of Applied Biologists, Warwick, pp. 311–316.

Hayward, C.F., Jackson, D.R. and Smith, K.A. (1993b) Nitrogen efficiency of autumn, winter and spring applications of organic manures on winter cereals and its effect on grain yield and quality. In: *Aspects of Applied Biology 36, Cereal Quality III.* Association of Applied Biologists, Warwick, pp. 301–310.

He, W.X., Zhang, Y.L., Sun, Y.Z. and Dong, Y.C. (1990) A study on the protein content of durum wheat varieties as related to ecological environment. *Crop Genetic Resources* No.4, 25–26.

Heaney, S.P., Martin, T.J. and Smith, J.M. (1988) Practical approaches to managing anti-resistance strategies with DMI fungicides. *Brighton Crop Protection Conference, Pests and Diseases – 1988.* British Crop Protection Council, Thornton Heath, UK, pp. 1097–1106.

Heenan, D.P., Taylor, A.C., Cullis, B.R. and Lill, W.J. (1994) Long term effects of rotation, tillage and stubble management on wheat production in southern N.S.W. *Australian Journal of Agricultural Research* 45, 93–117.

Hemantaranjan, A. and Garg, O.K. (1988) Iron and zinc fertilization with reference to the grain quality of *Triticum aestivum* L. *Journal of Plant Nutrition* 11, 1439–1450.

Henly, S.J., Dodge, A.D. and Stephens, R.J. (1985) Aspects of the selectivity of barley, *Bromus sterilis* and *Bromus willdenowii* to isoproturon. In: *Aspects of Applied Biology 9, The Biology and Control of Weeds in Cereals*. Association of Applied Biologists, Warwick, pp. 81–89.

Hera, C., Popescu, S., Vines, I., Stan, S. and Oproiu, E. (1986) Influence of some technological factors on wheat quality. *Productia Vegetala, Cereale si Plante Tehnice* 38 (7), 27–33.

Herold, A. and Walker, D.A. (1979) Transport across chloroplast envelopes. The role of phosphate. In: Tosteson, D.C. (ed.) *Membrane Transport in Biology II. Transport Across Single Biological Membranes*. Springer-Verlag, Berlin, pp. 411–439.

Hevia, H., Tollenaar, G.H., Serri, C.H. and Villegas, R. (1988) Effect of sowing date on baking quality of several wheat cultivars in south-central Chile. *Agro-Ciencia* 4, 117–123.

Hill, J. and Leaver J.D. (1990) Utilisation of whole-crop wheat by dairy cattle. In: Pollot, G.E. (ed.) *Milk and Meat from Forage Crops, Occasional Symposium No. 24*. The British Grassland Society, Reading, pp. 99–102.

Hill, J. and Leaver J.D. (1991a) Effect of stage of growth and urea addition on the preservation and nutritive value of whole crop wheat. *Animal Production* 52, 606.

Hill, J. and Leaver J.D. (1991b) Replacement of whole crop wheat by grass silage in the diet of dairy cows. *Animal Production* 52, 606.

Hoare, F.A. and Jordan, V.W.L. (1984) Interactions between stem base pathogens of wheat and their control by seed treatment and fungicide sprays. In: *Brighton Crop Protection Conference, Pests and Diseases 1984*. British Crop Protection Council, Thornton Heath, UK, pp. 77–82.

Hobbs, P.R., Woodhead, T. and Meisner, C. (1994) Soil physical factors limiting the productivity of the rice–wheat rotation and ways to reduce their impact through management. In: Saunders, D.A. and Hettel, G.P. (eds) *Wheat in Heat-Stressed Environments: Irrigated, Dry Areas and Rice–Wheat Farming Systems*. CIMMYT, Mexico D.F., pp. 276–289.

Hoffman, L.A., Harwood, J.L. and Leath, M.N. (1988) Economic effects of standardizing wheat protein reporting. *Staff-Report, Economic Research Service* No. AGES880826. US Department of Agriculture.

Hofmann, B. and Pallutt, B. (1989) Studies on the control of *Galium aparine* L. with SYS 67 Gebifan, SYS 67 Gebifan + Basagran as well as tank mixes of these herbicides with bercema-Bitosen N or ammonium nitrate with urea solution. *Nachrichtenblatt fur den Pflanzenschutz in der DDR* 43 (9), 180–183.

Holbrook, J.R., Byrne, W.R. and Ridgman, W.J. (1983) Assessment of results from wheat trials testing varieties and application of nitrogenous fertilizer. *Journal of Agricultural Science, Cambridge* 101, 447–452.

Holland, J.M. (1994) Progress towards integrated arable farming research in Western Europe. *Pesticide Outlook* (December), 17–23.

Holliday, R. (1956) Fodder production from winter-sown cereals and its effect upon grain yield. *Field Crop Abstracts* 9, 129–135, 207–213.

Holmes, M.R.J. (1980) Nitrogen – interaction with variety. In: *Winter Wheat: Proceedings of the 16th NIAB Crop Conference*. National Institute of Agricultural Botany, Cambridge, pp. 62–70.

Holzer, Z., Levy, D., Samuel, V. and Angemar, Y. (1985) Wheat silage for fattening young male cattle. *Animal Feed Science and Technology* 12, 253–266.

Honek, A. (1991) Nitrogen fertilization and abundance of the cereal aphids *Metopolophium dirhodum* and *Sitobium avenae* (Homoptera, Aphididae). *Zeitschrift fur Pflanzenkrankheiten und Pflanzenschutz* 98, 655–660.

Hook, S.C.W., Salmon, S.E. and Stewart, B.A. (1989) *Effects of Various Nitrogen-fertilizer Regimes on the Milling and Baking Qualities of Home-grown Bread-making Wheats. Project Report No. 14*. Home-Grown Cereals Authority, London.

Hopkins, J.W. (1968) Protein content of western Canadian hard red spring wheat in relation to some environmental factors. *Agricultural Meteorology* 5, 411–431.

Horn, F.P., Horn, G.W. and Crookshank, H.R. (1977) Effect of short-term fasting and re-feeding on weight and blood components of steers grazing wheat pasture. *Journal of Animal Science* 44, 288–294.

Hough, M.N. (1990) Weather patterns and cereal quality. In: *Aspects of Applied Biology 25, Cereal Quality II*. Association of Applied Biologists, Warwick, pp. 143–148.

Howard, D.D. (1986) Ammonium nitrate, urea and urea-ammonium nitrate solution as nitrogen sources for winter wheat. *Journal of Fertilizer Issues* 3, 25–29.

Hunshal, C.S., Chimmad, B.V., Hunshal, S.C., Viswanath, D.P. and Balikai, R.A. (1989) Effect of saline water irrigation on chapati quality. *Karnatka Journal of Agricultural Sciences* 2, 309–314.

Hunter, P.J. (1967) The effect of cultivations on slugs of arable ground. *Plant Pathology* 16, 153.

Hunter, A.S. (1974) Field comparisons of formamide, urea-ammonium nitrate solution and ammonium nitrate as nitrogen sources for grasses and wheat. *Agronomy Journal* 66, 540–543.

Hunter, P.J. and Symunds, P.V. (1970) The distribution of bait pellets for slug control. *Annals of Applied Biology* 65, 1.

Hurle, K. (1993) Integrated management of grass weeds in arable crops. In: *Brighton Crop Protection Conference, Weeds – 1993*. British Crop Protection Council, Thornton Heath, UK, pp. 81–88.

Hyde, M.B. and Galleymore, H.B. (1951) The subepidermal fungi of cereal grains II. The nature, identity and origin of the mycelium in wheat. *Annals of Applied Biology* 38, 348–356.

Ingram, J. (1990) The potential yield and varietal choice available for the major forage crops. In: Pollot, G.E. (ed.) *Milk and Meat from Forage Crops, Occasional Symposium No. 24*. The British Grassland Society, Reading, pp. 13–23.

International Food Policy Research Institute (1995) *Nutrient-enriched Food Crops could Save Millions of Poor People from Malnutrition and Disease and Benefit World Farmers through Higher Yields*. News Release. International Food Policy Research Institute, Washington, USA.

Ionescu, S., Patrascoiu, C., Nedelciuc, C., Nedelciuc, M., and Toma, M. (1988) Contributions to establishing fertilizer systems in maize and wheat crops under irrigation, on the south eastern Oltenian Plain. *Analele Institutului de Cercetari pentru Cereale si Plante Tehnice Fundulea* 56, 245–259.

Ivanic, J. and Vnuk, L. (1988) Effect of different nutrition on the crude and actual protein contents in grain DM of winter wheat cv. Kosutka. *Pol'nohospodarstvo* 34, 961–968.

Jackson, G.D., Kushnak, G.D., Benson, A.N., Skogley, E.O. and Lund, R.E. (1991) Potassium response in no-till small grain production. *Journal of Fertilizer Issues* 8 (3), 89–92.

Jaggi, I.K., Gorwantiwar, S.M. and Khanna, S.S. (1972) Effect of bulk density and aggregate size on wheat growth. *Journal of the Indian Society of Soil Science* 20, 421–423.

Jain, N.K., Maurya, D.P. and Singh, H.P. (1971) Effects of time and method of applying nitrogen to dwarf wheats. *Experimental Agriculture* 7, 21–26.

Jarman, R.J. and Pickett, A.A. (1994) *Botanical Descriptions of Cereal Varieties.* National Institute of Agricultural Botany, Cambridge.

Jedel, P.E. and Salmon, D.F. (1994) Date and rate of seeding winter cereals in central Alberta. *Canadian Journal of Plant Science* 74, 447–453.

Jeger, M.J., Jones, D.G. and Griffiths, E. (1981) Disease progress of non-specialised fungal pathogens in intra-specific mixed stands of cereal cultivars. II. Field experiments. *Annals of Applied Biology* 98, 199–210.

Jenkinson, D.S. (1986) Nitrogen in UK arable agriculture. *Journal of the Royal Agricultural Society of England* 147, 178–189.

Joelsson, A. and Petterson, O. (1984) Agriculture in Southern Halland; nutrient supply, leakage and counter measures. Rapport 1597, *Naturvardsverket*, pp. 25–28.

Johnson, V.A. and Mattern, P.J. (1987) Wheat, rye and triticale. In: *Nutritional Quality of Cereal Grains: Genetic and Agronomic Improvement. Agronomy Monograph No. 28.* ASA-CSSA-SSSA, Madison, WI, USA, pp. 135–150.

Johnson, V.A., Mattern, P.J. and Scmidt, J.W. (1967) Nitrogen relations during spring growth in varieties of *Triticum aestivum* L. differing in grain protein content. *Crop Science* 7, 664–667.

Johnson, V.A., Dreier, A.F. and Grabouski, P.H. (1973) Yield and protein responses to nitrogen fertilizer of two winter wheat varieties differing in inherent protein content of their grain. *Agronomy Journal* 65, 259–263.

Johnsson, L. (1992) Climate factors influencing attack of common bunt (*Tilletia caries* (DC) Tul.) in winter wheat in 1940–1988 in Sweden. *Zeitschrift fur Pflanzenkrankheiten und Pflanzenschutz* 99, 21–28.

Jones, R.G. (1987a) Quality requirements for wheat starch and gluten extraction. In: *Aspects of Applied Biology 15, Cereal Quality.* Association of Applied Biologists, Warwick.

Jones, D.R. (1987b) The effects of early season fungicides on quality of winter wheat. In: *Aspects of Applied Biology 15, Cereal Quality.* Association of Applied Biologists, Warwick, pp. 395–402.

Jones, D.R. (1994) Evaluation of fungicides for control of eyespot disease and yield loss relationships in winter wheat. *Plant Pathology* 43, 831–846.

Jones, D.G. and Clifford, B.C. (1983) *Cereal Diseases, Their Pathology and Control,* 2nd edn. John Wiley and Sons, Chichester, UK.

Jones, F.G.W. and Jones, M.G. (1973) *Pests of Field Crops.* Edward Arnold, London.

Jones, P., Twoney, U. and Dhitaphichit, P. (1995) Effect of loose smut infection on wheat and barley plant development: impact of dwarfing genes on disease epidemiology. In: *Aspects of Applied Biology 42, Physiological Responses of Plants to Pathogens.* Association of Applied Biologists, Warwick, pp. 33–41.

Jordan, V.W.L. (1992) *Nitrogen and Fungicide Interactions in Breadmaking Wheat.* Home-Grown Cereal Authority, London.

Jordan, V.W.L. and Hutcheon, J.A. (1994) Less intensive integrated farming systems for arable crop production and environmental protection. *Proceedings of the Fertilizer Society* No. 346. Fertilizer Society, London.

Jordan, V.W.L. and Hutcheon, J.A. (1995) Strategies for disease control to meet the needs of current and future cereal production systems. In: Hewitt, H.G., Tyson, D., Hollomon, D.W., Smith, J.M., Davies, W.P. and Dixon K.R. (eds) *A Vital Role for Fungicides in Cereal Production.* BIOS, Oxford.

Jordan, V.W.L., Hutcheon, J.A. and Glen, D.M. (1993) *Studies in Technology Transfer of Integrated Farming Systems: Considerations and Principles for Development.* Long Ashton Research Station, Bristol.

Justus, N. and Thurman, R.L. (1955) The effect of clipping and grazing on the subsequent growth of winter oats. *Agronomy Journal* 47, 82–83.

Kahurananga, J. (1991) Intercropping Ethiopian *Trifolium* species with wheat. *Experimental Agriculture* 27, 391–395.

Kalid, A.W. and Wali, S.B. (1988) Effect of seeding rate and nitrogen fertilization on quality of two wheat cultivars grown under rain fed condition in north Iraq. 3. Effects on quality. *Iraqi Journal of Agricultural Sciences, ZANCO* 6, 147–157.

Kalinin, N.I. (1986) Change of spring wheat grain quality under the effect of air temperature. *Soviet Agricultural Sciences* No. 1, 16–18.

Kalinin, N.I. (1988) Protein content of spring wheat grain as a function of hydrothermic conditions. *Soviet Agricultural Sciences* No. 2, 20–33.

Karjalainen, R. and Salovaara, H. (1988) Effects of severe infection with *Septoria nodorum* on spring wheat quality. *Acta Agriculturae Scandinavica* 38, 183–188.

Karvonen, T., Peltonen, J. and Kivi, E. (1991) The effect of northern climate conditions on sprouting damage of wheat grains. *Acta Agriculturae Scandinavica* 41, 55–64.

Kasatikov, V.A. and Runik, V.E. (1989) Amino acid composition of grain when using municipal sewage sludge as fertilizer. *Soviet Agricultural Sciences* No. 2, 15–18.

Katyal, J.C. and Friesen, D.K. (1988) Deficiencies of micronutrients and sulfur in wheat. In: Klatt, A.R. (ed.) *Wheat Production Constraints in Tropical Environments.* CIMMYT, Mexico D.F., pp. 99–129.

Keen, B.W. (1991) Weed control in small grain cereals in a low profit situation. The short term view. In: *Brighton Crop Protection Conference, Weeds – 1991.* British Crop Protection Council, Thornton Heath, UK, pp. 113–120.

Kelman, W.M. and Qualset, C.O. (1993) Responses of recombinant inbred lines of wheat to saline irrigation: Milling and baking qualities. *Crop Science* 33, 1223–1228.

Kent, N.L. and Evers, A.D. (1994) *Kent's Technology of Cereals.* Pergamon Press, Oxford.

Kernan, J.A., Coxworth, E.C., Crowle, W.L. and Spurr, D.T. (1984) The nutritional value of crop residue components from several wheat cultivars grown at different fertilizer levels. *Animal Feed Science and Technology* 11, 301–311.

Kettlewell, P.S. (1989) Breadmaking quality in wheat. *Agricultural Progress* 64, 30–45.

Kettlewell, P.S. (1993) Pre-harvest prediction of Hagberg falling number of wheat: some possible methods for the United Kingdom. In: *Aspects of Applied Biology 36, Cereal Quality III*. Association of Applied Biologists, Warwick, pp. 257–266.

Khatkar, B.S., Bell, A.E. and Schofield, J.D. (1995) The dynamic rheological properties of glutens and gluten subfractions from wheat of good and poor breadmaking quality. *Journal of Cereal Science* 22, 29–44.

Khodabandeh, N. (1985) Study of the effect of planting date on the productivity and protein content of Omid wheat. *Iranian Journal of Agricultural Sciences* 16, 27–33.

King, J.E. (1973) Cereal foliar disease surveys. *Proceedings of the 7th British Insecticide and Fungicide Conference, Brighton 1973*. BCPC Publications, Croydon, pp. 771–780.

King, J.E. (1976) Relationship between yield loss and severity of yellow rust recorded on a large number of single stems of winter wheat. *Plant Pathology* 25, 172–177.

King, R.W. and Richards, R.A. (1984) Water uptake in relation to pre-harvest sprouting damage in wheat: ear characteristics. *Australian Journal of Agricultural Research* 35, 327–336.

King, J.E., Jenkins, J.E.E. and Morgan, W.A. (1983) The estimation of yield losses in wheat from severity of infection by *Septoria* species. *Plant Pathology* 32, 239–249.

Kivisaari, S. (1985) Sowing date and its effect on yield and some quality characters in wheat barley and oats in 3 different soil types in south Finland. *Nordisk-Jordbrugsforskning* 67, 183.

Kiyomoto, R.K. (1986) Wheat performance in Connecticut (1983–1984) *Bulletin, Connecticut Agricultural Experiment Station, New Haven No. 835*. Connecticut Agricultural Experiment Station, New Haven.

Klapp, E. (1967) *Lehrbuch des Acker- und Pllanzenbaues 6*. Auflage, Berlin.

Knapp, J.S. and Harms, C.L. (1988) Nitrogen fertilization and plant growth regulator effects on yield and quality of four wheat cultivars. *Journal of Production Agriculture* 1, 94–98.

Kolderup, F. (1979) Application of different temperatures in three growth phases of wheat. III. Effects on protein content and composition. *Acta Agriculturae Scandinavica* 29, 379–384.

Kolenbrander, I.G.J. (1983) Fertilisers, farming practice and water quality. *Fertiliser Society Proceedings* 135, 1–52.

Kosmolak, F.G. and Crowle, W.L. (1980) An effect of nitrogen fertilization on the agronomic traits and dough mixing strength of five Canadian hard red spring wheat cultivars. *Canadian Journal of Plant Science* 60, 1071–1076.

Kosmolak, F.G., Dexter, J.E., Matsuo, R.R. and Marchylo, B.A. (1980) A relationship between durum wheat quality and gliadin electrophorgrams. *Canadian Journal of Plant Science* 60, 427–432.

Kostowska, B., Gabinska, K., Rola, J., Szymczak, J., Sykut, A. and Wybieralska, A. (1984) Influence of herbicides on the biological value of the grain of some wheat varieties. *Roczniki Nauk Rolniczych, E Ochrona Roslin* 14, 209–221.

Kovacs, M.I.P., Noll, J.S., Dahlke, G. and Leisle, D. (1995) Pasta viscoelasticity: Its usefulness in the Canadian durum wheat breeding program. *Journal of Cereal Science* 22, 115–121.

Kreyger, J. (1963) General considerations concerning the drying of seeds. *Journal of the International Seed Testing Association* 28, 753–784.

Kristenson, V.F. (1991) Production and feeding of whole-crop cereals and legumes in Denmark. In: *Proceedings of the Whole Crop Cereals Conference, The Royal Agricultural College, 18th April 1991.* The Maize Growers Association, Reading.

Kruez, E. and Zabel, S. (1989) The effect of interrupting a long term grain rotation with two years of lucerne production on the yield and baking quality of four winter wheat cultivars. *Archiv fur Acker und Pflanzenbau und Bodenkunde* 33, 539–544.

Kulik, M.F., Demchenko, V.E., Khimich, V.V. and Yurchenko, V.K. (1983) Deacylation of cellulose in the rumen of cattle. *Sel'skokhozyyaistvennaya Biologiya* 6, 108–113.

Kuroyanagi, T. and Paulsen, G.M. (1988) Mediation of high-temperature injury by roots and shoots during reproductive growth of wheat. *Plant, Cell and Environment* 11, 517–523.

Lacey, J. (1990) Mycotoxins in UK cereals and their control. In: *Aspects of Applied Biology 25, Cereal Quality II.* Association of Applied Biologists, Warwick, pp. 395–405.

Lagudah, E.S., O'Brien, L. and Halloran, G.M. (1988) Influence of gliadin composition and high molecular wheat subunits of glutenin on dough properties in an F_3 population of a bread wheat cross. *Journal of Cereal Science* 7, 33–42.

Lake, P.A. (1987) Quality requirements for the feed compounder. In: *Aspects of Applied Biology 15, Cereal Quality.* Association of Applied Biologists, Warwick, pp. 32–37.

Lal, R. (1991) Tillage practices and soil degradation in the wheat cropping systems of the warmer areas of Africa and Asia. In: Saunders, D.A. (ed.) *Wheat for the Nontraditional, Warm Areas.* CIMMYT, Mexico D.F., pp. 257–265.

Languasco, L., Orsi, C. and Rossi, V. (1993) Forecasting blackpoint of wheat using meteorological and fungal isolation data. *Proceedings of Workshop on Computer-based DSS on Crop Protection.* Parma, Italy.

Lapis, D.B. (1985) Chemical control of wheat diseases in the Philippines. In: *Wheats for More Tropical Environments, A Proceedings of an International Symposium.* CIMMYT, Mexico D.F., pp. 204–208.

Large, E.C. (1954) Growth stages in cereals. Illustrations of Feekes scale. *Plant Pathology* 3, 128–129.

Large, E.C. and Doling, D.A. (1962) The measurement of cereal mildew and its effect on yield. *Plant Pathology* 25, 63–73.

Larson, I. and Nielson, I.D. (1966) The effect of varying nitrogen supply on the content of amino-acid in wheat grains. *Plant and Soil* 24, 299–308.

Last, F.T. (1954) The effect of time of application of nitrogenous fertilizer on powdery mildew of winter wheat. *Annals of Applied Biology* 41, 381–392.

Lathia, D.N. and Koch, M. (1989) Comparative study of phytic acid content, *in-vitro* protein digestibility and amino acid composition of different types of flat breads. *Journal of the Science of Food and Agriculture* 47, 353–364.

Lawrence, G.J., MacRitchie, F. and Wrigley, C.W. (1988) Dough and baking quality of wheat lines deficient in glutenin subunits controlled by the *Glu-A1, Glu-B1* and *Glu-D1* loci. *Journal of Cereal Science* 7, 109–112.

Lawson, J.A. (1989) The effect of propiconazole on the physiology and pathology of winter wheat. PhD Thesis (CNAA), Harper Adams Agricultural College.

Leaver, J.D. and Hill, J. (1990) The nutritive value of whole-crop cereals for dairy cattle. In: Wilkinson, J.M. and Stark, B.A. (eds) *Whole-Crop Cereals: Making and Feeding Cereal Silage.* Chalcombe, Canterbury, pp. 21–26.

Lebidinskaya, V.N., Bober, L.V. and Kokhan, N.G. (1988) Systematic application of fertilizers and productivity of crops in a rotation. *Agrokhimiya* No. 7, 39–46.

Lee, G., Stevens, D.J., Stokes, S. and Wratten, S.D. (1981) Duration of cereal aphid populations on the effects on wheat yield and breadmaking quality. *Annals of Applied Biology* 98, 169–178.

Leggett, M.E. and Sivasithamparam, K. (1986) Interaction of fluorescent pseudo-monads, hyphae of the take-all fungus and growth of wheat roots in soil. In: *British Crop Protection Conference, Pest and Diseases.* British Crop Protection Council, Thornton Heath, UK, pp. 1177–1184.

Leitch, M. (1993) A view of wheat production in Shanxi province, China. *Journal of the University of Wales Agricultural Society* 73, 80–90.

Leprince, X. (1995) The market for fungicides in European wheat and barley crops. In: Hewitt, H.G., Tyson, D., Hollomon, D.W., Smith, J.M., Davies, W.P. and Dixon K.R. (eds) *A Vital Role for Fungicides in Cereal Production.* BIOS, Oxford, pp. 19–28.

Lerch, R.N., Barbarick, K.A., Westfall, D.G., Follett, R.H., McBride, T.M. and Owen, W.F. (1990) *Journal of Production Agriculture* 3, 66–71.

Lewis, O.A.M. (1986) *Plants and Nitrogen.* Edward Arnold, London.

Lewis, D.A. and Tatchell, J.A. (1979) Energy in UK agriculture. *Journal of the Science of Food and Agriculture* 30, 449–457.

Leygue, J.P. (1993) Cereals as industrial feedstock. In: *Aspects of Applied Biology 36, Cereal Quality III.* Association of Applied Biologists, Warwick, pp. 29–42.

Li, J.X., Cui, J.M., Gao, R.L. and Wang, X.C. (1991) Study on accumulation dynamic of grain protein in wheat. *Acta Agriculturae Universitatis Henanensis* 25, 366–371.

Lin, Z-J., Miskelly, D.M. and Moss, H.J. (1990) Suitability of various Australian wheats for Chinese-style steamed bread. *Journal of the Science of Food and Agriculture* 53, 203–213.

Lipsett, J. (1963) Factors affecting the occurrence of mottling in wheat. *Australian Journal of Agricultural Research* 14, 303–314.

Lisoval, A.P., Gudz, V.P., Dolya, N.N., Pravilov, N.V. and Malienko, N.V. (1991) Methods for increasing winter wheat yields. *Khimizatsiya Sel'skogo-Khozyaistva* No. 8, 64–66.

Littler, J.W. (1984) Effect of pasture on subsequent wheat crops on a black earth of the Darling Downs. I. The overall experiment. *Queensland Journal of Agricultural and Animal Sciences* 41, 1–12.

Liu, Z.Z. (1985) Recent advances in research on wheat scab in China. In: *Wheats for More Tropical Environments, A Proceedings of an International Symposium.* CIMMYT, Mexico D.F., pp. 174–181.

Lockhart, D.A.S. (1989) Cereal farming and harvesting. In: Palmer, G.H. (ed.) *Cereal Science and Technology.* Aberdeen University Press, Aberdeen, pp. 1–36.

Loewer, O.J., Bridges, T.C. and Bucklin, R.A. (1994) *On-Farm Drying and Storage Systems*. American Society of Agricultural Engineers, New York.

Loffler, C.M. and Busch, R.H. (1982) Selection of grain protein, grain yield, and nitrogen partitioning efficiency in hard red spring wheat. *Crop Science* 22, 591–595.

Lorenz, F. and Steffens, G. (1992) Agronomically efficient and environmentally careful application to arable crops. In: *Aspects of Applied Biology 30, Nitrate and Farming Systems*. Association of Applied Biologists, Warwick, pp. 109–116.

Lucas, J.A. (1995) Cereal management and the epidemiology of stem-base diseases. In: Hewitt, H.G., Tyson, D., Hollomon, D.W., Smith, J.M., Davies, W.P. and Dixon K.R. (eds) *A Vital Role for Fungicides in Cereal Production*. BIOS, Oxford, pp. 95–104.

Lui, M.Z., Zhang, H.Y. and Zhou, X.G. (1987) Effects of fertilizer application on the quality and yield of wheat. *Journal of Agricultural Sciences, Jiangsu Academy of Agricultural Sciences* 3 (3), 31–38.

Lukow, O.M., Payne, P.I. and Tkachuk, R. (1989) The HMW glutenin subunit composition of Canadian wheat cultivars and their association with breadmaking quality. *Journal of the Science of Food and Agriculture* 46, 451–460.

Lunn, G.D., Scott, R.K., Kettlewell, P.S. and Major, B.J. (1995) Effects of orange wheat blossom midge (*Sitodiplosis mosellana*) infection on pre-maturity sprouting and Hagberg falling number of wheat. In: *Aspects of Applied Biology 42, Physiological Responses of Plants to Pathogens*. Association of Applied Biologists, Warwick, pp. 355–358.

Lupton, F.G.H. (1972) Further experiments on photosynthesis and translocation in wheat. *Annals of Applied Biology* 71, 69–71.

Luzzardi, G.C. (1985) Wheat breeding for scab resistance. In: *Wheats for More Tropical Environments, A Proceedings of an International Symposium*. CIMMYT, Mexico D.F., pp. 158–168.

Lyon, T.L., Bizzell, J.A. and Wilson, B.D. (1929) *Journal of the American Society of Agronomy* 15, 457–467.

MacKenzie, D.R. (1982) Meeting world food needs: lessons of the green revolution. *Journal of the Northeastern Agricultural Economics Council* (fall), 107–111.

MacRitchie, F. (1973) Conversion of a weak flour to a strong one by increasing the proportion of its high molecular weight gluten protein. *Journal of the Science of Food and Agriculture* 24, 1325–1329.

Maga, J.P. (1975) *Critical Reviews of Food Technology* 5, 443–486.

Magan, N. and Lacey, J. (1984) Effect of water activity, temperature and substrate on interactions between field and storage fungi. *Transactions of the British Mycological Society* 82, 83–93.

Mahajan, A.R. and Nayeem, K.A. (1990) Effects of date of sowing on test weight, protein percent and yield in wheat and triticale genotypes. *Journal of Maharashtra Agricultural Universities* 15, 69–71.

Makepeace, R.J. (1982) A review of broad-leaved weed problems in spring cereals. In: *Aspects of Applied Biology 1, Broad-leaved Weeds and Their Control in Cereals*. Association of Applied Biologists, Warwick, pp. 103–108.

Manley, M. and Joubert, G.D. (1989) The effect of cultivar, location and sowing date on the protein content of wheat. *Applied Plant Science* 3, 15–17.

Marshall, E.J.P. (1985) Weed distributions associated with cereal field edges – some preliminary observations. In: *Aspects of Applied Biology 9, The Biology and Control of Weeds in Cereals.* Association of Applied Biologists, Warwick, pp. 49–58.

Martin, J.H., Leonard, W.H. and Stamp, D.L. (1976) *Principles of Field Crop Production.* Macmillan Publishing Co., New York.

Martin, D.A., Miller, S.D. and Alley, H.P. (1986) Small grain response to herbicides. *Proceedings of the Western Society of Weed Science* 39, 178–179.

Martin, D.A., Miller, S.D. and Alley, H.P. (1989) Winter wheat (*Triticum aestivum*) response to herbicides applied at three growth stages. *Weed Technology* 3 (1), 90–94.

Martin, D.A., Miller, S.D. and Alley, H.P. (1990) Spring wheat response to herbicides applied at three growth stages. *Agronomy Journal* 82, 95–97.

Marwat, A.Q., Karim, M., Khalil, S.K., Wazir, A.L. and Bakht, J. (1989) Effect of land preparation methods and seeding rates on weed populations and wheat yield. *Sarhad Journal of Agriculture* 5, 107–111.

Matsoukas, N.P. and Morrison, W.L. (1990) Breadmaking quality of ten Greek breadwheats – baking and storage tests on bread made by long fermentation and activated (chemical) dough development processes, and the effects of bug-damaged wheat. *Journal of the Science of Food and Agriculture* 53, 363–377.

Matsuo, R.R. and Dexter, J.E. (1980) Relationship between some durum wheat physical characteristics and semolina milling properties. *Canadian Journal of Plant Science* 60, 49–53.

Mavi, H.S., Jhorar, O.P., Sharma, I., Singh, G., Mahi, G.S., Mathauda, S.S., Aujla, S.S. and Singh, G. (1992) Forecasting Karnal bunt disease of wheat – a meteorological method. *Cereal Research Communications* 20, 67–74.

Mazurek, J. and Kus, J. (1991) Effect of nitrogen fertilizer application and sowing date and rate on yield and quality of grain of spring wheat cultivars grown after different preceding crops. Part 1. *Biuletyn Instytutu Hodowli i Aklimatyzacji Roslin* 177, 123–136.

McCaig, T.N. and DePauw, R.M. (1992) Breeding for preharvest sprouting tolerance in white-seed-coat spring wheat. *Crop Science* 32, 19–23.

McCance, R.A. and Widdowson, E.M. (1960) *The Composition of Foods*, 3rd edn. HMSO, London.

McClean, S.P. (1987) The management of milling wheat. In: *Aspects of Applied Biology 15, Cereal Quality.* Association of Applied Biologists, Warwick, pp. 125–136.

McCormack, G., Panozzo, J., Bekes, F. and MacRitchie, F. (1991) Contributions to breadmaking of inherent variations in lipid content and composition of wheat cultivars. I. Results of survey. *Journal of Cereal Science* 13, 255–261.

McDermott, E.E. and Pace, J. (1960) Comparison of the amino-acid composition of the protein in flour and endosperm from different types of wheat, with particular reference to variation in lysine content. *Journal of the Science of Food and Agriculture* 11, 109–115.

McDonald, G.K. (1989) The contribution of nitrogen fertiliser to the nitrogen nutrition of rainfed wheat crops in Australia: a review. *Australian Journal of Experimental Agriculture* 29, 455–481.

McDonald, H.G. and Vaidyanathan, L.V. (1987) Effects of nitrogen fertilizer on the Hagberg falling number of winter wheat. In: *Aspects of Applied Biology 15, Cereal Quality.* Association of Applied Biologists, Warwick, pp. 359–369.

McEwen, F.L. and Stephenson, G.R. (1979) *The Use and Significance of Pesticides in the Environment*. Wiley, New York.

McGrath, S.P. (1985) The effects of increasing yields on the macro- and micro-element concentration and offtakes in the grain of winter wheat. *Journal of the Science of Food and Agriculture* 36, 1073–1083.

McGrath, S.P., Zhao, F., Crosland, A.R. and Salmon, S.E. (1993) Sulphur status of British wheat grain and its relationship with quality parameters. In: *Aspects of Applied Biology 36, Cereal Quality III*. Association of Applied Biologists, Warwick, pp. 317–326.

McLean, K.A. (1987) Post-harvest manipulation and measurement of grain quality. In: *Aspects of Applied Biology 15, Cereal Quality*. Association of Applied Biologists, Warwick, pp. 483–494.

McLean, K.A. (1989) *Drying and Storing Combinable Crops*, 2nd edn. Farming Press, Ipswich, UK.

Mehta, Y.R. and Igarashi, S. (1985) Chemical control measures for the major diseases of wheat, with special attention to spot blotch. In: *Wheats for More Tropical Environments, A Proceedings of an International Symposium*. CIMMYT, Mexico D.F., pp. 196–200.

Melander, B. (1993) Population dynamics of *Apera spica-venti* as influenced by cultural methods. In: *Brighton Crop Protection Conference, Weeds – 1993*. British Crop Protection Council, Thornton Heath, UK, pp. 107–112.

Mercier, S. and Hyberg, B. (1995) Grain quality revisited. *Choices*, Vol. 1. American Agricultural Economics Association, pp. 35–38.

Meuser, F., Brummer, J.M. and Seibel, W. (1994) Bread varieties in Central Europe. *Cereal Foods World* 39, 224–230.

Micke, P. (1983) International research programmes for the genetic improvement of grain proteins. In: Gottshalk, W.E. and Muller, H.P. (eds) *Advances in Agricultural Biochemistry: Seed Proteins and Biochemistry*. Nijhoff Junk, the Hague, pp. 25–44.

Mielke, H. and Meyer, D. (1990) New investigations on control of head blight with reference to efficacy of fungicide treatment on yield and breadmaking quality in wheat. *Nachrichtenblatt des Deutschen Pflanzenshtzdienstes* 42 (11), 161–170.

Miller, T.E. (1987) Systematics and evolution. In: Lupton, F.G.H. (ed.) *Wheat Breeding, its Scientific Basis*. Chapman and Hall, London, pp. 1–30.

Miller, R.H. (1994) Integrated pest management of cereal aphids in the Nile Valley: Problems and potential. In: Saunders, D.A. and Hettel, G.P. (eds) *Wheat in Heat-Stressed Environments: Irrigated, Dry Areas and Rice–Wheat Farming Systems*. CIMMYT, Mexico D.F., pp. 109–120.

Miller, P.C.H. and Binns, T. (1994) *A Comparison of Methods of Applying Pesticides to Cereal Grains before Storage, Project Report No. 96*. Home-Grown Cereals Authority, London.

Mills, C.E. and Ryan, P.J. (1995) The control of herbicide-resistant *Alopecurus myosuroides* (blackgrass). In: *Brighton Crop Protection Conference, Weeds – 1995*. British Crop Protection Council, Thornton Heath, UK, pp. 1153–1160.

Milosev, D. (1989) Contents of nitrogen in grain and straw of wheat as affected by rate of nitrogen and temperature during grain formation and grain filling. *Agrochemija* No. 1–3, 33–46.

Ministry of Agriculture, Fisheries and Food (1986) *The Analysis of Agricultural Materials, MAFF Reference Book 427,* 3rd edn. HMSO, London.

Ministry of Agriculture, Fisheries and Food (1992) *Feed Composition: UK Tables of Feed Composition and Nutritive Value for Ruminants,* 2nd edn. Chalcombe, Canterbury.

Ministry of Agriculture, Fisheries and Food (1994) *Fertiliser Recommendations for Agricultural and Horticultural Crops, MAFF Reference Book 209,* 6th edn. HMSO, London.

Miskelly, D.M. (1984) Flour components affecting paste and noodle colour. *Journal of the Science of Food and Agriculture* 35, 463–471.

Miskelly, D.M. and Moss, H.J. (1985) Flour quality requirements for Chinese noodle manufacture. *Journal of Cereal Science* 3, 379–387.

Mitchell, G.A., Bingham, F.T. and Page, A.L. (1978) Yield and metal composition of lettuce and wheat grown on soils amended with sewage sludge enriched with cadmium, copper and zinc. *Journal of Environmental Quality* 9, 451–455.

Mitra, A.K. and Jana, P.K. (1991) Effect of doses and method of boron application on wheat in acid terai soils of north Bengal. *Indian Journal of Agronomy* 36, 72–74.

Mohamed, M.A., Steiner, J.J., Wright, S.D., Bhangoo, M.S. and Millhouse, D.E. (1990) Intensive crop management practices on wheat yield and quality. *Agronomy Journal* 82, 701–707.

Moonen, J.H.E. and Zeven, A.C. (1985) Association between high molecular weight subunits of glutenin and breadmaking quality in wheat lines derived from backcrosses between *Triticum aestivum* and *Triticum speltoides. Journal of Cereal Science* 3, 97–101.

Moore, A. (1985) Moulds and mycotoxins in animal feedstuffs for England and Wales 1981–1983. In: Moss, M.O. and Frank, M. (eds) *5th Meeting on Mycotoxins in Animal and Human Health.* University of Surrey, Guildford, pp. 117–126.

Morris, M.L., Belaid, A. and Byerlee, D. (1991) Wheat and barley production in rainfed marginal environments of the developing world. In: *Part I of CIMMYT World Wheat Facts and Trends.* CIMMYT, Mexico, D.F.

Morrison, I.N. and Bourgeois, L. (1995) Approaches to managing ACCase inhibitor resistance in wild oat on the Canadian prairies. In: *Brighton Crop Protection Conference, Weeds – 1995.* British Crop Protection Council, Thornton Heath, UK, pp. 567–576.

Moss, S.R. (1985) The effect of drilling date, pre-drilling cultivations and herbicides on *Alopecurus myosuroides* (black-grass) populations in winter cereals. In: *Aspects of Applied Biology 9, The Biology and Control of Weeds in Cereals.* Association of Applied Biologists, Warwick, pp. 31–39.

Moss, S.R. (1987) Competition between black-grass (*Alopecurus myosuroides*) and winter wheat. In: *Proceedings 1987 British Crop Protection Conference, Weeds.* British Crop Protection Council, Thornton Heath, UK, pp. 367–374.

Moss, H.J., Wrigley, C.W., MacRitchie, F. and Randall, P.J. (1981) Sulfur and nitrogen fertilizer effects on wheat II. Influence on grain quality. *Australian Journal of Agricultural Research* 32, 213–226.

Moss, H.J., Randall, P.J. and Wrigley, C.W. (1983) Alteration to grain, flour and dough quality in three wheat types with variation in soil sulfur supply. *Journal of Cereal Science* 1, 255–264.

Mount, M.S. and Ellingboe, A.H. (1969) ^{32}P and ^{35}S transfer from susceptible wheat to *Erysiphe graminis* f.sp. *hordei* during primary infection. *Phytopathology* 59, 235.

Muldoon, D.K. (1985) Simulation of the effect of forage cutting on subsequent grain yields in temperate cereals. *Agricultural Systems* 17, 231–242.

Murphy, M. (1991) A comparative perspective on the economic performance of organic and conventional farming systems in Great Britain. In: *Brighton Crop Protection Conference – Weeds*. British Crop Protection Council, Thornton Heath, UK, pp. 763–774.

Murray, A.W.A. and Nunn, P.A. (1987) A non-linear function to describe the response of per cent nitrogen in the grain to applied nitrogen fertiliser. In: *Aspects of Applied Biology 15, Cereal Quality*. Association of Applied Biologists, Warwick, pp. 219–226.

Mustafa, A.I., Makki, Y.M., Burhan, H.O. and Al-Tahir, O.A. (1987) Effect of sowing date and nitrogen fertilizer level on the quality of wheat grown in the eastern region of Saudi Arabia. *Arab Gulf Journal of Scientific Research, B Agricultural and Biological Sciences* 5, 349–365.

Myers, D.G. and Edsall, K.J. (1989) The application of image processing techniques to the identification of Australian wheat varieties. *Plant Varieties and Seeds* 2, 109–116.

Myram, C. (1984) Fungicides and their effect on milling quality. *Milling* (April), 40.

Nagao, S. (1995) Wheat Products in East Asia. *Cereal Foods World* 40, 482–487.

Nakhtore, C.L. and Kewat, M.L. (1989) Response of dwarf wheat to varying fertility levels under limited and adequate irrigation conditions. *Indian Journal of Agronomy* 34, 508–509.

Nambiar, K.K.M. (1991) Long-term fertility effects on wheat productivity. In: Saunders, D.A. (ed.) *Wheat for the Nontraditional, Warm Areas*. CIMMYT, Mexico D.F., pp. 516–521.

Narwal, S.S., Malik, D.S. and Malik, R.S. (1983) Studies in multiple cropping II. Effects of preceding grain legumes on the nitrogen requirement of wheat. *Experimental Agriculture* 19, 143–151.

Nash, J.N. (1985) *Crop Conservation and Storage in Cool Temperate Climates*. Pergamon Press, Oxford.

National Academy of Sciences (1978) *Postharvest Food Losses in Developing Countries*. National Academy of Sciences, Washington, DC.

National Institute of Agricultural Botany (1995) *Cereal Variety Handbook*. NIAB, Cambridge.

Neales, T.F., Anderson, M.J. and Wardlaw, I.F. (1963) The role of the leaves in the accumulation of nitrogen by wheat during ear development. *Australian Journal of Agricultural Research* 14, 725–736.

Nellist, M.E. (1986) *Drying Wheat in the UK. AFRC Institute of Engineering Research Report No. 50*. Silsoe, UK.

Nellist, M.E. and Bruce, D.M. (1987) Drying and cereal quality. In: *Aspects of Applied Biology 15, Cereal Quality*. Association of Applied Biologists, Warwick, pp. 439–456.

Netis, I.T. (1987) Fertilizer application to winter wheat with different sowing dates. *Agrokhimiya* No. 4, 46–50.

Newman, G. (1992) Future prospects for whole-crop cereals in the UK. In: Wilkinson, J.M. and Stark, B.A. (eds) *Whole-Crop Cereals: Making and Feeding Cereal Silage*, 2nd edn. Chalcombe, Canterbury, pp. 165–175.

Newman, G., East, J. and Kelly, P. (1992) Wholecrop cereals. *Technical Bulletin* No. 29. Maize Growers Association, Reading.

Nilsson-Ehle, H. (1914) Zur Kenntnis det mit der Keimungsphysiologie des Weizens in Zusammenhang stehenden inneren Faktoren. *Zeitschrift fur Pflanzenzuchtung* 2, 153–187.

Oakley, J.N., Ellis, S.A., Walters, K.F.A. and Watling, M. (1993) The effects of cereal aphid feeding on wheat quality. In: *Aspects of Applied Biology 36, Cereal Quality III*. Association of Applied Biologists, Warwick, pp. 383–390.

O'Brien, L., Brown, J.S., Panozzo, J.F. and Archer, M.J. (1990) The effect of stripe rust on the quality of Australian wheat varieties. *Australian Journal of Agricultural Research* 41, 827–833.

Oerke, E.C., Dehne, H.W., Shonbeck, F. and Weber, A. (1994) *Crop Production and Crop Protection – Estimated Losses in Major Food and Cash Crops*. Amsterdam, Elsevier.

Oglezneva, V.V. and Berkutova, N.S. (1987) Effect of growth retardants on grain technological qualities of winter wheat and rye. *Agrokhimiya* No. 11, 104–109.

Olered, R. (1967) Development of α-amylase and falling number in wheat and rye during ripening. *Vaxtodling* 23, 1–106.

Olered, R. and Jonsson, G. (1970) Electrophoretic studies of α-amylase in wheat. II. *Journal of the Science of Food and Agriculture* 21, 385–391.

Orson, J.H. (1988) The control of *Galium aparine* (cleavers) in winter cereals with herbicides: ADAS results, harvest years 1985 to 1987. In: *Aspects of Applied Biology 18, Weed Control in Cereals and the Impact of Legislation on Pesticide Application*. Association of Applied Biologists, Warwick, pp. 99–107.

Orson, J.H. (1995) Crop protection practices in wheat and barley crops. In: Hewitt, H.G., Tyson, D., Hollomon, D.W., Smith, J.M., Davies, W.P. and Dixon K.R. (eds) *A Vital Role for Fungicides in Cereal Production*. BIOS, Oxford, pp. 9–18.

Orth, R.A. and Moss, H.J. (1987) The sensitivity of various products to sprouted wheat. In: Mares, D.J. (ed.) *Fourth International Symposium on Pre-harvest Sprouting in Cereals*. Westview Press, Boulder, Colorado, pp. 167–175.

Orth, R.A. and Shellenberger, J.A. (1988) Origin, production, and utilization of wheat. In: Pomeranz, Y. (ed.) *Wheat Chemistry and Technology*. American Association of Cereal Chemists, Minnesota, pp. 1–14.

Osborne, B.G. and Fearn, T. (1983) Collaborative evaluation of near infrared reflectance analysis for the determination of protein, moisture and hardness in wheat. *Journal of the Science of Food and Agriculture* 34, 1011–1017.

Oskarsen, H. (1989) Fertilizers and growth regulation in spring wheat. *Norsk Landbruksforsking* 3, 177–183.

Oxley, T.A. and Hyde, M.B. (1955) Recent experiments on hermetic storage of wheat. *Proceedings, Third International Bread Congress*, Hamburg. pp. 179–182.

Pain, B.F., Gooding, M.J., Chadd, S.A. and Alliston, J.C. (1993) *Maize, Whole Crop Cereals and the Environment*. MAFF, London.

Palgrave, D.A. (1986) Focus on liquids. In: *Fertilizer Review 1986*. The Fertilizer Manufacturers Association Ltd, London, pp. 17–19.

Pandey, M. and Srivastava, G.P. (1985) Nitrate reductase activity in wheat leaves as influenced by herbicide application and its relationship with grain protein content. *Indian Journal of Agricultural Chemistry* 18, 239–245.

Parrish, S.K., Kaufmann, J.E., Croon, K.A., Ishida, Y., Ohta, K. and Itoh, S. (1995) MON 37500: A new selective herbicide to control annual and perennial weeds in wheat. In: *Brighton Crop Protection Conference, Weeds 1995*. British Crop Protection Council, Thornton Heath, UK, pp. 57–63.

Parry, D.W. (1990) *Plant Pathology in Agriculture*. Cambridge University Press, Cambridge.

Parry, D.W., Jenkinson, P. and McLeod, L. (1995) *Fusarium* ear blight (scab) in small grain cereals – a review. *Plant Pathology* 44, 207–238.

Partridge, J.R.D. and Shaykewich, C.F. (1972) Effects of nitrogen, temperature, and moisture regime on the yield and protein content of Neepawa wheat. *Canadian Journal of Soil Science* 52, 179–185.

Patterson, D.E. (1975) The development and assessment of reduced cultivation machinery. *Outlook on Agriculture* 8, 236–239.

Payne, P.I. (1987a) Genetics of wheat storage proteins and the effect of allelic variation on bread-making quality. *Annual Review of Plant Physiology* 38, 141–153.

Payne, P.I. (1987b) The genetical basis of bread-making quality in wheat. In: *Aspects of Applied Biology 15, Cereal Quality*. Association of Applied Biologists, Warwick, pp. 79–95.

Payne, P.I., Jackson, E.A. and Holt, L.M. (1984) The association between γ-gliadin 45 and gluten strength in durum wheat varieties: a direct causal effect or the result of genetic linkage? *Journal of Cereal Science* 2, 73–81.

Payne, P.I., Nightingale, M.A., Krattiger, A.F. and Holt, L.M. (1987) The relationship between HMW glutenin subunit composition and the bread-making quality of British-grown wheat varieties. *Journal of the Science of Food and Agriculture* 40, 51–65.

Payne, P.I., Holt, L.M., Krattiger, A.F. and Carrillo, J.M. (1988) Relationships between seed quality characteristics and HMW glutenin subunit composition determined using wheats grown in Spain. *Journal of Cereal Science* 7, 229–235.

Pedersen, B. and Eggum, B.O. (1983) The influence of milling on the nutritive value of flour from cereal grains. 2. Wheat. *Qualitas Plantarum – Plant Foods for Human Nutrition* 33, 51–61.

Pelikan, M. and Belan, F. (1987) Effect of some agrometeorological factors on quality and yield of winter wheat grain. *Acta Universitatis Agriculturae, Facultas Agronomica* 35, 83–89.

Peltonen, J. and Karjalainen, R. (1992) Effects of fungicide sprays on foliar diseases, yield, and quality of spring wheat in Finland. *Canadian Journal of Plant Science* 72, 955–963.

Pena, R.J. (1995) Wheat usage in Mexico and Central America. In: Faridi, H. and Faubion, J.M. (eds) *Wheat End Uses Around the World*. American Association of Cereal Chemists, St Paul, Minnesota, pp. 43–64.

Pena, R.J., Zarco-Hernandez, J., Amaya-Celis, A. and Mujeeb-Kazi, A. (1994) Relationships between 1B-encoded glutenin subunit compositions and bread-making quality characteristics of some durum wheat (*Triticum turgidum*) cultivars. *Journal of Cereal Science* 19, 243–249.

Pendleton, J.W. and Weibel, R.O. (1965) Shading studies in winter wheat. *Agronomy Journal* 57, 292–293.

Penning de Vries, F.W.T., Brunsting, A.H.M. and van Laar, H.H. (1974) Products, requirements and efficiency of biosynthesis: A quantitative approach. *Journal of Theoretical Biology* 45, 339–377.

Penny, A. and Freeman, C.R. (1974) Results from experiments with winter wheat, spring barley and grass, comparing a liquid N-fertilizer either alone or with added herbicide, and top-dressings of 'Nitro-Chalk' without or with herbicide sprayed alone. *Journal of Agricultural Science, Cambridge* 83, 511–529.

Penny, A. and Jenkyn, J.F. (1975) Results from two experiments with winter wheat, comparing top-dressings of a liquid N-fertilizer either alone or with added herbicide or mildew fungicide or both, and of 'Nitro-Chalk' without or with the herbicide or fungicide or both. *Journal of Agricultural Science, Cambridge* 85, 533–539.

Penny, A., Widdowson, F.N. and Jenkyn, J.E. (1978) Spring top-dressings of 'Nitro-Chalk' and late sprays of liquid N-fertilizer and a broad spectrum fungicide for consecutive crops of winter wheat at Saxmundham, Suffolk. *Journal of Agricultural Science, Cambridge* 90, 509–516.

Perten, H. (1964) Application of the falling number method for evaluating α-amylase activity. *Cereal Chemistry* 41, 127–140.

Peters, N.C.B. (1984) Competition between *Galium aparine* (cleavers) and cereals. In: *Understanding Cleavers (Galium aparine) and their Control in Cereals and Oilseed Rape – Abstracts of Papers*. Association of Applied Biologists, Warwick, pp. 3–4.

Petinov, N.S. and Pavlov, A.N. (1955) Increase of protein content of spring wheat grain (grown under irrigation) by means of spraying with nitrogenous supplements. *Fiziologiia Rastenii* 2, 113–122.

Petr, J. (1991) *Weather and Yield*. Elsevier, Amsterdam.

Petr, J., Cerny, V. and Hruska, L. (1988) *Yield Formation in the Main Field Crops*. Elsevier, Amsterdam.

Petrakova, L.V., Domanevskaia, N.A. and Bredikhin, V.N. (1974) Effect of nitrogen fertilizers on the fractional composition of protein and quality of gluten of spring wheat. *Agrokhimiya* No. 11, 3.

Petroczi, I.M. (1991) Effect of growth regulator (with ethephon as active agent) on the grain yield and baking quality of winter wheat. *Novenytermeles* 40, 241–249.

Pettigrew, D.E., Amos, H.E., Sisk, L.R. and McCormick, M.E. (1988) An evaluation of anhydrous ammonia treated sorghum and wheat silages and soybean hulls in total mixed rations. *Nutrition Reports International* 37, 607–621.

Phipps, R.H., Weller, R.F. and Siviter, J.W. (1990) Whole-crop cereals for dairy cows. In: Wilkinson, J.M. and Stark, B.A. (eds) *Whole-Crop Cereals: Making and Feeding Cereal Silage*, 2nd edn. Chalcombe, Canterbury, pp. 27–35.

Phipps, R.H., Weller, R.F., Clark, H. and Reeve, A. (1991) Preliminary observations on the production and quality of whole crop wheat. *Animal Production* 52, 604–605.

Pingali, P.L., Moya, P.F. and Velasco, L.E. (1990) The post-Green Revolution blues in Asian rice production: The diminished gap between experimental stations and farmers yields. Paper presented at the Annual Meeting of the American Agricultural Economics Association, Vancouver, Canada, August 1990.

Pixton, S.W. and Hill, S.T. (1967) Long-term storage of wheat – II. *Journal of the Science of Food and Agriculture* 18, 94–98.

Pixton, S.W., Warburton, S. and Hill, S.T. (1975) Long-term storage of wheat – III: Some changes in the quality of wheat observed during 16 years of storage. *Journal of Stored Products Research* 11, 177–185.

Plant Breeding Institute (1989) *Cereals: A Guide to Varieties*. Plant Breeding International, Cambridge.

Pogna, N.E., Boggini, G., Corbellini, M., Cattaneo, M. and Dal Belin Peruffo, A. (1982) Association between gliadin electrophoretic bands and quality in common wheat. *Canadian Journal of Plant Science* 62, 913–918.

Polley, R.W. and Clarkson, J.D.S. (1978) Forecasting cereal disease epidemics. In: Scott, P.R. and Bainbridge, A. (eds) *Plant Disease Epidemiology*. Blackwell Scientific Publications, Oxford, pp. 141–151.

Pomeranz, Y. (1982) Biochemical, functional and nutritive changes during storage. In: Christensen, C.M. (ed.) *Storage of Cereal Grain and their Products*. American Association of Cereal Chemists, St Paul, Minnesota. pp. 145–218.

Pomeranz, Y. (1987) *Modern Cereal Science and Technology*. VCH Publishers, New York.

Ponce, R.G., Lamela, A., Salas, M.L. and van Beusichem, M.L. (1990) Effects of nitrogen and herbicide on grain yield and protein content of a tall and a semi-dwarf wheat (*Triticum aestivum*) cultivar in semi-arid conditions. *Development in Plant and Soil Sciences* 41, 565–568.

Porceddu, E. and Srivastava, J.P. (1990) Evaluation, documentation and utilization of durum wheat germplasm at ICARDA and the University of Tuscia, Italy. In: Srivastava, J.P. and Damania, A.B. (eds) *Wheat Genetic Resources: Meeting Diverse Needs*. John Wiley and Sons, Chichester, pp. 3–8.

Porwal, M.K. and Gupta, O.P. (1993) Relationship between weed biomass and crude protein content in wheat as influenced by various methods of weed control. *Crop Research Hisar* 6, 377–382.

Powlson, D.S. (1987) Measuring and minimising losses of fertilizer nitrogen in arable agriculture. In: Jenkinson, D.S. and Smith, K.A. (eds) *Nitrogen Efficiency in Agricultural Soils*. London, Elsevier, pp. 231–245.

Powlson, D.S. and Jenkinson, D.S. (1981) A comparison of the organic matter, biomass, adenosine triphosphate and mineralizable nitrogen contents of ploughed and direct-drilled soils. *Journal of Agricultural Science, Cambridge* 97, 713–721.

Powlson, D.S., Pruden, G., Johnston, A.E. and Jenkinson, D.S. (1986) The nitrogen cycle in the Broadbalk wheat experiment: recovery and losses of ^{15}N-labelled fertilizer applied in spring and impact of nitrogen from the atmosphere. *Journal of Agricultural Science, Cambridge* 107, 591–609.

Powlson, D.S., Hart, P.B.S., Poulton, P.R., Johnston, A.E. and Jenkinson, D.S. (1992) Influence of soil type, crop management and weather on the recovery of ^{15}N-labelled fertilizer applied to winter wheat in spring. *Journal of Agricultural Science, Cambridge* 118, 83–100.

Pritchard, P.E. (1993) The glutenin fraction (gel-protein) of wheat protein – a new tool in the prediction of baking quality. In: *Aspects of Applied Biology 36, Cereal Quality III*. Association of Applied Biologists, Warwick, pp. 75–84.

Proven, M.J. and Dobson, S.C. (1987) Some observations on the effects of disease control on specific weight of winter wheat. In: *Aspects of Applied Biology 15, Cereal Quality*. Association of Applied Biologists, Warwick, pp. 425–431.

Pushman, F.M. and Bingham, J. (1975) Components of test weight of ten varieties of winter wheat grown with two rates of nitrogen fertilizer application. *Journal of Agricultural Science* 85, 559–563.

Pushman, F.M. and Bingham, J. (1976) The effects of a granular nitrogen fertilizer and a foliar spray of urea on yield and bread-making quality of ten winter wheats. *Journal of Agricultural Science* 87, 281–292.

Qarani, J., Ponte, J.G. and Posner, S. (1992) Flat breads of the World. *Cereal Foods World* 37, 863–865.

Radford, B.J., Gibson, G., Nielsen, R.G.H., Butler, D.G., Smith, G.D. and Orange, D.N. (1992) Fallowing practices, soil water storage, plant-available soil nitrogen accumulation and wheat performance in South West Queensland. *Soil and Tillage Research* 22, 73–93.

Raemaekers, R. (1985) Chemical control of *Helminthosporium sativum* on rainfed wheat in Zambia. In: *Wheats for More Tropical Environments, A Proceedings of an International Symposium*. CIMMYT, Mexico D.F., pp. 201–203.

Rao, A.C.S., Smith, J.L., Jandhyala, V.K., Papendick, R.I. and Parr, J.F. (1993) Cultivar and climatic effects on the protein content of soft white winter wheat. *Agronomy Journal* 85, 1023–1028.

Rasmussen, P.E. and Rohde, C.R. (1991) Tillage, soil depth, and precipitation effects on wheat response to nitrogen. *Soil Science Society of America Journal* 55, 121–124.

Rautapaa, J. (1966) The effect of the English grain aphid *Macrosiphum avenae* (F.) (Hom., Aphidaea) on the yield and quality of wheat. *Annales Agriculturae Fenniae* 5, 334–341.

Razzaque, M.A. and Hossain, A.B.S. (1991) The wheat development program in Bangladesh. In: Saunders, D.A. (ed.) *Wheat for the Nontraditional, Warm Areas*. CIMMYT, Mexico, D.F., pp. 44–54.

Redman, P.L and Knight, M. (1992) Machinery for whole-crop cereals. In: Wilkinson, J.M. and Stark, B.A. (eds) *Whole-Crop Cereals: Making and Feeding Cereal Silage*, 2nd edn. Chalcombe, Canterbury, pp. 97–106.

Redway, F.A., Vasil, V., Lu, D. and Vasil, I.K. (1990) Identification of callus types for long term maintenance and regeneration from commercial cultivars of wheat (*Triticum aestivum* L.). *Theoretical and Applied Genetics* 79, 609–617.

Reeves, T.G. and Brooke, H.D. (1977) The effect of genotype and phenotype on the competition between wheat and annual ryegrass. In: *Proceedings of the 6th Asian Pacific Weed Science Conference*. Asian Pacific Weed Science Society, Wellington, New Zealand, pp. 166–171.

Rerkasem, B., Lodkaew, S. and Jamjod, S. (1991) Assessment of grain set failure and diagnosis for boron deficiency in wheat. In: Saunders, D.A. (ed.) *Wheat for the Nontraditional, Warm Areas*. CIMMYT, Mexico D.F., pp. 500–504.

Reuter, D.J. and Dyson, C.B. (1990) *Farming Practices and Protein Levels in ASW Wheat. Technical Report No. 162*. Department of Agriculture, South Australia.

Rhoades, J.D., Bingham, F.T., Letey, J., Dedrick, A.R., Bean, M., Hoffman, G.J., Alves, W.J., Swain, R.V., Pacheco, P.G. and Lemert, R.D. (1988) Reuse of drainage water for irrigation: results of Imperial Valley study. *Hilgardia* 56 (5), 1–16.

Rifkin, H. (1989) *Utilization of Wheat in the Scotch Whisky Industry*. Institute of Brewing, Cambridge.

Rifkin, H.L., Bringhurst, T.A., McDonald, A.M.L. and Hands, E. (1990) Quality requirements of wheat for distilling. In: *Aspects of Applied Biology 25, Cereal Quality II*. Association of Applied Biologists, Warwick, pp. 29–40.

Rixhon, L. and Vandam, J. (1987) Study of some factors influencing technologic quality of bread wheat, in Belgium. In: Borghi, B. (ed.) *Hard Wheat: Agronomic, Technological, Biochemical and Genetic Aspects*. Commission of the European Communities, Luxembourg, pp. 85–92.

Roderuck, C.E. and Fox, H. (1987) Nutritional value of cereal grains. In: Olson, R.A. and Frey, K.J. (eds) *Nutritional Quality of Cereal Grains: Genetic and Agronomic Improvement. Agronomy Monograph No. 28*. ASA-CSSA-SSSA, Madison, WI, USA, pp. 1–9.

Rodriguez-Bores, F.J. and Bushuk, W. (1975) Factors affecting the Pelshenke test. *Abstracts of the 60th Annual Meeting of the AACC (Paper 115). Cereal Foods World* 20, 459.

Rola, J. and Rola, H. (1987) The influence of *Galium aparine* density, nitrogen fertilization and wheat sowing rate on yield. *Fragmenta Herbologica Jugoslavia* 16, 149–153.

Romig, R.W. and Calpouzos, L. (1970) The relationship between stem rust and loss in yield of spring wheat. *Phytopathology* 72, 1278–1280.

Rowlands, D.G., Pinniger, D.B. and Wilkin, D.R. (1989) *The Control of Pests in Stored Cereals. Research Review No. 17*. Home-Grown Cereals Authority, London.

Royle, D.J., Parker, S.R., Lovell, D.J. and Hunter, T. (1995) Interpreting trends and risks for better control of *Septoria* in winter wheat. In: Hewitt, H.G., Tyson, D., Hollomon, D.W., Smith, J.M., Davies, W.P. and Dixon K.R. (eds) *A Vital Role for Fungicides in Cereal Production*. BIOS, Oxford, pp. 105–115.

Ruegger, A., Winzeler, M. and Winzeler, H. (1993) The influence of different nitrogen levels and seeding rates on the dry matter production and nitrogen uptake of spelt (*Triticum spelta* L.) and wheat (*Triticum aestivum* L.) under field conditions. *Journal of Agronomy and Crop Science* 171, 124–132.

Rule, J.S. (1987) The effect of late-nitrogen on the grain quality of winter wheat varieties. In: *Aspects of Applied Biology 15, Cereal Quality*. Association of Applied Biologists, Warwick, pp. 249–253.

Russell, G.E. (1981) Disease and crop yield: The problems and prospects for agriculture. In: Ayers, P.G. (ed.) *Effects of Disease on the Physiology of the Growing Plant*. Cambridge University Press, Cambridge, pp. 14–28.

Russell, R.S., Cannell, R.Q. and Goss, M.J. (1975) Effects of direct drilling on soil conditions and root growth. *Outlook on Agriculture* 8, 227–232.

Saad, A.O.M., Ashour, N.I., Thalooth, A.T. and Nour, T.A. (1986) Growth, photosynthetic pigments content and yield of wheat plants as influenced by the time of potassium application. *Egyptian Journal of Agronomy* 11, 53–62.

Saari, E.E. (1985) Distribution and importance of root rot diseases of wheat, barley and triticale in south and southeast Asia. In: *Wheats for More Tropical Environments, A Proceedings of an International Symposium*. CIMMYT, Mexico D.F., pp. 189–195.

Sadler, R. and Scott, K.J. (1974) Nitrogen assimilation and metabolism in barley leaves infected with powdery mildew fungus. *Physiological Plant Pathology* 4, 353–380.

Salmon, S.E. and Cook, R.J. (1987) Effects of fungicides on the milling and baking quality of wheat. In: *Aspects of Applied Biology 15, Cereal Quality*. Association of Applied Biologists, Warwick, pp. 373–384.

Salmon, S.E., Greenwell, P. and Dampney, P.M.R. (1990) The effect of rate and timing of late nitrogen applications to breadmaking wheats as ammonium nitrate or foliar urea-N, and the effect of foliar sulphur application. In: *Aspects of Applied Biology 25, Cereal Quality II*. Association of Applied Biologists, Warwick, pp. 242–253.

Salomonsson, L., Jonsson, A., Salomonsson, A.C. and Nilsson, G. (1994) Effects of organic fertilizers and urea when applied to spring wheat. *Acta Agriculturae Scandinavica. Section B, Soil and Plant Science* 44, 170–178.

Salunkhe, D.K., Chavan, J.K. and Kadam, S.S. (1985) *Postharvest Biotechnology of Cereals*. CRC Press, Boca Ratan, Florida.

Samuel, A.M. and East, J. (1990) Organically grown wheat – the effect of crop husbandry on grain quality. In: *Aspects of Applied Biology 36, Cereal Quality III*. Association of Applied Biologists, Warwick, pp. 199–208.

Sander, D.H., Allaway, W.H. and Olson, R.A. (1987) Modification of nutritional quality by environmental and production practices. In: Olson, R.A. and Frey, K.J. (eds) *Nutritional Quality of Cereal Grains: Genetic and Agronomic Improvement. Agronomy Monograph No. 28*. ASA-CSSA-SSSA, Madison, WI, USA, pp. 45–81.

Sansoucy, R. (1981) Silage from whole-crop forage cereals for cattle feeding. *World Animal Review* 37, 25–30.

Sarrafi, A., Ecochard, R. and Grignac, P. (1989) Genetic variability for some grain quality characters in tetraploid wheats. *Plant Varieties and Seeds* 2, 163–169.

Saunders, D.A. (1988) Characterisation of tropical wheat environments: Identification of production constraints and progress achieved. In: Klatt, A.R. (ed.) *Wheat Production Constraints in Tropical Environments*. CIMMYT, Mexico D.F., pp. 12–26.

Schipper, A. (1991) Modifications of the dough physical properties of various wheat cultivars by environmental influences. *Agribiological Research* 44, 114–132.

Schulthess, U., Tedla, A., Mohamed-Saleem, M.A. and Said, A.N. (1995) Effects of variety, altitude, and undersowing with legumes on the nutritive value of wheat straw. *Experimental Agriculture* 31, 169–176.

Schwarz, F.J., Reimann, W. and Kirchgessner, M. (1990) Fodder intake and milk production of cows fed wheat whole crop silage with and without undersown ryegrass. *Bayerisches Landwirtschaftliches Jahrbuch* 67, 3–13.

Scrimshaw, N.S. and Young, V.R. (1976) The requirements of human nutrition. *Scientific American* 235 (3), 51–65.

Scudamore, K.A. (1993) Occurrence and significance of mycotoxins in cereals grown and stored in the United Kingdom. In: *Aspects of Applied Biology 36, Cereal Quality III*. Association of Applied Biologists, Warwick, pp. 361–373.

Seaton, J.C. (1987) Malt types and beers. *European Brewery Convention Proceedings*, Spain, pp. 177–188.

Sen, A. and Misra, N.M. (1987) Effect of presowing seed treatments and phosphate doses on yield, quality and nutrient uptake of wheat. *Agricultural Science Digest, India* 7, 208–212.

Sharma, B.K., Sirdhi, A. and Sirdhi, A. (1994) Present status of Karnal bunt of wheat in foothills of Himachal Pradesh. *Plant Disease Research* 9, 77–79.

Sharrow, S.H. (1990) Defoliation effects on biomass yield components of winter wheat. *Canadian Journal of Plant Science* 70, 1191–1194.

Sharrow, S.H. and Motazedian, I. (1987) Spring grazing effects on components of winter wheat yield. *Agronomy Journal* 79, 502–504.

Shatilov, I.S. and Sharov, A.F. (1992) Productivity of field crops and indications of soil fertility with application of different rates of organic fertilizers. *Izvestiya Timiryazevskoi Sel'skokhozyaistvennoi Akademii* No. 1, 3–11.

Sherrott, A.P. (1988) Control of *Poa* spp. in winter cereals ADAS trials 1985-87. In: *Aspects of Applied Biology 18, Weed Control in Cereals and the Impact of Legislation on Pesticide Application.* Association of Applied Biologists, Warwick, pp. 207–227.

Shewry, P.R., Tatham, A.S., Barro, F., Barcelo, P. and Lazzeri, P. (1995) Biotechnology of breadmaking: Unraveling and manipulating the multi-protein gluten complex. *Bio/tecnology* 13, 1185–1190.

Shipton, W.A., Boyd, W.R.J., Rosielle, A.A. and Shearer, B.I. (1971) The common septoria diseases of wheat. *Botanical Reviews* 37, 231–262.

Sidhu, D.S. and Byerlee, D. (1992) *Technical Change and Wheat Productivity in the Indian Punjab in the Post-Green Revolution Period. CIMMYT Economics Working Paper 92/02.* CIMMYT, Mexico, D.F.

Silsbury, J.H. (1990) Grain yield of wheat in rotation with pea, vetch or medic grown with three systems of management. *Australian Journal of Experimental Agriculture* 30, 645–649.

Singh, S.B. (1985) Production, storage and marketing of wheat seed in India. In: *Wheats for More Tropical Environments, A Proceedings of an International Symposium.* CIMMYT, Mexico D.F., pp. 297–302.

Singh, R.P. (1988) Macro-element requirements and fertility management issues in rice–wheat rotation areas. In: Klatt, A.R. (ed.) *Wheat Production Constraints in Tropical Environments.* CIMMYT, Mexico D.F., pp. 78–98.

Singh, A.J., Singh, B. and Bal, H.S. (1987) Indiscriminate fertiliser use vis-a-vis groundwater pollution in Central Punjab. *Indian Journal of Agricultural Economics* 42, 404–409.

Singh, K., Nema, D.P., Verma, H.D. and Singh, K. (1991) Effect of graded levels of NPK on yield and quality of irrigated wheat. *Orissa Journal of Agricultural Research* 4, 133–136.

Singh, N., Mehta, S.C., Singh, M., Mittal, S.B., Singh, N. and Singh, M. (1992) Effect of potassium and magnesium application on quality and grain yield of wheat. *Journal of Potassium Research* 8, 231–238.

Singhal, N.C., Srivastava, K.N. and Mehta, S.L. (1989) Pattern of dry matter and protein accumulation in developing wheat seeds and their relationship. *Indian Journal of Genetics* 49, 95–102.

Slovic, P. (1984) Perception of risk. *Science* 236, 280–285.

Smika, D.E. and Greb, B.W. (1973) Protein content of winter wheat grain as related to soil and climatic factors in the semiarid central great plains. *Agronomy Journal* 65, 433–436.

Smith, K.A., and Chambers, B.J. (1993) Utilising the nitrogen content of organic manures on farms. Problems and practical solutions. *Soil Use and Management* 9, 20–27.

Smith, S.P. and Davies, W.P. (1990) Response of wheat to high and low nitrogen and fungicide inputs on shallow limestone soil. *BCPC Monograph No. 45 Organic and Low Input Agriculture.* BCPC Publications, Croydon, pp. 227–230.

Smith, G.P. and Gooding, M.J. (1996a) Models describing wheat crude protein content and Hagberg falling number in regions of England between 1974 and 1993. *Aspects of Applied Biology 46, Modelling in Applied Biology: Spatial Aspects.* Association of Applied Biologists, Warwick, pp. 131–134.

Smith, G.P. and Gooding, M.J. (1996b) Relationships of wheat quality with climate and nitrogen application in regions of England (1974–1993). *Annals of Applied Biology* 129, 97–108.

Smith, S.P., Davies, W.P., Bulman, C. and Evans, E.J. (1990) Responses of grain protein in winter wheat to nitrogen. In: *Aspects of Applied Biology 25, Cereal Quality II.* Association of Applied Biologists, Warwick, pp. 255–260.

Smolik, J.D. and Dobbs, T.L. (1991) Crop yield and economic returns accompanying the transition to alternative farming systems. *Journal of Production Agriculture* 4, 153–161.

Sofield, I., Wardlaw, I.F., Evans, L.T. and Zee, S.Y. (1977) Nitrogen, phosphorus and water contents during grain development and maturation in wheat. *Australian Journal of Plant Physiology* 4, 799–810.

Sosulski, F.W., Lin, D.M. and Paul, E.A. (1966) Effect of moisture, temperature, and nitrogen on yield and protein quality of Thatcher wheat. *Canadian Journal of Plant Science* 46, 583–588.

Soulaka, A.B. and Morrison, W.R. (1985) The bread baking quality of six wheat starches differing in composition and physical properties. *Journal of the Science of Food and Agriculture* 36, 719–727.

Spiertz, J.H.J. (1977) The influence of temperature and light intensity on grain growth in relation to the carbohydrate and nitrogen economy of the wheat plant. *Netherlands Journal of Agricultural Science* 25, 182–197.

Spiertz, J.H.J. (1978) Grain production and assimilate utilization of wheat in relation to cultivar characteristics, climatic factors and nitrogen supply. *Agricultural Research Reports (Wageningen)* No. 881.

Spiertz, J.H.J, Hag, B.A. and Kupers, L.J.P. (1971) Relation between green area duration and grain yield in some varieties of spring wheat. *Netherlands Journal of Agricultural Science* 19.

Spink, J.H., Clare, R.W. and Kilpatrick, J.B. (1993) Grain quality of milling wheats at eight different sowing dates. In: *Aspects of Applied Biology 36, Cereal Quality III.* Association of Applied Biologists, Warwick, pp. 231–240.

Sprague, M.A. (1986) Overview. In: Sprague, M.A. and Triplett, G.B (eds) *No-tillage and Surface-tillage Agriculture.* John Wiley and Sons, New York, pp. 1–18.

Stapper, M. and Fischer, R.A. (1990) Genotype, sowing date and plant spacing influence on high-yielding irrigated wheat in southern New South Wales. II. Growth, yield and nitrogen use. *Australian Journal of Agricultural Research* 41, 1021–1041.

Stenning, B.C. and Channa, K.S. (1987) Sources of inaccuracy in grain moisture measurement. In: *Aspects of Applied Biology 15, Cereal Quality.* Association of Applied Biologists, Warwick, pp. 457–468.

Stenvert, N.L. (1974) Grinding resistance: a simple measure of wheat hardness. *Journal of Flour and Animal Feed Milling* 56, 24–27.

Stewart, B.A. and Dyke, G.V. (1993) Factors affecting the grain yield, milling and breadmaking quality of wheat, 1969–72. II. Breadmaking quality in relation to variety and nitrogen fertilizer. *Plant Varieties and Seeds* 6, 169–178.

Stoddard, F.L. and Marshall, D.R. (1990) Variability in grain protein in Australian hexaploid wheats. *Australian Journal of Agricultural Research* 41, 277–288.

Stoskopf, N.C. (1992) *Cereal Grain Crops.* Reston Publishing Company, Virginia.

Strange, R.N., Deano, A. and Smith, H. (1978) Virulence enhancement of *Fusarium graminearum* by choline and betain and of *Botrytis cinerea* by other constituents of wheat germ. *Transactions of the British Mycological Society* 70, 201–207.

Strong, W.M. (1982) Effect of late-application of nitrogen on the yield and protein content of wheat. *Australian Journal of Experimental Agriculture and Animal Husbandry* 22, 54–61.

Summers, R.W. (1990) The effect on winter wheat of grazing by brent geese *Branta bernicla. Journal of Applied Ecology* 27, 821–833.

Sundstol, F. (1988) Straw and other fibrous by-products. *Livestock Production Science* 19, 137–158.

Sutton, J.C. (1982) Epidemiology of wheat head blight and maize ear rot caused by *Fusarium graminearum. Canadian Journal of Plant Pathology* 4, 195–209.

Svetov, V.G. (1990) Effect of agroecological factors on the black germ of wheat. *Mikologiya i Fitopatologiya* 23, 471–473.

Sward, R.J. and Lister, R.M. (1987) The incidence of barley yellow dwarf viruses in wheat in Victoria. *Australian Journal of Agricultural Research* 38, 821–828.

Syers, J.K., Skinner, R.J. and Curtin, D. (1987) Soil and fertiliser sulphur in UK agriculture. *The Fertiliser Society, Proceedings No. 264.* The Fertiliser Society, London.

Sylvester-Bradley, R. (1990) Does extra nitrogen applied to breadmaking wheat benefit the baker? In: *Aspects of Applied Biology 25, Cereal Quality II.* Association of Applied Biologists, Warwick, pp. 217–228.

Sylvester-Bradley, R. and George, B.J. (1987) Effects of quality payments on the economics of applying nitrogen to wheat. In: *Aspects of Applied Biology 15, Cereal Quality.* Association of Applied Biologists, Warwick, pp. 303–335.

Sylvester-Bradley, R., Dampney, P.M.R. and Murray, A.W.A. (1984) The response of winter wheat to nitrogen. In: *The Nitrogen Requirement of Cereals. Ministry of Agriculture, Fisheries and Food. Reference Book 385.* HMSO, London, pp. 151–174.

Sylvester-Bradley, R., Addiscott, T.M., Vaidyanathan, L.V., Murray, A.W.A. and Whitmore, A.P. (1987a) Nitrogen advice for cereals. *Proceedings of the Fertiliser Society No. 263.* The Fertiliser Society, London.

Sylvester-Bradley, R., Marriot, N.J., Haywood, C.F. and Hook, S.C.W. (1987b) Effect of urea sprays during ripening on nitrogen content and breadmaking quality of winter wheat. In: *Aspects of Applied Biology 15, Cereal Quality.* Association of Applied Biologists, Warwick, pp. 283–287.

Szunics, L., Szunics, L. and Stehli, L. (1987) Microorganisms isolated from wheat and damage caused by them. IV. Data on damage due to *Fusarium. Noveny-termeles* 36, 421–430.

Tabl, M.M. and Kiss, A. (1983) Chemical and quality characters of triticale and wheat grown at two plant densities and different levels of nitrogen fertilizer. *Cereal Research Communications* 11, 275–281.

Takruri, H.R., Humeid, M.A. and Umari, M.A.H. (1990) Protein quality of parched immature durum wheat (*Frekeh*). *Journal of the Science of Food and Agriculture* 50, 319–327.

Tatham, A.S., Field, J.M. and Shewry, P.R. (1987) Model studies of wheat gluten elasticity. In: *Aspects of Applied Biology 15, Cereal Quality.* Association of Applied Biologists, Warwick, pp. 91–95.

Taylor, A.C. and Gilmour, A.R. (1971) Wheat protein prediction from climatic factors in southern New South Wales. *Australian Journal of Experimental Agriculture and Animal Husbandry* 11, 546–549.

Taylor, A.C. and Lill, W.J. (1986) Wheat crop surveys in southern New South Wales. 4. The response of grain yield and other wheat attributes to weeds. *Australian Journal of Experimental Agriculture* 26, 709–715.

Taylor, B.R. and Roscrow, J.C. (1990) The quality of winter wheat varieties for distilling from Scottish sites. In: *Aspects of Applied Biology 25, Cereal Quality II.* Association of Applied Biologists, Warwick, pp. 183–191.

Taylor, B.R., Cranstoun, D.A.S. and Roscrow, J.C. (1993) The quality of winter wheat varieties for distilling from Scottish sites. In: *Aspects of Applied Biology 36, Cereal Quality III.* Association of Applied Biologists, Warwick, pp. 481–489.

Tessemma, T. (1987) Durum wheat breeding in Ethiopia. In: van Ginke, M. and Tanner, D.G. (eds) *The Fifth Regional Wheat Workshop for Eastern, Central Southern Africa and the Indian Ocean.* CIMMYT, Mexico D.F., pp. 18–22.

Tester, R.F., Morrison, W.R., Ellis, R.H., Piggott, J.R., Batts, G.R., Wheeler, T.R., Morison, W.R., Hadley, P. and Ledward, D.A. (1995) Effects of elevated growth temperature and carbon dioxide levels on some physicochemical properties of wheat starch. *Journal of Cereal Science* 22, 63–71.

Tetlow, R.M. (1990) Whole-crop cereal for beef cattle. In: Wilkinson, J.M. and Stark, B.A. (eds) *Whole-Crop Cereals: Making and Feeding Cereal Silage.* Chalcombe, Canterbury, pp. 55–66.

Tetlow, R.M. (1992) A decade of research into whole-crop cereals at Hurley. In: Wilkinson, J.M. and Stark, B.A. (eds) *Whole-Crop Cereals: Making and Feeding Cereal Silage*, 2nd edn. Chalcombe, Canterbury, pp. 1–19.

Tetlow, R.M. and Mason, V.C. (1987) Treatment of whole-crop cereals with alkali. 1. The influence of sodium hydroxide and ensiling on the chemical composition and *in vitro* digestibility of rye, barley and wheat crops harvested at increasing maturity and dry matter content. *Animal Feed Science and Technology* 18, 257–269.

Tetlow, R.M., Mason, V.C. and Deschard, G. (1987) Treatment of whole-crop cereals with alkali. 2. Voluntary intake and digestibility by sheep of rye, barley and wheat crops ensiled with sodium hydroxide. *Animal Feed Science and Technology* 18, 271–281.

Thirakhupt, V. and Araya, J.E. (1992) Effects of cereal aphid feeding and barley yellow dwarf virus on 'Abe' wheat in the laboratory. *Zeitschrift fur Pflanzenkrankheiten und Pflanzenshutz* 99, 420–425.

Thompson, A.J. (1995) The comparative performance of wheat cultivars and genotypes in different organic systems of production. PhD Thesis, Royal Agricultural College/University of Reading.

Thompson, A.J., Gooding, M.J. and Davies, W.P. (1992) Shading ability, grain yield and grain quality of organically grown cultivars of winter wheat. *Tests of*

Agrochemicals and Cultivars 13, Annals of Applied Biology 120 (Supplement), pp. 86–87.

Thompson, A.J., Gooding, M.J. and Davies, W.P. (1993a) Blackpoint on the grain of organically grown cultivars of winter wheat. *Tests of Agrochemicals and Cultivars 14, Annals of Applied Biology* 122 (Supplement), pp. 164–165.

Thompson, A.J., Gooding, M.J. and Davies, W.P. (1993b) The effect of season and management on the grain yield and breadmaking quality of organically grown wheat cultivars. In: *Aspects of Applied Biology 36, Cereal Quality III.* Association of Applied Biologists, Warwick, pp. 179–188.

Thompson, A.J., Gooding, M.J. and Davies, W.P. (1993c) Implications of wheat husbandry for the success of undersown clover and grass. *Forward With Grass into Europe. Occasional Symposium No. 27.* British Grassland Society, Reading, pp. 173–175.

Timms, M.F., Bottomly, R.C., Ellis, J.R.S and Schofield, J.D. (1981) The baking quality and protein characteristics of a winter wheat grown at different levels of nitrogen fertilisation. *Journal of the Science of Food and Agriculture* 32, 684–698.

Tipples, K.H., Dubetz, S. and Irvine, G.N. (1977) Effects of high rates of nitrogen on Neepawa wheat grown under irrigation II. Milling and baking quality. *Canadian Journal of Plant Science* 57, 337–350.

Tivy, J. (1990) *Agricultural Ecology.* Longman, Harlow.

Todorov, N.A. (1988) Cereals, pulses and oilseeds. *Livestock Production Science* 19, 47–95.

Tollenaar, H. and Houston, B.R. (1967) A study of the epidemiology of stripe rust, *Puccinia striiformis* West, in California. *Canadian Journal of Botany* 45, 291–307.

Tompkins, D.K. (1992) Effects of agronomic treatment on yield, pattern of soil water use and foliar diseases in winter wheat in northeast Saskatchewan. *Dissertation Abstracts International. B, Sciences and Engineering* 52 (11), 5592B.

Tompkins, D.K., Fowler, D.B. and Wright, A.T. (1991) Water-use by no-till winter wheat: Influence of seed rate and row spacing. *Agronomy Journal* 83, 766–769.

Tonkin, J.H.B. (1987) Seed impurities in samples of cereal seed and feed grain. In: *Aspects of Applied Biology 15, Cereal Quality.* Association of Applied Biologists, Warwick, pp. 473–482.

Tottman, D.R., Steer, P.M. and Martin, T.D. (1988) The tolerance of cereals to several foliage-applied broad-leaved weed herbicides at different growth stages. In: *Aspects of Applied Biology 18, Weed Control in Cereals and the Impact of Legislation on Pesticide Application.* Association of Applied Biologists, Warwick, pp. 145–156.

Triboi, E., Branlard, G. and Landry, J. (1990) Environmental and husbandry effects on the content and composition of proteins in wheat. In: *Aspects of Applied Biology 25, Cereal Quality II.* Association of Applied Biologists, Warwick, pp. 149–158.

UKROFS (1992) *Standards for Organic Food Production.* United Kingdom Register of Organic Food Standards, London.

Vaidyanathan, L.V. (1987) Precision and reliability of measuring Hagberg falling number of wheat including variability associated with crop husbandry and grain handling. In: *Aspects of Applied Biology 15, Cereal Quality.* Association of Applied Biologists, Warwick, pp. 495–513.

Vaidyanathan, L.V. and Davies, D.B. (1980) Response of winter wheat to fertiliser nitrogen in undisturbed and cultivated soil. *Journal of the Science of Food and Agriculture* 31, 414–419.

Vaidyanathan, L.V., Sylvester-Bradley, R., Bloom, T.M. and Murray, A.W.A. (1987) Effects of previous cropping and applied nitrogen on grain nitrogen content in winter wheat. In: *Aspects of Applied Biology 15, Cereal Quality*. Association of Applied Biologists, Warwick, pp. 227–237.

Vasil, V., Brown, S.M., Re, D., Fromm, M.E. and Vasil, I.K. (1991) Stably transformed callus lines from microprojectile bombardment of cell suspension cultures of wheat. *Bio/technology* 9, 743–747.

Vasil, V., Castillo, A.M., Fromm, M.E. and Vasil, I.K. (1992) Herbicide resistant fertile transgenic wheat plants obtained by microprojectile bombardment of embryogenic callus. *Bio/technology* 10, 667–674.

Vedrov, N.G. and Frolov, I.N. (1990) The role of sowing depth in the formation of spring wheat yield. *Sibirskii Vestnik Sel'skokhozyaistvennoi Nauki* No. 1, 3–6.

Verma, H.S., Singh, A. and Agarwal, V.K. (1986) Environmental factors in relation to loose smut infection in wheat seeds. *Indian Phytopathology* 39, 423–426.

Villareal, R.L., Rajaram, S. and Nelson, W. (1985) Breeding wheat for more tropical environments at CIMMYT. In: *Wheats for More Tropical Environments, A Proceedings of an International Symposium*. CIMMYT, Mexico D.F., pp. 89–99.

Vos, J. (1981) Effects of temperature and nitrogen supply on post-floral growth of wheat; measurements and simulations. *Agricultural Research Reports (Wageningen)* No. 911.

Wagstaff, H. (1987) Husbandry methods and farm systems in industrialised countries which use lower levels of external inputs: a review. *Agriculture, Ecosystems and the Environment* 19, 1–27.

Wainwright, A. (1995) The control of seed-borne pathogens. In: Hewitt, H.G., Tyson, D., Hollomon, D.W., Smith, J.M., Davies, W.P. and Dixon K.R. (eds) *A Vital Role for Fungicides in Cereal Production*. BIOS, Oxford, pp. 83–94.

Walker, N.H. (1992) How other crops can complement grass to cope with seasonal and annual fluctuations. In: Hopkins, A. (ed.) *Grass on the Move: A Positive Way Forward for the Grassland Farmer, Occasional Symposium No. 26*. British Grassland Society, Reading, pp. 113–119.

Wall, J.S. (1971) Disulfide bonds: determination, location and influence on molecular properties of proteins. *Agricultural and Food Chemistry* 19, 619–625.

Wall, P.C., Hobbs, P.R., Saunders, D.A., Sayre, K.D. and Tanner, D.G. (1991) Wheat crop management for the warmer areas: A review of issues and advances. In: Saunders, D.A. (ed.) *Wheat for the Nontraditional, Warm Areas*. CIMMYT, Mexico D.F., pp. 225–241.

Wallwork, H. and Spooner, B. (1988) *Tapesia yallundae* – the teleomorph of *Pseudocercosporella herpotrichoides*. *Transactions of the British Mycological Society* 91, 703–705.

Wardlaw, I.F., Carr, D.J. and Anderson, M.J. (1965) The relative supply of carbohydrate and nitrogen to wheat grain, and an assessment of the shading and defoliation techniques used for these determinations. *Australian Journal of Agricultural Research* 16, 893–901.

Wardlaw, I.F., Dawson, I.A. and Munibi, P. (1989) The tolerance of wheat to high temperatures during reproductive growth. II. Grain development. *Australian Journal of Agricultural Research* 40, 15–24.

Wasowicz, E. (1991) Changes of chemical grain components, especially lipids, during their deterioration by fungi. In: Chelkowski, J. (ed.) *Cereal Grain: Mycotoxins, Fungi and Quality in Drying and Storage.* Elsevier, Amsterdam, pp. 259–280.

Wassermann, L., Muhlbauer, W. and Schreiber, H. (1983) Influence of drying on wheat quality 3). Effect of grain moisture content on drying behaviour and baking properties. *Getreide Mehl und Brot* 9, 268–274.

Webb, J. and Sylvester-Bradley, R. (1995) A comparison of the responses of two cultivars of late-autumn-sown wheat to applied nitrogen. *Journal of Agricultural Science, Cambridge* 125, 11–24.

Webb, A., Haley, S.L. and Leetma, S. (1995) Enhancing US wheat export performance: the implications of wheat cleaning. *Agribusiness* 11, 317–332.

Wedin, W.F. and Hoveland, C. (1987) Cereal vegetation as a source of forage. In: Olson, R.A. and Frey, K.J. (eds) *Nutritional Quality of Cereal Grains: Genetic and Agronomic Improvement. Agronomy Monograph No. 28.* ASA-CSSA-SSSA, Madison, WI, USA, pp. 83–99.

Weegels, P.L., Marseille, J.P., Bosveld, P. and Hamer, R.J. (1994) Large-scale separation of gliadins and their bread-making quality. *Journal of Cereal Science* 20, 253–264.

Weegels, P.L., Hamer, R.J. and Schofield, J.D. (1996) Critical review: Functional properties of wheat glutenin. *Journal of Cereal Science* 23, 1–18.

Weeks, J.T., Anderson, O.D. and Bechl, A.E. (1993) Rapid production of multiple independent lines of fertile transgenic wheat (*Triticum aestivum*). *Plant Physiology* 102, 1077–1084.

Weibel, R.O. and Pendleton, J.W. (1964) Effect of artificial lodging on winter wheat grain yield and quality. *Agronomy Journal* 56, 487–488.

Weller, R.F. (1991) The national whole crop cereals survey (summary). *Proceedings of the Whole Crop Cereals Conference, The Royal Agricultural College, 18th April 1991.* The Maize Growers Association, Reading.

Weller, R.F. (1992) The national whole-crop cereals survey. In: Wilkinson, J.M. and Stark, B.A. (eds) *Whole-Crop Cereals: Making and Feeding Cereal Silage,* 2nd edn. Chalcombe, Canterbury, pp. 137–156.

Weller, R.F., Cooper, A. and Dhanoa, M.S. (1995) The selection of winter wheat varieties for whole crop cereal conservation. *Grass and Forage Science* 50, 172–177.

Wheeler, B.E.J. (1969) *An Introduction to Plant Diseases.* John Wiley and Sons, London.

White, D.J. (1979) Support energy use in forage conservation. In: Thomas, C. (ed.) *Forage Conservation in the 80s, Occasional Symposium No. 11, British Grassland Society.* The British Grassland Society, Reading, pp. 33–45.

White, D.J. (1981) Energy in agriculture. *Proceedings of the Fertiliser Society No. 203.* Fertiliser Society, London.

White, L.M., Hartman, G.P. and Bergman, J.W. (1981) *In vitro* digestibility, crude protein, and phosphorus content of straw of winter wheat, barley and oat cultivars in eastern Montana. *Agronomy Journal* 73, 117–121.

Whitehead, D.C., Pain, B.F. and Ryden, J.C. (1986) Nitrogen in UK grassland agriculture. *Journal of the Royal Agricultural Society of England* 147, 190–201.

Wibberley, E.J. (1989) *Cereal Husbandry*. Farming Press, Ipswich, 258 pp.

Wicks, G.A. (1986) Substitutes for tillage on the Great Plains. In: Sprague, M.A. and Triplett, G.B (eds) *No-tillage and Surface-tillage Agriculture*. John Wiley and Sons, New York, pp. 183–196.

Widdowson, F.V., Penny, A., Darby, R.J., Bird, E. and Hewitt, M.V. (1987) Amounts of NO_3-N and NH_4-N in soil, from autumn to spring, under winter wheat and their relationship to soil type, sowing date, previous crop and N uptake at Rothamsted, Woburn and Saxmundham. *Journal of Agricultural Science, Cambridge* 108, 73–95.

Wiese, M.V. (1987) *Compendium of Wheat Diseases*, 2nd edn. APS Press, Minnesota.

Wilkin, D.R. and Rowlands, D.G. (1988) *The Biodeterioration of Stored Cereals. Research Review No. 3*. Home-Grown Cereals Authority, London.

Wilkinson, B. (1975) Soil types and direct drilling – a provisional assessment. *Outlook on Agriculture* 8, 233–235.

Willey, R.W. (1979) Intercropping – its importance and research needs. I. Competition and yield advantages. *Field Crop Abstracts* 79, 517–529.

Williams, P.E.V. and Chesson, A. (1989) Cereal raw materials and animal production. In: Palmer, G.H. (ed.) *Cereal Science and Technology*. Aberdeen University Press, Aberdeen, pp. 413–463.

Wilson, B.J. (1984) Competition between *Galium aparine* and cereals. In: *Understanding Cleavers (*Galium aparine*) and their Control in Cereals and Oilseed Rape – Abstracts of Papers*. Association of Applied Biologists, Warwick, pp. 5–6.

Wilson, W.W. (1989) Differentiation and implicit prices in export wheat markets. *Western Journal of Agricultural Economics* 14, 67–77.

Wilson, P.R. (1990) A new instrument concept for nitrogen/protein analysis. A challenge to the Kjeldahl method. In: *Aspects of Applied Biology 25, Cereal Quality II*. Association of Applied Biologists, Warwick, pp. 443–446.

Wilson, B.J., Thornton, M.E. and Lutman, P.J.W. (1985) Yields of winter cereals in relation to the timing of control of black-grass (*Alopecurus myosuroides* Huds.) and broad-leaved weeds. In: *Aspects of Applied Biology 9, The Biology and Control of Weeds in Cereals*. Association of Applied Biologists, Warwick, pp. 41–48.

Wimschneider, W., Bachtaler, G. and Fischbeck, G. (1990) Competitive effects of *Avena fatua* L. (wild oats) on wheat (*Triticum aestivum* L.) as a basis for effective weed control. *Weed Research, Oxford* 30, 43–52.

Wiseman, J. (1990) Quality requirements of wheat for poultry feed. In: *Aspects of Applied Biology 25, Cereal Quality II*. Association of Applied Biologists, Warwick, pp. 41–51.

Wiseman, J. (1993) The nutritional value of wheat in pig and poultry rations. *The Agronomist* No. 2, 12–13.

Wiseman, J. and McNab, J. (1995) *Nutritive Value of Wheat Varieties Fed to Non-ruminants. Project Report No. 111*. Home-Grown Cereals Authority, London.

Wiseman, J., Nicol, N. and Norton, G. (1993) Biochemical and nutritional characterisation of wheat varieties for young broilers. In: *Aspects of Applied Biology 34, Physiology of Varieties*. Association of Applied Biologists, Warwick, pp. 329–334.

Withers, P.J.A. (1987) Effect of applied nitrogen on specific weight of winter wheat. In: *Aspects of Applied Biology 15, Cereal Quality*. Association of Applied Biologists, Warwick, pp. 289–292.

Wood, H.L. and Fox, W.E. (1965) The interaction of nitrogen and water on yield, protein and mottling in wheat grown on black earth in Queensland. *Experimental Agriculture* 1, 107–112.

Woolford, M.K., Bolsen, K.K. and Peart, L.A. (1982) Studies on the aerobic deterioration of whole-crop cereal silages. *Journal of Agricultural Science, Cambridge* 98, 529–535.

Woolley, E.W. and Sherrott, A.F. (1993) Determination of economic threshold populations of *Poa annua* in winter wheat. In: *Brighton Crop Protection Conference, Weeds – 1993*. British Crop Protection Council, Thornton Heath, UK, pp. 95–101.

Wright, A.T. (1990) Quality effects of pulses on subsequent cereal crops in the northern prairies. *Canadian Journal of Plant Sciences* 70, 1013–1021.

Wright, K.J. and Wilson, B.J. (1992) Effects of nitrogen fertiliser on competition and seed production of *Avena fatua* and *Galium aparine* in winter wheat. In: *Aspects of Applied Biology 30, Nitrate and Farming Systems*. Association of Applied Biologists, Warwick, pp. 381–386.

Wrigley, C.W., Robinson, P.J. and Williams, W.T. (1981) Association between electrophoretic patterns of gliadin proteins and quality characteristics of wheat cultivars. *Journal of the Science of Food and Agriculture* 32, 433–442.

Wrigley, C.W., du Cros, D.L., Moss, H.J., Randall, P.J., Fullington, J.G. and Kasarda, D.D. (1984) Effect of sulphur deficiency on wheat quality. *Sulphur in Agriculture* 8, 2–7.

Yarham, D.J. (1975) The effect of non-ploughing on cereal diseases. *Outlook on Agriculture* 8, 245–247.

Yarham, D.J. (1980) Ripening diseases – do they matter? *Proceedings of the 16th NIAB Crop Conference 'Winter Wheat'*. National Institute of Agricultural Botany, Cambridge, pp. 89–94.

Yarham, D.J. (1993a) *Ergot of Cereals: A Literature Review and Survey of Incidence in Traded Grain. HGCA Research Review No. 25*. Home-Grown Cereals Authority, London.

Yarham, D.J. (1993b) Soil-borne spores as a source of inoculum for wheat bunt (*Tilletia caries*). *Plant Pathology* 42, 654–656.

Yarham, D.J. (1995) Soil-borne diseases of cereals. In: Hewitt, H.G., Tyson, D., Hollomon, D.W., Smith, J.M., Davies, W.P. and Dixon K.R. (eds) *A Vital Role for Fungicides in Cereal Production*. BIOS, Oxford, pp. 83–94.

Yarham, D.J. and Turner, J. (1992) ADAS organic wheat survey. *New Farmer and Grower* (Spring), 31–33.

Zadoks, J.C., Chang, T.T. and Konzak, C.F. (1974) A decimal code for the growth stages of cereals. *Weed Research, Oxford* 14, 415–421.

Zeleny, F. (1986) Direct and residual effects of molybdenum on winter wheat yield and grain quality. *Annals of the Research Institute for Crop Production, Prague* 24, 309–321.

Zentner, R.P., Bowren, K.E., Edwards, W. and Campbell, C.A. (1990) Effects of crop rotations and fertilization on yields and quality of spring wheat grown on a Black Chernozem in north-central Saskatchewan. *Canadian Journal of Plant Science* 70, 383–397.

Zhang, T.Y., Wang, H.L. and Xu, F.L. (1990) Effects of grain blackpoint disease of wheat and the pathogenic fungi. *Acta Phytophylacia Sinicia* 17, 313–316.

Zhong, W.Y. (1988) Screening techniques and sources of resistance to *Fusarium* head blight. In: Klatt, A.R. (ed.) *Wheat Production Constraints in Tropical Environments.* CIMMYT, Mexico D.F., pp. 239–250.

Zhmurko, N.G. (1992) Effect of trace elements under a regime of different mineral fertilizer doses on the yield and quality of grain in different varieties of winter wheat. *Fiziologiya i Biokhimiya Kul'turnykh Rastenii* 24, 583–587.

Zhou, X. and Carter, N. (1991) The effects of nitrogen and fungicide on cereal aphid population development and the consequences for the aphid–yield relationship in winter wheat. *Annals of Applied Biology* 119.

Zhou, J.B. and Li, C.W. (1991) Effect of applying N fertilizer with P fertilizer or manure on wheat yield. *Shaanxi Journal of Agricultural Sciences* No. 1, 12–13.

Zohary, D., Harlan, J.R. and Vardi, A. (1969) The wild diploid progenitors of wheat and their breeding value. *Euphytica* 18, 58–65.

Zuniga, E. (1991) Integrated pest management: Aphid control in South America. In: Saunders, D.A. (ed.) *Wheat for the Nontraditional, Warm Areas.* CIMMYT, Mexico D.F., pp. 214–224.

Index